Structure Formation in the Universe

NATO Science Series

A Series presenting the results of activities sponsored by the NATO Science Committee. The Series is published by IOS Press and Kluwer Academic Publishers, in conjunction with the NATO Scientific Affairs Division.

A. Life Sciences	IOS Press
B. Physics	Kluwer Academic Publishers
C. Mathematical and Physical Sciences	Kluwer Academic Publishers
D. Behavioural and Social Sciences	Kluwer Academic Publishers
E. Applied Sciences	Kluwer Academic Publishers
F. Computer and Systems Sciences	IOS Press

1. Disarmament Technologies	Kluwer Academic Publishers
2. Environmental Security	Kluwer Academic Publishers
3. High Technology	Kluwer Academic Publishers
4. Science and Technology Policy	IOS Press
5. Computer Networking	IOS Press

As a consequence of the restructuring of the NATO Science Programme in 1999, the NATO Science Series has been re-organized and new volumes will be incorporated into the following revised sub-series structure:

I. Life and Behavioural Sciences	IOS Press
II. Mathematics, Physics and Chemistry	Kluwer Academic Publishers
III. Computer and Systems Sciences	IOS Press
IV. Earth and Environmental Sciences	Kluwer Academic Publishers
V. Science and Technology Policy	IOS Press

NATO-PCO-DATA BASE

The NATO Science Series continues the series of books published formerly in the NATO ASI Series. An electronic index to the NATO ASI Series provides full bibliographical references (with keywords and/or abstracts) to more than 50000 contributions from international scientists published in all sections of the NATO ASI Series.
Access to the NATO-PCO-DATA BASE is possible via CD-ROM "NATO-PCO-DATA BASE" with user-friendly retrieval software in English, French and German (WTV GmbH and DATAWARE Technologies Inc. 1989).

The CD-ROM of the NATO ASI Series can be ordered from: PCO, Overijse, Belgium

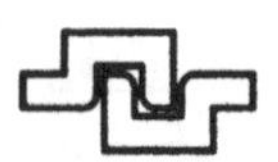

Series C: Mathematical and Physical Sciences – Vol. 565

Structure Formation in the Universe

edited by

Robert G. Crittenden

and

Neil G. Turok

University of Cambridge,
Cambridge, U.K.

Kluwer Academic Publishers

Dordrecht / Boston / London

Published in cooperation with NATO Scientific Affairs Division

Proceedings of the NATO Advanced Study Institute on
Structure Formation in the Universe
Cambridge, U.K.
26 July - 6 August 1999

A C.I.P. Catalogue record for this book is available from the Library of Congress.

ISBN 1-4020-0155-X (HB)
ISBN 1-4020-0156-8 (PB)

Published by Kluwer Academic Publishers,
P.O. Box 17, 3300 AA Dordrecht, The Netherlands.

Sold and distributed in North, Central and South America
by Kluwer Academic Publishers,
101 Philip Drive, Norwell, MA 02061, U.S.A.

In all other countries, sold and distributed
by Kluwer Academic Publishers,
P.O. Box 322, 3300 AH Dordrecht, The Netherlands.

Printed on acid-free paper

Printed in the Netherlands.

CONTENTS

PREFACE

This book contains a series of lectures given at the NATO Advanced Study Institute (ASI) "Structure Formation in the Universe", held at the Isaac Newton Institute in Cambridge in August, 1999. The ASI was held at a critical juncture in the development of physical cosmology, when a flood of new data concerning the large scale structure of the Universe was just becoming available. There was an air of excitement and anticipation: would the standard theories fit the data, or would new ideas and models be required?

Cosmology has long been a field of common interest between East and West, with many seminal contributions made by scientists working in the former Soviet Union and Eastern bloc. A major aim of the ASI was to bring together scientists from across the world to discuss exciting recent developments and strengthen links. However, a few months before the meeting it appeared that it might have to be cancelled. The war in the former Yugoslavia escalated and NATO began a protracted bombing campaign against targets in Kosovo and Serbia. Many scientists felt uneasy about participating in a NATO-funded meeting in this situation.

After a great deal of discussion, it was agreed that the developing East-West conflict only heightened the need for further communication and that the school should go ahead as planned, but with a special session devoted to discussion of the legitimacy of NATO's actions. In the end, the bombings were halted well before the conference so the issue became less urgent. Nevertheless, an evening meeting was held at which a wide spectrum of views were expressed in an informative and constructive dialog.

If the political scene was exciting, the scientific scene was even more so. Over the previous year a consensus was emerging clearly favouring one theoretical model, a nearly flat universe with cold dark matter (CDM) and a cosmological constant, over its rivals, including standard CDM, open CDM, topological defects o r mixed dark matter. For the first time the jigsaw puzzle of different data from galaxies, Lyman alpha clouds, gravitational lensing and the cosmic microwave sky appeared to fit together in a consistent way.

Observationally, the cosmic microwave measurements were already threatening to steal the show. The TOCO results had recently appeared suggesting a peak in the angular power spectrum, and exciting new results were rumoured from the Boomerang and Maxima collaborations (although they only a ppeared in final form many months later). But many lectures focussed on the signatures in the cosmic microwave sky of various theoretical models.

In addition, two new surveys of galaxies and large scale structure were just commencing. The 2-degree field survey and the Sloan Digital Sky survey promised an order of magnitude more data on galaxies and how they were distributed in the universe. These surveys and their potential impact on cosmology were addressed in detail at the meeting, a long with other possible probes of the matter distribution such as gravitational lensing.

The emergence of a tentative 'concordance model' was exciting but it was also recognised that the model itself left much to explain theoretically, especially the remarkable degree of fine tuning needed to obtain the desired cosmological constant, as well as the coincidences in the densities of other forms of matter. How could these be explained? There was much discussion of 'quintessence' models and how these might be made more natural.

Whilst the concordance model was certainly doing well in broad brush terms, cracks were also beginning to appear. The first was a hint from gravitational lensing that the cosmological constant might be smaller than that required to explain the high redshift supernovae data. The second was a topic of great interest at the meeting, the fact that cold dark matter models produce dwarf galaxy halos which are too numerous and too dense to be compatible with the observations. This discourse led to a flurry of theoretical papers after the meeting.

There were also a series of lectures on inflation and quantum cosmological theories of the initial conditions. These led to several lengthy debates revealing weak points in various approaches. Again these discussions had a profound impact on work performed by many researchers after the meeting and their influence is still being felt in the field.

The ASI was only the first in a series of events during a six month program at the Newton Institute, two of which deserve special mention. The first was a workshop on statistical tools in cosmology where the Boomerang test flight results were presented, and the second was an exciting summer school on brane cosmology, a topic then in its infancy. The workshop at the Newton Institute was critical in stimulating a burst of activity internationally in this field.

It should be noted that this book documents only the morning pedagogical lectures at the ASI, and so contains only a fraction of what was covered at the meeting. Of equal importance were the afternoon sessions, where younger scientists were able to present their recent research to their international colleagues. But as with any such school, the most valuable time was that spent in informal dialogs and in forming the lasting relationships which will span the scientific careers of those involved.

We would like to thank all those individuals who made this school a success and NATO for generously sponsoring this ASI. In particular, the staff of the Isaac Newton Institute were crucial in helping the school run smoothly. We also want to thank the other INI program coordinators, Valery Rubakov and Paul Steinhardt. Finally, the school and this book would not have been possible without the lecturers who volunteered their time (though in some cases, very reluctantly) in preparing, presenting and writing up the contributions contained in this volume.

Robert Crittenden
Neil Turok
July, 2001

INTRODUCTORY LECTURE

MARTIN J. REES
Institute of Astronomy
Madingley Road, Cambridge, CB3 0HA

Abstract. The agenda for this meeting is broad. In this brief introduction there is no time to address all the key topics that will be covered. Some comments are offered on current controversies, on the limits and prospects of cosmology in the coming decade, and on how the 'sociology' of our subject may change in the decades beyond that.

1. Preamble

First, a historical perspective:

What would a conference on 'cosmology' have been like in earlier decades? In the first half of the century the agenda would have been almost solely theoretical. The standard models date from the 1920s and 1930s; the Hubble expansion was first claimed in 1929, but not until the 1950s was there any prospect of discriminating among the various models, Indeed, there was even then little quantitative data on how closely any isotropic homogeneous model fitted the actual universe.

A cosmology meeting in the 1950s would have focussed on the question: is there evidence for evolution, or is the universe in a steady state? Key protagonists on the theoretical side would have included Hoyle and Bondi. Ryle would have been arguing that counts of radio sources – objects so powerful that many lay beyond the range of optical telescopes – already offered evidence for cosmic evolution; and Sandage would have advocated the potential of the Mount Palomar 200 inch telescope for extending the Hubble diagram far enough to probe the deceleration. Intimations from radio counts that the universe was indeed evolving, were strengthened after 1963 by the redshift data on quasars.

The modern era of physical cosmology of course began in 1965, when the discovery of the microwave background brought the early 'fireball phase'

R.G. Crittenden and N.G. Turok (eds.), Structure Formation in the Universe, 1–14.

into the realm of empirical science, and the basic physics of the 'hot big bang' was worked out. (The far earlier contributions by Gamow's associates, Alpher and Herman, continued, however, to be under-appreciated). There was also substantial theoretical work on anisotropic models, etc. Throughout the 1970s this evidence for the 'hot big bang' firmed up, as did the data on light elements, and their interpretation as primordial relics.

Theoretical advances in the 1980s gave momentum to the study of the ultra-early universe, and fostered the 'particle physics connection': the sociological mix of cosmologists changed. There was intense discussion of inflationary models, non-baryonic matter, and so forth.

Here we are in the late 1990s with a still larger and more diverse community. The pace of cosmology has never been faster. We are witnessing a crescendo of discoveries that promises to continue into the next millennium. This is because of a confluence of developments:

1. The microwave background fluctuations : these are now being probed with enough sensitivity to provide crucial tests of inflation and discrimination among different models.

2. The high-redshift universe: the Hubble Space Telescope (HST) has fulfilled its potential; two Keck Telescopes have come on line, along with the first VLT telescope, Subaru, and the first Gemini telescope. These have opened up the study of 'ordinary' galaxies right back to large redshifts, and to epochs when they were newly formed. In the coming year, three new X-ray telescopes will offer higher resolution and higher sensitivity for the study of distant galaxies and clusters.

3. Large scale clustering and dynamics: big surveys currently in progress are leading to far larger and more reliable samples, from which we will be able to infer the quantitative details of how galaxies of different types are clustered, and how their motions deviate from Hubble flow. Simultaneously with this progress, there have been dramatic advances in computer simulations. These now incorporate realistic gas dynamics as well as gravity.

4. Developments in fundamental physics offer important new speculative insights, which will certainly figure prominently in our discussions of the ultra-early universe.

It is something of a coincidence – of technology and funding – that the impetus on all these fronts has been more or less concurrent.

Max Planck claimed that theories are never abandoned until their proponents are all dead: that is too cynical, even in cosmology! Some debates have been settled; some earlier issues are no longer controversial; some of us change our minds (quite frequently, sometimes). And as the consensus advances, new questions which could not even have been posed in earlier decades are now being debated. This conference's agenda is therefore a 'snapshot' of evolving knowledge, opinion and speculation.

Consider the following set of assertions – a typical utterance of the r-m-s cosmologist whom you might encounter on a Cambridge street:

Our universe is expanding
from a hot big bang
in which the light elements were synthesised.
There was a period of inflation,
which led to a 'flat' universe today.
Structure was 'seeded' by Gaussian irregularities,
which are the relics of quantum fluctuations,
and the large-scale dynamics is dominated by 'cold' dark matter.
but Λ (or quintessence) is dynamically dominant.

I've written it like nine lines of 'free verse' to highlight how some claims are now quite firm, but others are fragile and tentative. Line one is now a quite ancient belief: it would have rated 99 percent confidence for several decades. Line two represents more recent history, but would now be believed almost equally strongly. So, probably, would line three, owing to the improved observations of abundances, together with refinements in the theory of cosmic nucleosynthesis. The concept of inflation is now 20 years old; most cosmologists suspect that it is was indeed a crucial formative process in the ultra-early universe, and this conference testifies to the intense and sophisticated theorising that it still stimulates.

Lower down in my list of statements, the confidence level drops below 50 percent. The 'stock' in some items – for instance CDM models , which have had several 'deaths' and as many 'resurrections' – is volatile, fluctuating year by year! The most spectacular 'growth stock' now (but for how long?) is the cosmological constant, lambda.

2. The Cosmological Numbers

Traditionally, cosmology was the quest for a few numbers. The first were H, q, and Λ. Since 1965 we have had another : the baryon/photon ratio. This is believed to result from a small favouritism for matter over antimatter in the early universe – something that was addressed in the context of 'grand unified theories' in the 1970s. (Indeed, baryon non-conservation seems a prerequisite for any plausible inflationary model. Our entire observable universe, containing at least 10^{79} baryons, could not have inflated from something microscopic if baryon number were strictly conserved)

In the 1980s non-baryonic matter became almost a natural expectation, and $\Omega_b/\Omega_{\rm CDM}$ is another fundamental number .

Another specially important dimensionless number, Q, tells us how smooth the universe is. It's measured by

— The Sachs-Wolfe fluctuations in the microwave background

— the gravitational binding energy of clusters as a fraction of their rest mass

- or by the square of the typical scale of mass- clustering as a fraction of the Hubble scale.

It's of course oversimplified to represent this by a single number Q, but insofar as one can, its value is pinned down to be 10^{-5}. (Detailed discussions introduce further numbers: the ratio of scalar and tensor amplitudes, and quantities such as the 'tilt', which measure the deviation from a pure scale-independent Harrison-Zeldovich spectrum.)

What is crucial is that Q is small. Numbers like Ω and H are only well-defined insofar as the universe possesses 'broad brush' homogeneity – so that our observational horizon encompasses many independent patches each big enough to be a fair sample. This would not be so, and the simple Friedmann models would not be useful approximations, if Q were not much less than unity. Q's smallness is necessary if the universe is to look homogeneous. But it is not, strictly speaking, a sufficient condition – a luminous tracer that didn't weigh much could be correlated on much larger scales without perturbing the metric. Simple fractal models for the luminous matter are nonetheless, as Lahav will discuss, strongly constrained by other observations such as the isotropy of the X-ray background, and of the radio sources detected in deep surveys.

3. How confident can we be of our models?

If our universe has indeed expanded, Friedmann-style, from an exceedingly high density, then after the first 10^{-12} seconds energies are within the range of accelerators. After the first millisecond – after the quark-hadron transition – conditions are so firmly within the realm of laboratory tests that there are no crucial uncertainties in the microphysics (though we should maybe leave our minds at least ajar to the possibility that the constants may still be time-dependent). And everything is still fairly uniform – perturbations are still in the linear regime.

It's easy to make quantitative predictions that pertain to this intermediate era, stretching from a millisecond to a million years. And we have now got high-quality data to confront them with. The marvellous COBE 'black body' pins down the microwave background spectrum to a part in 10,000. The 'hot big bang' has lived dangerously for thirty years: it could have been shot down by (for instance) the discovery of a nebula with zero helium, or of a stable neutrino with keV mass; but nothing like this has happened. The debate (concurrence or crisis?) now focuses on 1 per cent effects in helium spectroscopy, and on traces of deuterium at very high redshifts. The case for extrapolating back to a millisecond is now compelling and battle-tested. Insofar as there is a 'standard model' in cosmology, this is now surely part of it.

When the primordial plasma recombined, after half a million years, the black body radiation shifted into the infrared, and the universe entered, literally, a dark age. This lasted until the first stars lit it up again. The basic microphysics remains, of course, straightforward. But once non-linearities develop and bound systems form, gravity, gas dynamics, and the physics in every volume of Landau and Lifshitz, combine to unfold the complexities we see around us and are part of.

Gravity is crucial in two ways. It first amplifies 'linear' density contrasts in an expanding universe; it then provides a negative specific heat so that dissipative bound systems heat up further as they radiate. There is no thermodynamic paradox in evolving from an almost structureless fireball to the present cosmos, with huge temperature differences between the 3 degrees of the night sky, and the blazing surfaces of stars.

It is feasible to calculate all the key cosmic processes that occurred between (say) a millisecond and a few million years: the basic physics is 'standard' and (according at least to the favoured models) everything is linear. The later universe, after the dark age is over, is difficult for the same reason that all environmental sciences are difficult.

The whole evolving tapestry is, however, the outcome of initial conditions (and fundamental numbers) imprinted in the first microsecond – during the era of inflation and baryogenesis, and perhaps even on the Planck scale . This is the intellectual habitat of specialists in quantum gravity, superstrings, unified theories, and the rest.

So cosmology is a sort of hybrid science. It's a 'fundamental' science, just as particle physics is. But it's also the grandest of the environmental sciences. This distinction is useful, because it signals to us what levels of explanation we can reasonably expect. The first million years is described by a few parameters: these numbers (plus of course the basic physical laws) determine all that comes later. It's a realistic goal to pin down their values. But the cosmic environment of galaxies and clusters is now messy and complex – the observational data are very rich, but we can aspire only to an approximate, statistical, or even qualitative 'scenario', rather like in geology and paleontology.

The relativist Werner Israel likened this dichotomy to the contrast between chess and mudwrestling. The participants in this meeting would seem to him, perhaps. an ill-assorted mix of extreme refinement and extreme brutishness (just in intellectual style, of course!).

4. Complexities of Structure, and Dark Matter

4.1. PREHISTORY OF IDEAS ON STRUCTURE FORMATION

Since we are meeting in the Isaac Newton Institute, it's fitting to recall that Newton himself envisaged structures forming via 'gravitational instability'. In an often-quoted letter to Richard Bentley, the Master of Trinity College, he wrote:

"If all the matter of the universe were evenly scattered throughout all the heavens, and every particle had an innate gravity towards all the rest, and ... if the matter were evenly dispersed throughout an infinite space, it could never convene into one mass, but some of it would convene into one mass and some into another, so as to make an infinite number of great masses, scattered at great distances from one another throughout all that infinite space. And thus might the sun and fixed stars be formed. ... supposing the matter to be of a lucent nature."

(It would of course be wishful thinking to interpret his last remark as a premonition of dark matter!)

4.2. THE ROLE OF SIMULATIONS

Our view of cosmic evolution is, like Darwinism, a compelling general scheme. As with Darwinism, how the whole process got started is still a mystery. But cosmology is simpler because, once the starting point is given, the gross features are predictable. The whole course of evolution is not, as in biology, sensitive to 'accidents'. All large patches that start off the same way, end up statistically similar.

That's why simulations of structure formation are so important. These have achieved higher resolution, and incorporate gas dynamics and radiative effects as well as gravity. They show how density contrasts grow from small-amplitude beginnings; these lead, eventually, to bound gas clouds and to internal dissipation.

Things are then more problematical. We are baffled by the details of star formation now, even in the Orion Nebula. What chance is there, then, of understanding the first generation of stars, and the associated feedback effects? In CDM-type models, the very first stars form at redshifts of 10-20 when the background radiation provides a heat bath of up to 50 degrees, and there are no heavy elements. There may be no magnetic fields, and this also may affect the initial mass function. We also need to know the efficiency of star formation, and how it depends on the depth of the potential wells in the first structures.

Because these problems are too daunting to simulate *ab initio*, we depend on parameter-fitting guided by observations. And the spectacular recent progress from 10-metre class ground based telescopes and the HST has been extraordinarily important here.

4.3. OBSERVING HIGH REDSHIFTS

We are used to quasars at very high redshifts. But quasars are rare and atypical – we would really like to know the history of matter in general. One of the most important advances in recent years has been the detection of many hundreds of galaxies at redshifts up to (and even beyond) 5. Absorption due to the hundreds of clouds along the line of sight to quasars probes the history of cosmic gas in exquisite detail, just as a core in the Greenland ice-sheet probes the history of Earth's climate.

Quasar activity reaches a peak at around $z = 2.5$. The rate of star formation may peak at somewhat smaller redshifts (even though the very first starlight appeared much earlier) But for at least the last half of its history, our universe has been getting dimmer. Gas gets incorporated in galaxies and 'used up' in stars – galaxies mature, black holes in their centres undergo fewer mergers and are starved of fuel, so AGN activity diminishes.

That, at least, is the scenario that most cosmologists accept. To fill in the details will need better simulations. But, even more, it will need better observations. I don't think there is much hope of 'predicting' or modelling the huge dynamic range and intricate feedback processes involved in star formation. A decade from now, when the Next Generation Space Telescope (NGST) flies, we may know the main cosmological parameters, and have exact simulations of how the dark matter clusters. But reliable knowledge of how stars form, when the intergalactic gas is reheated, and how bright the first 'pregalaxies' are will still depend on observations. The aim is get a consistent model that matches not only all we know about galaxies at the present cosmic epoch, but also the increasingly detailed snapshots of what they looked like, and how they were clustered, at all earlier times.

But don't be too gloomy about the messiness of the 'recent' universe. There are some 'cleaner' tests. Simulations can reliably predict the present clustering and large-scale distribution of non-dissipative dark matter. This can be observationally probed by weak lensing, large scale streaming, and so forth, and checked for consistency with the CMB fluctuations, which probe the linear precursors of these structures.

4.4. DARK MATTER: WHAT, AND HOW MUCH?

The nature of the dark matter – how much there is and what it is – still eludes us. It's embarrassing that 90 percent of the universe remains unaccounted for.

This key question may yield to a three-pronged attack:

1. Direct detection. Astronomical searches are underway for 'machos' in the Galactic Halo; and several groups are developing cryogenic detectors for supersymmetric particles and axions.

2. Progress in particle physics. Important recent measurements suggest that neutrinos have non-zero masses; this result has crucially important implications for physics beyond the standard model; however the inferred masses seem too low to be cosmologically important. If theorists could pin down the properties of supersymmetric particles, the number of particles that survive from the big bang could be calculated just as we now calculate the helium and deuterium made in the first three minutes. Optimists may hope for progress on still more exotic options.

3. Simulations of galaxy formation and large-scale structure. When and how galaxies form, the way they are clustered, and the density profiles within individual systems, depend on what their gravitationally-dominant constituent is, and are now severely constraining the options.

5. Steps Beyond the Simplest Universe: Open Models, Λ, etc.

5.1. THE CASE FOR $\Omega < 1$

Everyone agrees that the 'simplest' universe would be a flat Einstein-de Sitter model. But we shall hear several claims during the present meeting that this model is now hard to reconcile with the data. Several lines of evidence suggest that gravitating CDM contributes substantially less than $\Omega_{\rm CDM} = 1$. The main lines of evidence are

(i) The baryon fraction in clusters is 0.15-0.2, On the other hand, the baryon contribution to omega is now pinned down by deuterium measurements to be around $\Omega_b = 0.015h^2$, where h is the Hubble constant in units of 100 km/sec/Mpc. If clusters are a fair sample of the universe, then this is incompatible with a dark matter density high enough to make $\Omega = 1$.

(ii) The presence of clusters of galaxies with $z = 1$ is hard to reconcile with the rapid recent growth of structure that would be expected if $\Omega_{\rm CDM}$ were unity.

(iii) The Supernova Hubble diagram (even though the case for actual acceleration may not be compelling) seems hard to reconcile with the large deceleration implied by an Einstein-de Sitter model.

(iv) The inferred ages of the oldest stars are only barely consistent with an Einstein-de Sitter model, for the favoured choices of Hubble constant.

5.2. OPEN UNIVERSE, OR VACUUM ENERGY?

The two currently-favoured options seem to be:

(A) an open model, or else

(B) a flat model where vacuum energy (or some negative-pressure component that didn't participate in clustering) makes up the balance.

If the universe is a more complicated place than some people hoped, which of these options is the more palatable? Opinions here may differ: How 'contrived' are the open-inflation models? Is it even more contrived that the vacuum-energy should have the specific small value that leads it to start dominating just at the present epoch?

Either of these models involves a specific large number. In case (A) this is the ratio of the Robertson Walker curvature scale to the Planck scale; in (B) it is the ratio of vacuum energy to some other (much higher) energy density. At present, (A) seems to accord less well than (B) with the data. In particular, the angular scale of the 'doppler peaks' in the CMB angular fluctuations seems to favour a flat universe; and the supernova Hubble diagram indicates an actual acceleration, rather than merely a slight deceleration (as would be expected in the open model).

We will certainly hear a great deal about the mounting evidence for Λ (or one of its time-dependent generalisations): the claimed best fit to all current data suggests a non-zero energy in the vacuum. However we should be mindful of the current large scatter in all CMB measurements relevant to the doppler peak, and the various uncertainties (especially those that depend on composition, etc.) in the supernovae from which a cosmic acceleration has been inferred. I think the jury is still out. However, CMB experiments are developing fast, and the high-z supernova sample is expanding fast too; so within two years we should know whether there is a vacuum energy, or whether systematic intrinsic differences between high-z and low-z supernovae are large enough to render the claims spurious. (On the same timescale we should learn whether the Universe actually is flat).

5.3. THE HISTORY OF Λ

I wouldn't venture bets on the final status of Λ. It is nonetheless interesting to recall its history. Λ was of course introduced by Einstein in 1917 to permit a static unbounded universe. After 1929, the cosmic expansion rendered Einstein's motivation irrelevant. However, by that time de Sitter had already proposed his expanding Λ-dominated model. In the 1930s, Eddington and Lemaitre proposed that the universe had expanded (under the action of the Λ-induced repulsion) from an initial Einstein state. Λ fell from favour after the 1930s: relativists disliked it as a 'field' acting on everything but acted on by nothing. A brief resurgence in the late 1960s was triggered by a (now discredited) claim for a pile-up in the redshifts of quasars at a value of z slightly below 2. (The CMB had already convinced most people that the universe emerged from a dense state, rather than from an Einstein static model, but it could have gone through a coasting or loitering phase where the expansion almost halted. A large range of affine distance would then correspond to a small range of redshifts, thereby accounting for a 'pile up' at a particular redshift. It was also noted that this model offered more opportunity for small-amplitude perturbations to grow.)

The 'modern' interest in Λ stems from its interpretation as an vacuum energy. This leads to the reverse problem: Why is $\Lambda \sim 120$ powers of 10 smaller than its 'natural' value, even though the effective vacuum density must have been very high in order to drive inflation? The interest has of course been hugely boosted recently, through the claims that the Hubble diagram for Type 1A supernovae indicates an acceleration.

(If Λ is fully resurrected, it will be a great 'coup' for de Sitter. His model, dating from the 1920s, not only describes the dynamics throughout the huge number of 'e-foldings' during inflation, but also describes future aeons of our cosmos with increasing accuracy. Only for the 50–odd decades of logarithmic time between the end of inflation and the present does it need modification!).

6. Inflation and the Very Early Universe

Numbers like Ω, $(\Omega_b/\Omega_{\rm CDM})$, Λ and Q are determined by physics as surely as the He and D abundances – it's just that the conditions at the ultra-early eras when these numbers were fixed are far beyond anything we can experiment on, so the relevant physics is itself still conjectural.

The inflation concept is the most important single idea. It suggests why the universe is so large and uniform – indeed, it suggests why it is expanding. It was compellingly attractive when first proposed, and most cosmologists (with a few eminent exceptions like Roger Penrose) would bet that it is, in some form, part of the grand cosmic scheme. The details are still unsettled. Indeed, cynics may feel that, since the early 1980s, there have been so many transmogrifications of inflation – old, new, chaotic, eternal, and open – that its predictive power is much eroded. (But here again extreme cynicism is unfair.)

We'll be hearing some discussion of whether inflationary models can 'naturally' account for the fluctuation amplitude $Q = 10^{-5}$; and, more controversially, whether it's plausible to have a non-flat universe, or a present-day vacuum energy in the permissible range. It's important to be clear about the methodology and scientific status of such discussion. I comment with great diffidence, because I'm not an expert here.

This strand of cosmology may still have unsure foundations, but it isn't just metaphysics: one can test particular variants of inflation. For instance, definite assumptions about the physics of the inflationary era have calculable consequences for the fluctuations –whether they're Gaussian, the ratio of scalar and tensor modes, and so forth – which can be probed by observing large scale structure and, even better, by microwave background observations. Cosmologists observe, stretched across the sky, giant protostructures that are the outcome of quantum fluctuations imprinted when the temperature was 10^{15} GeV or above. Measurements with the MAP and Planck/Surveyor spacecraft will surely tell us things about 'grand unified' physics that can't be directly inferred from ordinary-energy experiments.

7. The Agenda 10 Years From Now: a Bifurcated Community?

7.1. THE NEXT FIVE YEARS

The current pace of advance is such that within five years we'll surely have made substantial further progress. We will not only agree that the value of H is known to 10 percent – we'll agree what that value is.

We'll know the key parameters (from high-z supernovae, from the CMB, from high-z observations, and from improved statistics on large scale clustering and streaming. I'd even bet (though maybe I'm being a bit rash here) that we'll know what the dominant dark matter is.

7.2. TEN YEARS AHEAD?

If we were to reconvene 10 years from now, what would be the 'hot topics' on the agenda? The key numbers specifying our universe – its geometry, fluctuations and content – may by then have been pinned down. I've heard people claim that cosmology will thereafter be less interesting– that the most important issues will be settled, leaving only the secondary drudgery of clearing up some details. I'd like to spend a moment trying to counter that view.

It may turn out, of course, that the new data don't fit at all into the parameter-space that these numbers are derived from. (I was tempted to describe this view as 'pessimistic' but of course some people may prefer to live in a more complicated and challenging universe!). But maybe everything will fit the framework, and we will pin down the contributions to from baryons, CDM, and the vacuum, along with the amplitude and tilt of the fluctuations, and so forth. If that happens, it will signal a great triumph for cosmology – we will know the 'measure of our universe' just as, over the last few centuries, we've learnt the size and shape of our Earth and Sun.

Our focus will then be redirected towards new challenges, as great as the earlier ones. But the character and 'sociology' of our subject will change: it will bifurcate into two sub-disciplines. This bifurcation would be analogous to what actually happened in the field of general relativity 20-30 years ago. The 'heroic age' of general relativity – leading to the rigourous understanding of gravitational waves, black holes, and singularities – occurred the 1960s and early 1970s. Thereafter, the number of active researchers in 'classical' relativity declined (except maybe in computational aspects of the subject): most of the leading researchers shifted either towards astrophysically-motivated problems, or towards quantum gravity and 'fundamental' physics.

What will be the foci of the two divergent branches of 'post classical' cosmology we'll be pursuing a decade from now? One will be 'environmental cosmology' – understanding the evolution of structure, stars and galaxies. The other will focus on the fundamental physics of the ultra-early universe (pre-inflation, m-branes, multiverses, etc). A few words about each of these:

7.3. ENVIRONMENTAL COSMOLOGY: LONG RANGE PROSPECTS

One continuing challenge will be to explore the emergence of structure. This is a tractable problem until the first star (or other collapsed system) forms. But the huge dynamic range and uncertain feedback thereafter renders the phenomena too complex for any feasible simulation.

To illustrate the uncertainty, consider a basic question such as when the intergalactic gas was first photoionized.

There have been many detailed models, but essentially this requires one photon for each baryon (somewhat more, in fact, to compensate for recombinations). A hot (O or B) star produces, over its lifetime, $10^4 - 10^5$ photons for each of its constituent baryons; if a black hole forms via efficient accretion of baryons, the corresponding number is several times 10^6. Thus, only a small amount of material need collapse into such objects in order to provide enough to ionize all the remaining baryons. But the key questions, of course, are how efficiently O-B stars or black holes can form. This depends on the so-called 'initial mass function' (IMF), which determines how much mass goes into high mass stars (or black holes) compared with the amount going concurrently into lower-mass stars? The challenge of calculating the IMF – involving gas-dynamical and radiative transfer calculations over an enormous dynamic range – may not have been met even ten years from now. But even if we assume that it has the same form as now, there is the issue of feedback: do the first stars provide a heat input (via radiation, stellar winds and supernovae) that inhibits later ones from forming? More specifically, we can imagine two options; either (a) all the gas that falls into gravitationally bound clumps of CDM turns into stars; or (b) one percent turns into stars, whose winds and supernovae provide enough momentum and energy to expel the other 99 percent. In the first case, the 3-sigma peaks would suffice; in case (b) more typical (1.5 sigma) peaks would be needed, or else larger and deeper potential wells more able to retain the gas.

Even if the clustering of the CDM under gravity could be exactly modelled, along with the gas dynamics, then as soon as the first stars form we face major uncertainties that will still be a challenge to the petaflop simulations being carried out a decade from now.

7.4. PROBING THE PLANCK ERA AND 'BEYOND'

The second challenge would be to firm up the physics of the ultra-early universe. Perhaps the most 'modest' expectation would be a better understanding of the candidate dark matter particles: if the masses and cross-sections of supersymmetric particles were known, it should be possible to predict how many survive, and their contribution to Ω, with the same confidence as that with which we can compute primordial nucleosynthesis. Associated with such progress, we might expect a better understanding of how the baryon-antibaryon asymmetry arose, and the consequence for Ω_b.

A somewhat more ambitious goal would be to pin down the physics of inflation. Knowing parameters like Q, the tilt, and the scalar/tensor ratio will narrow down the range of options. The hope must be to make this physics as well established as the physics that prevails after the first millisecond.

One question that interests me specially is whether there are multiple big bangs, and which features of our actual universe are contingent rather than necessary. Could the others have different values of Q, or different Robertson-Walker curvature? Furthermore, will the 'final theory' determine uniquely what we call the fundamental constants of physics – particle masses and coupling constants? Are these 'constants' uniquely specified by some equation that we can eventually write down? Or are they in some sense accidental features of a phase transition as our universe cooled – secondary manifestations of some still deeper laws governing a whole ensemble of universes?

This might seem arcane stuff, disjoint from 'traditional' cosmology – or even from serious science. But my prejudice is to be open minded about ensembles of universe and suchlike. This makes a real difference to how I weigh the evidence and place my bets on rival models.

Rocky Kolb's highly readable history 'Blind Watchers of the Sky' reminds us of some fascinating debates that occurred 400 years ago. Kepler was upset to find that planetary orbits were elliptical. Circles were more beautiful – and simpler, with one parameter not two. But Newton later explained all orbits in terms of a universal law with just one parameter, G.

Had Kepler still been alive then, he'd surely have been joyfully reconciled to ellipses.

The parallel is obvious. The Einstein-de Sitter model seems to have fewer free parameters than any other. Models with low Ω, non-zero Λ, two kinds of dark matter, and the rest may seem ugly. But maybe this is our limited vision. Just as Earth follows an orbit that is no more special than it needed to be to make it habitable, so we may realise that our universe is just one of the anthropically-allowed members of a grander ensemble. So maybe we should go easy with Occam's razor and be wary of arguments that $\Omega = 1$ and $\Lambda = 0$ are *a priori* more natural and less ad hoc.

There's fortunately no time to sink further into these murky waters, so I'll briefly conclude.

A recent cosmology book (not written by anyone at this conference) was praised, in the publisher's blurb, for 'its thorough coverage of the inflammatory universe'. That was a misprint, of course. But maybe enough sparks will fly here in the next few days to make it seem a not inapt description.

The organisers have chosen a set of fascinating open questions. I suspect they'll still seem open at the end of this meeting, but we'll look forward to learning the balance of current opinion, and what bets people are prepared to place on the various options.

Part I
EARLY UNIVERSE PHYSICS

COSMIC INFLATION

ANDREAS ALBRECHT
U. C. Davis Department of Physics
One Shields Avenue, Davis, CA, 95616

Abstract. I review the current status of the theory of Cosmic Inflation. My discussion covers the motivation and implementation of the idea, as well as an analysis of recent successes and open questions. There is a special discussion of the physics of "cosmic coherence" in the primordial perturbations. The issue of coherence is at the heart of much of the success inflation theory has achieved at the hands of the new microwave background data. While much of this review should be useful to anyone seeking to update their knowledge of inflation theory, I have also made a point of including basic material directed at students just starting research.

1. Introduction

It has been almost 20 years since Guth's original paper fuelled considerable excitement in the cosmology community [1]. It was through this paper that many cosmologists saw the first glimmer of hope that certain deep mysteries of the Universe could actually be resolved through the idea of cosmic Inflation. Today Inflation theory has come a long way. While at first the driving forces were deep theoretical questions, now the most exciting developments have come at the hands of new cosmological observations. In the intervening period we have learned a lot about the predictions inflation makes for the state of the Universe today, and new technology is allowing these predictions to be tested to an ever growing degree of precision. The continual confrontation with new observational data has produced a string of striking successes for inflation which have increased our confidence that we are on the right track. Even so, there are key unanswered questions about the foundations of inflation which become ever more compelling in the face of growing observational successes.

R.G. Crittenden and N.G. Turok (eds.), Structure Formation in the Universe, 17–42.

2. Motivating Inflation: The Cosmological Features/Problems

It turns out that most of the cosmological "problems" that are usually introduced as a motivation for inflation are actually only "problems" if you take a very special perspective. I will take extra care to be clear about this by first presenting these issues simply as "features" of the standard Big Bang, and then discussing circumstances under which these features can be regarded as "problems".

2.1. THE STANDARD BIG BANG

To start with, I will introduce the basic tools and ideas of the standard "Big Bang" (SBB) cosmology. Students needing further background should consult a text such as Kolb and Turner [2]. An extensive discussion of inflation can be found in [3]. The SBB treats a nearly perfectly homogeneous and isotropic universe, which gives a good fit to present observations. The single dynamical parameter describing the broad features of the SBB is the "scale factor" a, which obeys the "Friedmann equation"

$$\left(\frac{\dot{a}}{a}\right)^2 \equiv H^2 = \frac{8\pi}{3}\rho - \frac{k}{a^2} \tag{1}$$

in units where $M_P = \hbar = c = 1$, ρ is the energy density and k is the curvature. The Friedmann equation can be solved for $a(t)$ once $\rho(a)$ is determined. This can be done using local energy conservation, which for the SBB cosmology reads

$$\frac{d}{da}\left(\rho a^3\right) = -3p\frac{d}{da}a^3. \tag{2}$$

Here (in the comoving frame) the stress energy tensor of the matter is given by

$$T^\mu_\nu = Diag(\rho, p, p, p). \tag{3}$$

In the SBB, the Universe is first dominated by relativistic matter ("radiation dominated") with $w = 1/3$ which gives $\rho \propto a^{-4}$ and $a \propto t^{1/2}$. Later the Universe is dominated by non-relativistic matter ("matter dominated") with $w = 0$ which gives $\rho \propto a^{-3}$ and $a \propto t^{2/3}$.

The scale factor measures the overall expansion of the Universe (it doubles in size as the separation between distant objects doubles). Current data suggests an additional "Cosmological Constant" term $\equiv \Lambda/3$ might be present on the right hand side of the Friedmann equation with a size similar to the other terms. However with the ρ and k terms evolving as negative powers of a these terms completely dominate over Λ at the earlier epochs we are discussing here. We set $\Lambda = 0$ for the rest of this article.

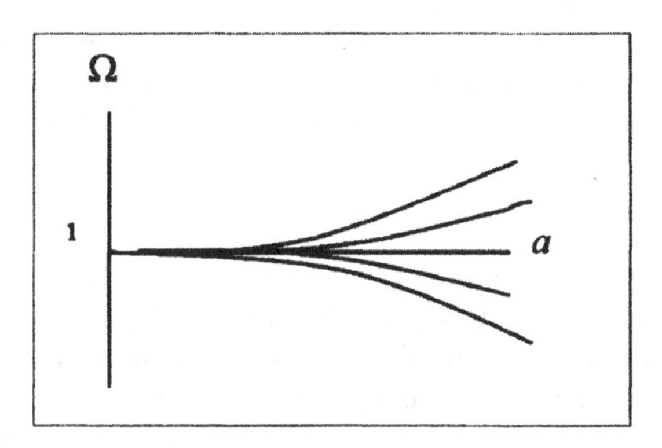

Figure 1. In the SBB $\Omega(a)$ tends to evolve away from unity as the Universe expands

2.2. THE FLATNESS FEATURE

The "critical density", ρ_c, is defined by

$$H^2 = \left(\frac{\dot{a}}{a}\right)^2 = \frac{8\pi}{3}\rho_c. \tag{4}$$

A universe with $k = 0$ has $\rho = \rho_c$ and is said to be "flat". It is useful to define the dimensionless density parameter

$$\Omega \equiv \frac{\rho}{\rho_c}. \tag{5}$$

If Ω is close to unity the ρ term dominates in the Friedmann equation and the Universe is nearly flat. If Ω deviates significantly from unity the k term (the "curvature") is dominant.

The Flatness feature stems from the fact that $\Omega = 1$ is an unstable point in the evolution of the Universe. Because $\rho \propto a^{-3}$ or a^{-4} throughout the history of the Universe, the ρ term in the Friedmann equation falls away much more quickly than the k/a^2 term as the Universe expands, and the k/a^2 comes to dominate. This behavior is illustrated in Fig. 1.

Despite the strong tendency for the equations to drive the Universe away from critical density, the value of Ω today is remarkably close to unity even after 15 Billion years of evolution. Today the value of Ω is within an order of magnitude of unity, and that means that at early times ρ must have taken values that were set extremely closely to ρ_c. For example at the epoch of Grand Unified Theories (GUTs) ($T \approx 10^{16} GeV$), ρ has to equal ρ_c to around 55 decimal places. This is simply an important property ρ must have for the SBB to fit the current observations, and at this point we merely take it as a feature (the "Flatness Feature") of the SBB.

2.3. THE HOMOGENEITY FEATURE

The Universe today is strikingly homogeneous on large scales. Observations of the Cosmic Microwave Background (CMB) anisotropies show that the Universe was even more inhomogeneous in the distant past. Because the processes of gravitational collapse causes inhomogeneities to *grow* with time, an even greater degree of homogeneity must have been present at earlier epochs.

As with the flatness feature, the equations are driving the Universe into a state (very inhomogeneous) that is very different from the way it is observed today. This difference is present despite the fact that the equations have had a very long time to act. To arrange this situation, very special (highly homogeneous) initial conditions must be chosen, so that excessive inhomogeneities are not produced.

One might note that at early times the Universe was a hot relativistic plasma in local equilibrium and ask: Why can't equilibration processes smooth out any fluctuations and produce the necessary homogeneity? This idea does not work because equilibration and equilibrium are only possible on scales smaller than the "Jeans Length" (R_J). On larger scales gravitational collapse wins over pressure and drives the system into an inhomogeneous non-equilibrium state.

At early epochs the region of the Universe we observe today occupied many Jeans volumes. For example, at the epoch of Grand Unification the observed Universe contained around 10^{80} Jeans volumes (each of which would correspond to a scale of only several *cm* today). So the tendency toward gravitational collapse has had a chance to have a great impact on the Universe over the course of its history. Only through extremely precisely determined initial conditions can the amplitudes of the resulting inhomogeneities be brought down to acceptable levels. At the GUT epoch, for example, initial density contrasts

$$\delta(x) \equiv (\rho(x) - \bar{\rho})/\bar{\rho} \tag{6}$$

that are non-zero only in the 20th, 40th or even later decimal places, depending on the scale, are required to arrange the correct degree of homogeneity. So here we have another special feature, the "Homogeneity Feature" which is required to make the SBB consistent with the Universe that we observe.

2.4. THE HORIZON FEATURE

One can take any physical system and choose an "initial time" t_i. The state of the system at any later time is affected both by the state at t_i (the initial conditions) and by the subsequent evolution. At any finite time after t_i there are causal limits on how large a scale can be affected by the subsequent evolution (limited ultimately by the speed of light, but really by the actual propagation speeds in that particularly system, which can be much slower). So there are always sufficiently large scales which have not been affected by the subsequent evolution, and on which the state of the system is simply a reflection of the initial conditions.

This same concept can equally well be applied to the SBB. However, there is one very interesting difference: The SBB starts with an initial singularity where $a = 0$ and $\rho = \infty$ at a finite time in the past. This is a serious enough singularity that the equations cannot propagate the Universe through it. In the SBB the Universe must simply start at this initial singularity with chosen initial conditions. So while in a laboratory situation the definition of t_i might be arbitrary in many cases, there is an absolutely defined t_i in the SBB.

As with any system, the causal "horizon" of the Universe grows with time. Today, the region with which we are just coming into causal contact (by observing distant points in the Universe) is one "causal radius" in size, which means objects we see in opposite directions are two causal radii apart and have not yet come into causal contact. One can calculate the number of causal regions that filled the currently observed Universe at other times. In the early Universe the size of a causally connected region was roughly the Jeans length, and so at the Grand Unification epoch there were around 10^{80} causally disconnected regions in the volume that would evolve into the part of the Universe we currently can observe. This is the "Horizon Feature" of the SBB, and is depicted in Fig. 2

2.5. WHEN THE FEATURES BECOME PROBLEMS

I have discussed three features of the SBB that are usually presented as "problems". What can make a feature a problem? The Flatness and Homogeneity features describe the need to give the SBB very special initial conditions in order to match current data. Who cares? If you look at the typical laboratory comparison between theory and experiment, success is usually measured by whether or not theoretical equations of motion correctly describe the evolution of the system. The choice of initial conditions is usually made only in order to facilitate the comparison. By these standards, the SBB does not have any problems. The equations, with suitably

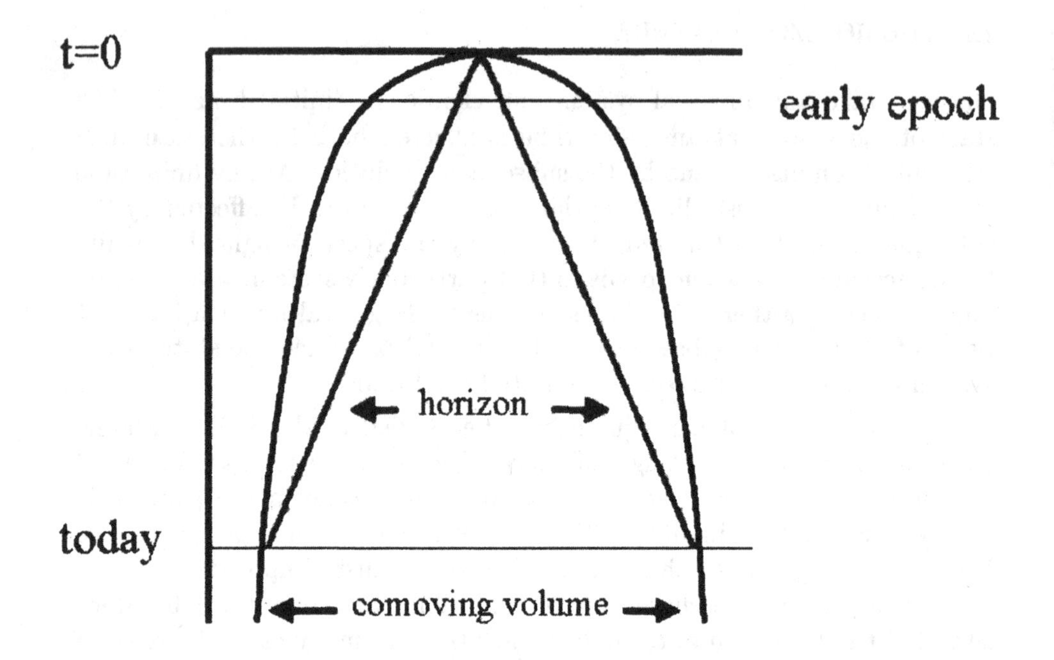

Figure 2. The region of the Universe we see today was composed of many causally disconnected regions in the past.

chosen initial conditions, do a perfectly good job of describing the evolution of the Universe.

However, there is something extremely strange about the initial conditions required by the SBB that makes us unable to simply accept the Flatness and Homogeneity features: The required initial conditions place the Universe very far away from where its equations of motion wish to take it. So much so that even today, 15 Billion years later, the Universe still has features (flatness and homogeneity) that the equations are trying to destroy.

I think a simple analogy is useful: Imagine that you come into my office and see a pencil balanced on its point. You notice it is still there the next day, the next week, and even a year later. You finally ask me "what is going on with the pencil?" to which I reply: "yes, that's pretty interesting, but I don't have to explain it because it was that way when I moved into my office". You would find such a response completely unreasonable. On the other hand, if the pencil was simply lying on my desk, you probably would not care at all how it got there.

The fact is that, like the fictitious balanced pencil, the initial conditions of the real Universe are so amazing that we cannot tear ourselves away from trying to understand what created them. It is at this point that the Flatness and Homogeneity features become "problems".

It is also at this point that the Horizon feature becomes a problem. The hope of a cosmologist who is trying to explain the special initial conditions is that some new physical processes can be identified which can "set up" these initial conditions. A natural place to hope for this new physics to occur is the Grand Unification (GUT) scale, where particle physics already indicates that something new will happen. However, once one starts thinking that way, one faces the fact that the Universe is compose of 10^{80} causally disconnected regions at the GUT epoch, and is seems impossible for physical processes to extend across these many disconnected regions and set up the special homogeneous initial conditions.

2.6. THE MONOPOLE PROBLEM

There is no way the Monopole problem can be regarded as merely a feature. In 1979 Preskill [4] showed how quite generally a significant number of magnetic monopoles would be produced during the cosmic phase transition required by Grand Unification. Being non relativistic ($\rho \propto a^{-3}$), they would rapidly come to dominate over the ordinary matter (which was then relativistic, $\rho \propto a^{-4}$). Thus it appeared that a Universe that is nearly 100% magnetic monopoles today was a robust prediction from GUT's, and thus the idea of Grand Unification was easily ruled out.

3. The Machinery of Inflation and what it can do

Over the years numerous inflationary "scenarios" have been proposed. My goal here is to introduce the basic ideas that underlie essentially all the scenarios.

3.1. THE POTENTIAL DOMINATED STATE

Consider adding an additional scalar matter field ϕ (the "inflaton") to the cosmological model. In general the stress-energy tensor of this field will depend on the space and time derivatives of the field as well as the potential $V(\phi)$. In the limit where the potential term dominates over the derivative terms the stress-energy tensor of ϕ takes the form

$$T^{\mu}_{\nu} = Diag(V, -V, -V, -V) \tag{7}$$

which leads (by comparison with Eqn. 3) to $\rho = -p = V$.

Solving Eqn. 2 gives $\rho(a) = V$. Because we are assuming $\dot{\phi}$ is negligible we also have $V =$ const.. Under these circumstances $H =$ const. and the

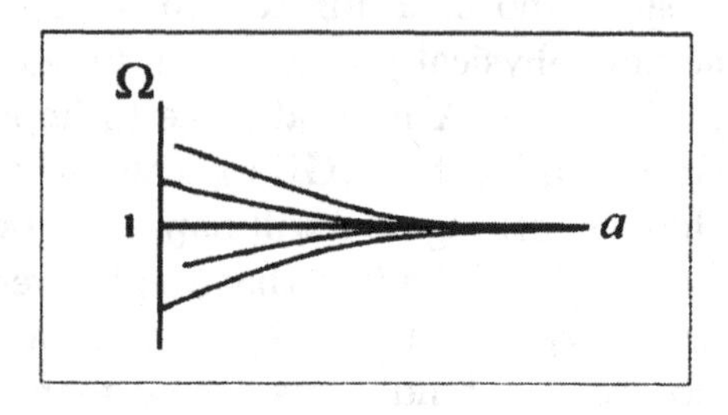

Figure 3. During inflation $\Omega = 1$ is an attractor

Friedmann equation gives exponential expansion:

$$a(t) \propto e^{Ht}, \tag{8}$$

very different from the power law behavior in the SBB.

3.2. A SIMPLE INFLATIONARY COSMOLOGY: SOLVING THE PROBLEMS

In the early days the inflationary cosmology was seen as a small modification of the SBB: One simply supposed that there was a period (typically around the GUT epoch) where the Universe temporarily entered a potential dominated state and "inflated" exponentially for a period of time. Things were arranged so that afterwards all the energy coupled out of $V(\phi)$ and into ordinary matter ("reheating") and the SBB continued normally.

This modification on the SBB has a profound effect on the cosmological problems. I will address each in turn:

Flatness: The flatness problem was phrased in terms of the tendency for the k term to dominate the ρ term in the Friedmann equation (Eqn. 1). During inflation with $\rho =$ const. the reverse is true and the k term becomes negligible, driving the Universe *toward* critical density and $\Omega = 1$. This process is illustrated in Fig. 3.

Horizon: A period of inflation radically changes the causality structure of the Universe. During inflation, each causally connected region is expanded exponentially. A suitable amount of inflation allows the entire

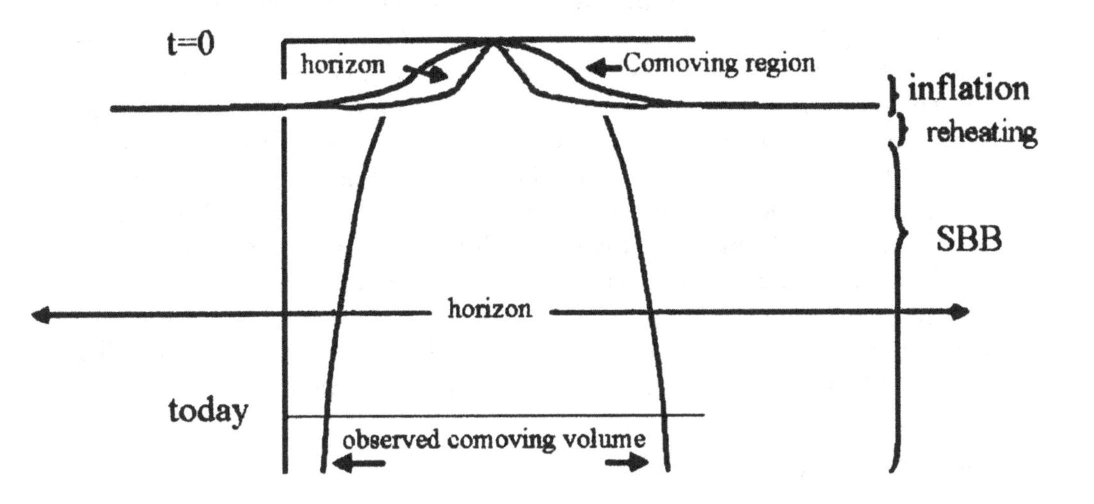

Figure 4. During inflation one horizon volume is exponentially expanded. Today, the observe region can occupy a small fraction of a causal volume.

observed Universe to come from a region that was causally connected before inflation, as depicted in Fig. 4. Typically one needs the Universe to expand by a factor of around e^{60} to solve the Horizon and Flatness problems with GUT epoch inflation.

Homogeneity: Of course "solving" the Horizon problem does not guarantee a successful picture. Bringing the entire observable Universe inside one causally connected domain in the distant past can give one *hope* that the initial conditions of the SBB can be explained by physical processes. Still, one must determine exactly what the relevant physical processes manage to accomplish. While the result of inflation is clearcut in terms of the flatness, it is much less so in terms of the homogeneity. The good news is that a given inflation model can actually make *predictions* for the spectrum of inhomogeneities that are present at the end of inflation. However, there is nothing intrinsic to inflation that predicts that these inhomogeneities are small. None the less, the mechanics of inflation can be further adjusted to give inhomogeneities of the right amplitude. Interestingly, inflation has a lot to say about other aspects of the early inhomogeneities, and these other aspects are testing out remarkably well in the face of new data. There will be more discussion of these matters in Sections 3.5 and 4.

Monopoles: The original Monopole problem occurs because GUTs produce magnetic monopoles at sufficiently high temperatures. In the SBB these monopoles "freeze out" in such high numbers as the Universe cools that they rapidly dominate over other matter (which is relativistic at that time). Inflation can get around this problem if the reheating after inflation does not reach temperatures high enough to produce the monopoles. During inflation, all the other matter (including any monopoles that may be present) is diluted to completely negligible densities and the overall density completely is dominated by $V(\phi)$. The ordinary matter of the SBB is created by the reheating process at the end of inflation. The key difference is that in a simple SBB the matter we see has in the past existed at all temperatures, right up to infinity at the initial singularity. With the introduction of an inflationary epoch, the matter around us has only existed up to a finite maximum temperature in the past (the reheating temperature). If this temperature is on the low side of the GUT temperature, monopoles will not be produced and the Monopole Problem is evaded.

3.3. MECHANISMS FOR PRODUCING INFLATION

Scalar fields play a key role in the standard mechanism of spontaneous symmetry breaking, which is widely regarded in particle physics as the fundamental origin of all particle masses. Essentially all attempts to establish a deeper picture of fundamental particles have introduced additional particles to the ones observed (typically to round out representations of larger symmetry groups used to unify the fundamental forces). Conflicts with observations are then typically avoided by giving the extra particles sufficiently large masses that they could not be produced in existing accelerators. Of course, this brings scalar fields into play.

The upshot is that additional scalar fields abound, at least in the imaginations of particle theorists, and if anything the problem for cosmologists has been that there are too many different models. It is difficult to put forward any one of them as the most compelling. This situation has caused the world of cosmology to regard the "inflaton" in a phenomenological way, simply investigating the behaviors of different inflaton potentials, and leaving the question of foundations to a time when the particle physics situation becomes clearer.

The challenge then is to account for at least 60 e-folding of inflation. Looking at the GUT scale (characterized by $M_G \approx M_P/1,000$) we can estimate $H = \sqrt{(8\pi/3)M_P^{-2}M_G^4}$, and a characteristic GUT timescale would be $t_g \approx M_G^{-1}$. So one could estimate a "natural" number of e-foldings for GUT-scale inflation to be $t_g \times H \approx \sqrt{8\pi/3}M_G/M_P$ where I have explicitly shown the Planck Mass for clarity. The upshot is that the "natural" number of e-foldings is a small fraction of unity, and there is the question of how to get sufficient inflation.

The "sufficient inflation" issue was easily addressed in the original Guth paper [1] because the end of inflation was brought on by an exponentially

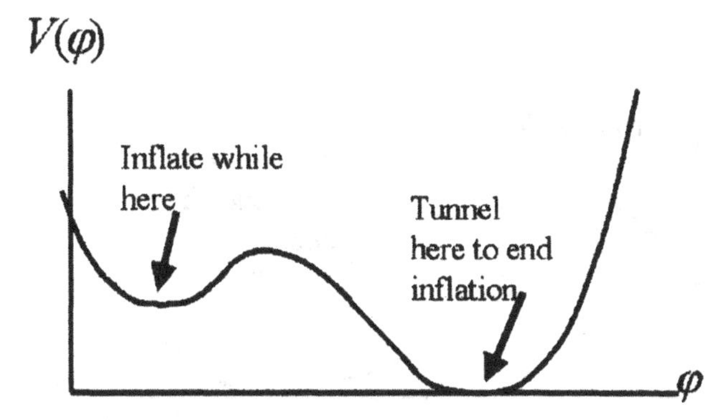

Figure 5. In the original version of inflation proposed by Guth, the potential dominated state was achieved in a local minimum of the inflaton potential. Long periods of inflation were easily produced because the timescale for inflation was set by a tunnelling process. However, these models had a "graceful exit problem" because the tunnelling process produced a very inhomogeneous energy distribution.

long tunnelling process with a timescale of the form $t \approx M_g^{-1} exp(B)$ where the dimensionless quantity B need take on only a modestly large value to provide sufficient inflation. In the Guth picture, the inflaton was trapped inside a classically stable local minimum of the inflation potential, and only quantum tunnelling processes could end inflation. This type of potential is depicted in Fig. 5. The original picture proposed by Guth had serious problems however, because the tunnelling operated via a process of bubble nucleation. This process was analyzed carefully by Guth and Weinberg [5] and it was show that reheating was problematic in these models: The bubbles formed with all the energy in their walls, and bubble collisions could not occur sufficiently rapidly to dissipate the energy in a more homogeneous way. The original model of inflation had a "graceful exit" problem that made its prediction completely incompatible with observations.

The graceful exit problem was resolved by the idea of "slow roll" inflation [6, 7]. In the slow roll picture inflation is *classically* unstable, as depicted in Fig. 6. Slow roll models are much easer to reheat and match onto the SBB. However, because the timescale of inflation is set by a classical instability, there is no exponential working to your advantage. Fine tuning of potential parameters is generally required to produce sufficient inflation in slow roll models. Essentially all current models of inflation use the slow roll mechanism. Interestingly, the constraints on the inflaton potential which are required to give the inhomogeneities a reasonable amplitude tend to exceed the constraints required to produce *sufficient* inflation. Thus the requirement of achieving sufficient inflation ends up in practice not providing any additional constraint on model building.

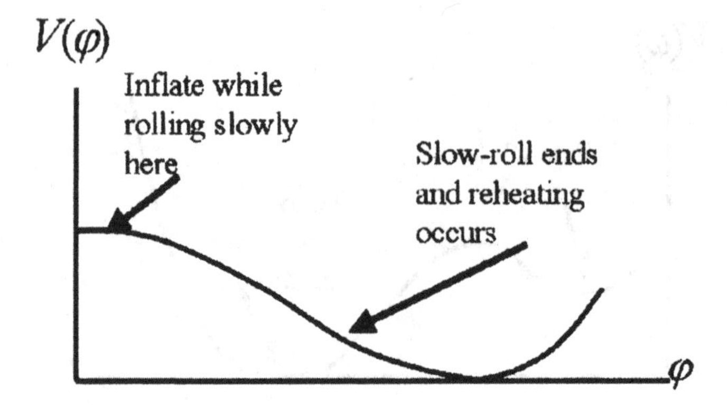

Figure 6. In "new" inflation a classical instability ends inflation. This allows for a graceful exit, but makes it more of a challenge to get sufficiently long inflation times.

I have outlined the tunnelling and slow roll pictures. While the slow roll remains the fundamental tool in all models, there are some variations that incorporate aspects of both. In some "open inflation" models [8] there is inflation with a tunnelling instability *and* a brief period of further inflation inside the bubble after tunnelling. This arrangement can produce values of Ω which are not necessarily close to unity. Also, inflation has been studied in a multi-dimensional inflaton space where there are some barriers and some slow roll directions.

3.4. THE START OF INFLATION

In the original basic pictures of inflation, where a period of inflation was inserted into a SBB model, various arguments were made about why the Universe *had* to enter an inflationary phase based on symmetry restoration at high temperatures [1, 6, 7]. However, starting with a SBB *before* inflation assumes a great deal about initial conditions (which after all we are trying to explain).

Currently cosmologists that think about these things take one of two perspectives. One group tries to treat "the most random possible" initial conditions [9], and discuss how inflation can emerge from such a state of matter. The other group tries to construct the "wavefunction of the Universe" based on deep principles [10]. In some cases the wavefunction of the Universe indicates that inflation is the most likely starting point for regions that take on SBB-like behavior. I have a personal prejudice against "principles" of initial conditions, because I do not see how that gets your further than simply postulating the initial conditions of the SBB.

Coming from the random initial conditions point of view, a popular argument is that even if the probability of starting inflation is extremely small in the primordial chaos, once inflation starts it takes over the Universe. When measured in terms of volume, the exponential expansion really does appear to take over. In fact, even though there is a classical instability, the small quantum probability of remaining on the flat inflationary part of the potential is amplified by the exponential inflationary growth, and in many models most of the Universe inflates forever (so-called "eternal inflation"). So perhaps as long as inflation is possible, one does not even have to think too hard about what went before.

All of these issues are plagued by key unresolved questions (for example to do with putting a measure on the large spaces one is contemplating). I will discuss some of the open questions associated with these issues in Section 6.

3.5. PERTURBATIONS

The technology for calculating the production of inhomogeneities during inflation is by now very well understood (For an excellent review see [11]. The natural variables to follow are the Fourier transformed density contrasts $\delta(k)$ in *comoving coordinates* ($k_{co} = ak_{phys}$). A given mode $\delta(k)$ behaves very differently depending on its relationship to the Hubble radius $R_H \equiv 1/H$. During inflation $H = const.$ while comoving scales are growing exponentially. The modes of interest start inside R_H ($k \gg R_H^{-1}$) and evolve outside R_H ($k < R_H^{-1}$) during inflation. After inflation, during the SBB, $R_H \propto t$ and $a \propto t^{1/2}$ or $t^{2/3}$ so R_H catches up with comoving scales and modes are said to "fall inside" the Hubble radius.

It is important to note that R_H is often called the "horizon". This is because in the SBB R_H is roughly equal to the distance light has travelled since the big bang. Inflation changes all that, but "Hubble radius" and "horizon" are still often (confusingly) used interchangeably.

Modes that are relevant to observed structures in the Universe started out on extremely small scales before inflation. By comparison with what we see in nature today, the only excitations one expects in these modes are zero-point fluctuations in the quantum fields. Even if they start out in an excited state, one might expect that on length-scales much smaller (or energies-scales much higher) than the scale ($\rho^{1/4}$) set by the energy density the modes would rapidly "equilibrate" to the ground state. These zero-point fluctuations are expanded to cosmic scales during inflation, and form the "initial conditions" that are used to calculate the perturbations today. Because the perturbations on cosmologically relevant scales must have been very small over most of the history of the Universe, linear perturbation theory in GR may be used. Thus the question of perturbations from inflation can be treated by a tractable calculation using well-defined initial conditions.

One can get a feeling for some key features by examining this result which holds for most inflationary models:

$$\delta_H = \frac{32}{75} \frac{16\pi}{M_p^6} \frac{V_*^3}{(V_*')^2}. \tag{9}$$

Here δ_H^2 is roughly the mean squared value of δ when the mode falls below R_H during the SBB. The "$*$" means evaluate the quantity when the mode in question goes outside R_H during inflation, and V is the inflaton potential. Since typically the inflaton is barely moving during inflation, δ_H^2 is nearly the same for all modes, resulting in a spectrum of perturbations that is called "nearly scale invariant". The cosmological data require $\delta_H \approx 10^{-5}$.

In typical models the "slow roll" condition,

$$V' M_P / V \ll 1 \tag{10}$$

is obeyed, and inflation ends when $V' M_P / V \approx 1$. If V is dominated by a single power of ϕ during the relevant period $V' M_P / V \approx 1$ means $\phi \approx M_P$ at the end of inflation. Since ϕ typically does not vary much during inflation $\phi \approx M_P$ would apply throughout. With these values of ϕ, and assuming simple power law forms for V, one might guess that the right side of Eqn. 9 is around unity (as opposed to 10^{-5}!) Clearly some subtlety is involved in constructing a successful model.

I have noticed people sometimes make rough dimensional arguments by inserting $V \approx M_G^4$ and $V' \approx M_G^3$ into Eqn. 9 and conclude $\delta_H \approx (M_G/M_P)^3$. This argument suggests that δ_H is naturally small for GUT inflation, and that the adjustments required to get $\delta_H \approx 10^{-5}$ could be modest. However that argument neglects the fact that the slow roll condition must be met. When this constraint is factored in, the challenge of achieving the right amplitude is much greater than the simplest dimensional argument would indicate.

3.6. FURTHER REMARKS

Inflation provides the first known mechanism for producing the flat, slightly perturbed initial conditions of the SBB. Despite the fact that there are many deep questions about inflation that have yet to be answered, it is still tremendous progress to know that nature has a way around the cosmological problems posed in Section 2. Whether nature has chosen this route is another question. Fortunately, inflation offers us a number of observational signatures that can help give an answer. These signatures are the subject of Sections 4 and 5.

4. Cosmic Coherence

In a typical inflation model the inflaton is completely out of the picture by the end of reheating and one is left with a SBB model set up with particular initial conditions. In realistic models the inhomogeneities are small at early times and can be treated in perturbation theory. Linear perturbation theory in the SBB has special properties that lead to a certain type of phase coherence in the perturbations. This coherence can be observed, especially in the microwave background anisotropies. Models where the perturbations are subject to only linear evolution at early times are call "passive" models

4.1. KEY INGREDIENTS

At sufficiently early times the Universe is hot and dense enough that photons interact frequently and are thus tightly coupled to the Baryons. Fourier modes of the density contrast $\delta_\gamma(k)$ in the resulting "photon-baryon fluid" have the following properties: On scales above the Jeans length the fluid experiences gravitational collapse. This is manifested by the presence of one growing and one decaying solution for δ. One can think of the growing solution corresponding to gravitational collapse and the decaying solution corresponding to the evolution of an expanding overdense region which sees its expansion slowed by gravity. During the radiation era Jeans length $R_J = R_H/\sqrt{3}$ and modes "fall inside" R_J just as they fall inside R_H. For wavelengths smaller than R_J the pressure counteracts gravity and instead of collapse the perturbations undergo oscillations, in the form of pressure (or "acoustic") waves. The experience of a given mode that starts outside R_J is to first experience gravitational collapse and then oscillatory behavior.

There is a later stage when the photons and Baryons decouple (which we can think of here in the instantaneous approximation). After decoupling the photons "free stream", interacting only gravitationally right up to the present day. The process of first undergoing collapse, followed by oscillation, is what creates the phase coherence. I will illustrate this mechanism first with a toy model.

4.2. A TOY MODEL

During the "outside R_J" regime, there is one growing and one decaying solution. The simplest system which has this qualitative behavior is the upside-down harmonic oscillator which obeys:

$$\ddot{q} = q. \tag{11}$$

The phase space trajectories are show on the left panel in Fig. 7. The system is unstable against the runaway (or growing) solution where $|q|$ and

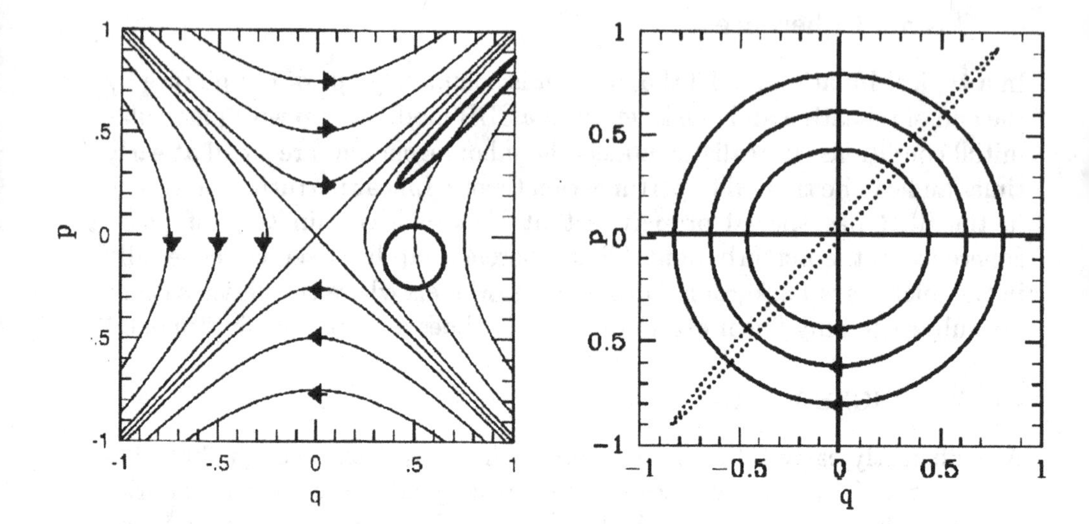

Figure 7. The phase space trajectories for an upside-down harmonic oscillator are depicted in the left panel. Any region of phase space will be squeezed along the diagonal line as the system evolves (i.e. the circle gets squeezed into the ellipse) . For a right-side-up harmonic oscillator paths in the phase space are circles, and angular position on the circle gives the phase of oscillation. Perturbations in the early universe exhibit first squeezing and then oscillatory behavior, and any initial phase space region will emerge into the oscillatory epoch in a form something like the dotted "cigar" due to the earlier squeezing. In this way the early period of squeezing fixes the phase of oscillation.

$|p|$ get arbitrarily large (and p and q have the same sign). This behavior "squeezes" any initial region in phase space toward the diagonal line with unit slope. The squeezing effect is illustrated by the circle which evolves, after a period of time, into the ellipse in Fig. 7.

The simplest system showing oscillatory behavior is the normal harmonic oscillator obeying

$$\ddot{q} = -q. \tag{12}$$

This phase space trajectories for this system are circles, as shown in the right panel of Fig. 7. The angular position around the circle corresponds to the phase of the oscillation. The effect of having first squeezing and then oscillation is to have just about any phase space region evolve into something like the dotted "cigar" in the right panel. The cigar then undergoes rotation in phase space, but the entire distribution has a fixed phase of oscillation (up to a sign). The degree of phase coherence (or inverse "cigar thickness") is extremely high in the real cosmological case because the relevant modes spend a long time in the squeezing regime.

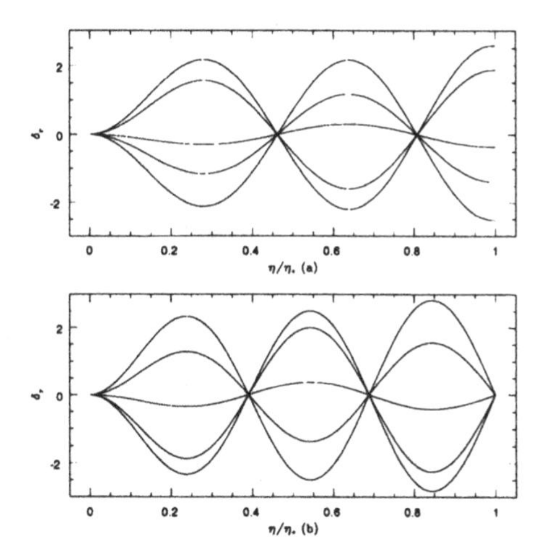

Figure 8. The evolution of δ_γ for two different wavelengths (upper and lower panels) as a function of time for an ensemble of initial conditions. Each wavelength shows an initial period of growth (squeezing) followed by oscillations. The initial squeezing fixes the phase of oscillation across the entire ensemble, but different wavelengths will have different phases. The two chosen wavelengths are maximally out of phase at η_* the time of last scattering.

4.3. THE RESULTS OF COHERENCE

Having presented the toy model, let as look at the behavior of the real cosmological variables. Figure 8 shows δ_γ (the photon fluctuations) as a function of (conformal) time η measured in units of η_*, the time at last scattering. Two different wavelengths are shown (top and bottom panels), and each panel shows several members of an ensemble of initial conditions. Each curve shows an early period of growth (squeezing) followed by oscillation. The onset of oscillation appears at different times for the two panels, as each mode "enters" R_J at a different time. As promised, because of the initial squeezing epoch all curves match onto the oscillatory behavior at the same phase of oscillation (up to a sign). Phases can be different for different wavelengths, as can be seen by comparing the two panels.

To a zeroth approximation the event of last scattering simply releases a snapshot of the photons at that moment and sends them free-streaming across nearly empty space. The left panel of Fig. 9 (solid curve) shows the mean squared photon perturbations at the time of last scattering in a standard inflationary model, vs k. Note how some wavenumbers have been caught at the nodes of their oscillations, while others have been caught at maxima. This feature is present despite the fact that the curve represents an *ensemble average* because the same phase is locked in for each member of the ensemble.

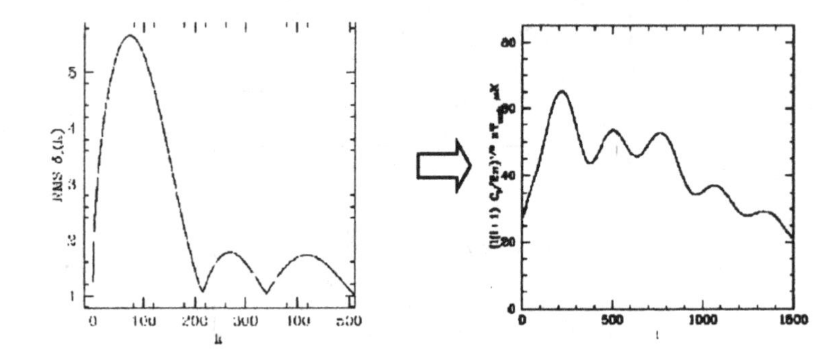

Figure 9. The left panel shows how temporal phase coherence manifests itself in the power spectrum for δ in Fourier space, shown here at one moment in time. Even after ensemble averaging, some wavelengths are caught at zeros of their oscillations while others are caught at maximum amplitudes. These features in the radiation density contrast δ are the root of the wiggles in the CMB angular power shown in the right panel.

The right panel of Fig. 9 shows a typical angular power spectrum of CMB anisotropies produced in an inflationary scenario. While the right hand plot is not exactly the same as the left one, it is closely related. The CMB anisotropy power is plotted vs. angular scale instead of Fourier mode, so the x-axis is "l" from spherical harmonics rather than k. The transition from k to l space, and the fact that other quantities besides δ_γ affect the anisotropies both serve to wash out the oscillations to some degree (there are no zeros on the right plot, for example). Still the extent to which there *are* oscillations in the CMB power is due to the coherence effects just discussed.

As our understanding of the inflationary predictions has developed, the defect models of cosmic structure formation have served as a useful contrast [12, 13]. In cosmic defect models there is an added matter component (the defects) that behaves in a highly nonlinear way, starting typically all the way back at the GUT epoch. This effectively adds a "random driving term" to the equations that is constantly driving the other perturbations. These models are called "active" models, in contrast to the passive models where

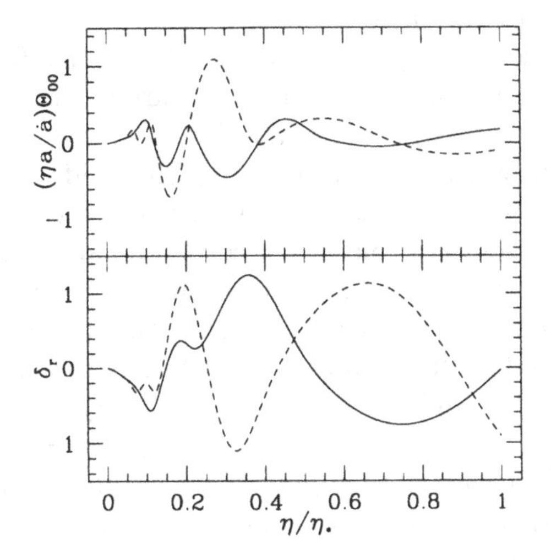

Figure 10. Decoherence in the defect models: The two curves in the upper panel show two different realizations of the effectively random driving force for a particular mode of δ_γ. The corresponding evolution of δ_γ is shown in the lower panel. There is no phase coherence from one member of the ensemble to the next.

all matter evolves in a linear way at early times. Figure 10 shows how despite the clear tendency to oscillate, the phase of oscillation is randomized by the driving force. In all known defect models the randomizing effect wins completely and there are no visible oscillations in the CMB power. This comparison will be discussed further in the next section.

5. Predictions

Given a region that has started inflating, and given the inflaton potential and the coupling of the inflaton to other matter, very detailed and unambiguous predictions can be made about the initial conditions of the SBB epoch that follows inflation. The catch is that at the moment we do not have a unique candidate for the inflaton and furthermore we are pretty confused in general about physics at the relevant energy scales.

Despite these uncertainties inflation makes a pretty powerful set of predictions. I will discuss these predictions here, but first it is worth discussing just what one might mean by "powerful" predictions. From time to time informal discussions of inflation degenerate into a debate about whether inflation will ever *really* be tested. The uncertainties about the inflaton mentioned above are the reason such debates can exist. As long as there is no single theory being tested, there is "wiggle room" for the theory to bend and adapt to pressure from the data.

Today there are certain broad expectations of what the inflaton sector might look like, and despite their breadth, these expectations lead to a set of highly testable predictions. However, radical departures from these broad expectations are also possible, and some will argue that such departures mean that one is never really testing inflation.

I have little patience for these arguments. The fundamental currency of scientific progress is the confirmation or rejections of specific ideas. There is no question that the standard picture of cosmic inflation commits itself to very specific predictions. To the extent that any one of these predictions is falsified, a real revolution will have to occur in our thinking. Furthermore, since most of these tests will be realized in the foreseeable future, the testing of inflation is not a philosophical question. It is a very real part of the field of cosmology.

To be absolutely clear on this matter, I have indicated in each topic below both the predictions of the standard picture and the possible exotic variations.

5.1. FLATNESS

The Standard Picture: In standard inflaton potentials a high degree of adjustment is required to achieve a suitably small amplitude for the density perturbations. It turns out that this adjustment also drives the overall time of inflation far (often exponentially) beyond the "minimal" 60 e-foldings. As a result, the total curvature of the Universe is precisely zero today in inflationary cosmologies, at least as foreseeable observations are concerned. Any evidence for a non-zero curvature at the precisions of realistic experiments will overturn the standard inflationary picture. The current data is consistent with a flat Universe as illustrated in Fig. 11.

Exotic variations: The possibility that an open Universe could emerge at the end of inflation has been considered for some time [15, 16] There exist very interesting models that combine ideas from old and new inflation to produce an open SBB at the end of inflation [8]. The old inflation ends by bubble nucleation, and there is subsequently a brief period of new inflation inside the bubble. The degree of openness is controlled by the shape of the inflaton potential, which determines the initial field values inside the bubble and the duration of inflation afterwards. These models suggest an interesting direction of study if the universe turns out to be open, but they also represent a picture that is radically different from the standard inflationary one.

5.2. COHERENCE

The standard picture The features in the CMB power spectrum produced by cosmic coherence (as illustrated in Fig 12) are very special signa-

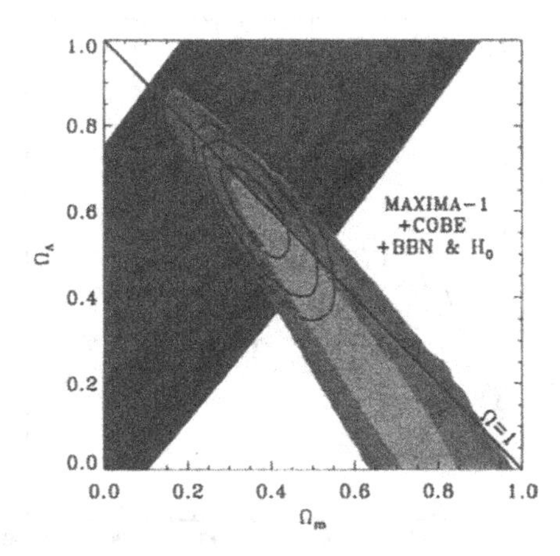

Figure 11. This is Fig. 4 from [14]. This shows that the Maxima and Supernova data consistent with a flat Universe (indicated by the diagonal $\Omega_m + \Omega_\Lambda = 1$ line).

tures of inflation (and other passive models). Similar features are predicted in the polarization anisotropy power spectrum and in the polarization-temperature cross-correlations. The absence of such features would overturn this standard picture. In light of this point, the presence of the so-called "first Doppler peak" in current data (Fig. 12) [17] reflects resounding support for the standard inflationary picture. Figure 12 also shows a cosmic defect based model [18, 19] that, through some pretty exotic physics has been manipulated to attempt the creation of a first Doppler peak. Although the broad shape is reproduced, the sharp peak and corresponding valley are not.

Exotic variations: It has been shown that under very extreme conditions, active perturbations could mimic the features produced by coherence in the inflationary models [20]. These "mimic models" achieved a similar effect by having the active sources produced impulses in the matter that are highly coherent and sharply peaked in time. Despite their efforts to mimic inflationary perturbations, these models have strikingly different signatures in the CMB polarization, and thus can be discriminated from inflationary models using future experiments. Also, there is no known physical mechanism that could produce the active sources required for the mimic models to work.

Another possibility is that the perturbations *are* produced by inflation but the oscillatory features in the CMB are hidden due to features introduced by a wiggly inflaton potential. It has been shown [21] that this sort of (extreme) effect would eventually turn up in CMB polarization and temperature-polarization cross-correlation measurements.

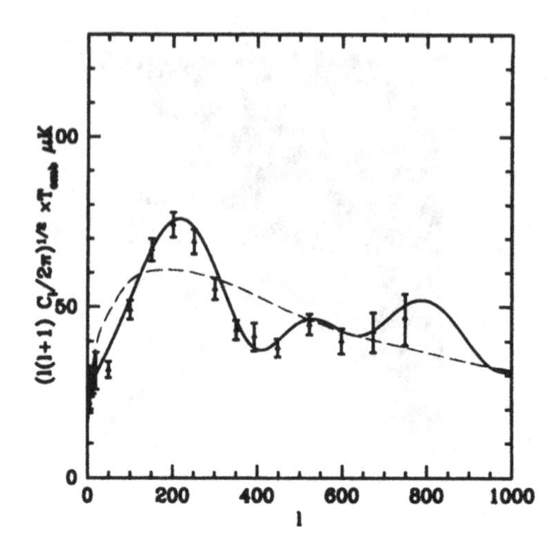

Figure 12. A current compilation of CMB anisotropy data with curves from an inflationary model (solid) and an active model (dashed). The sharp peak, a result of the phase coherence in the inflation model is impossible for realistic active models to produce.

Another possibility is that inflation solves the flatness, homogeneity and horizon problems, but generates perturbations at an unobservably low amplitude. The perturbations then must be generated by some other (presumably "active") mechanism. Currently there is no good candidate active mechanism to play this role [19], but if there was, it would certainly generate clear signatures in the CMB polarization and temperature-polarization cross-correlation measurements.

I also note here that a time varying speed of light (VSL) has been proposed as an alternative to inflation [22, 23]. So far all know realizations of this idea have produced a highly homogeneous universe which also must be perturbed by some subsequent mechanism. The building evidence against active mechanisms of structure formation are also weakening the case for the VSL idea.

5.3. PERTURBATION SPECTRUM AND OTHER SPECIFICS

One clear prediction from the standard picture is the "nearly scale invariant" spectrum ($\delta_H(k) \approx$ const.) discussed in section 3.5. Strong deviations (more than several percent) from a scale invariant spectrum would destroy our standard picture. Fortunately for inflation, there current CMB and other cosmological data gives substantial support to the idea that the primordial spectrum was indeed close to scale invariant. Current data is consistent with a nearly scale invariant spectrum, as illustrated for example in Fig 13.

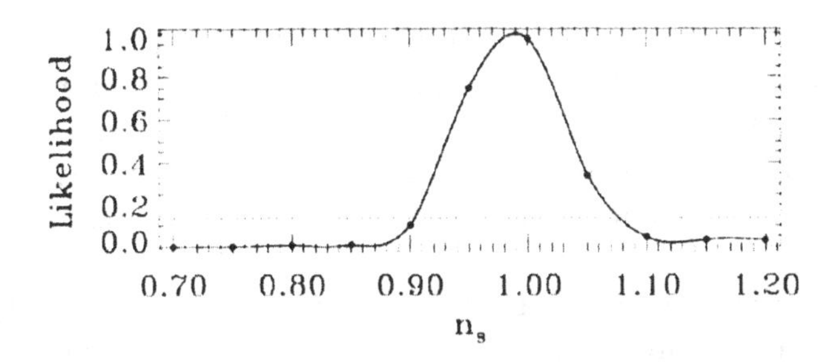

Figure 13. This is Fig. 2 from [14]. This shows that the Maxima data are consistent with a scale invariant spectrum ($n = 1$). Other data show a similar level of consistency.

Also, the creation of gravitational waves during inflation is a unique effect [24]. While gravity waves from inflation are only observable with foreseeable experiments for certain inflation models [25], we know of no other source of a similar spectrum of gravity waves. Thus, if we do observe the right gravity wave spectrum this would be strong evidence for inflation.

There are a host of specific details of the perturbations that are not uniquely specified by all "Standard Picture" models. Some models have significant contributions form tensor perturbations (gravity waves) while others do not. The inflaton potential specifies the ratio of scalar to tensor amplitudes, but this ratio can be different for different models.

6. The Big Picture and Open Questions

There are a number of interesting open questions connected with inflation:

The origin of the Inflaton: It is far from clear what the inflaton actually is and where its potential comes from. This is intimately connected with the question of why the perturbations have the amplitude and spectrum they do. Currently, there is much confusion about physics at the relevant energy scales, and thus there is much speculation about different possible classes of inflaton potentials. One can hope that a clearer picture will eventually appear as some deeper theory (such as string/M theory) emerges to dictate the fundamental laws of physics at inflation scale.

Physics of the inflaton: Having chosen an inflaton potential, one can calculate the perturbations produced during inflation *assuming* the relevant field modes for wavelengths much smaller than the Hubble radius are in their ground states. This seems plausible, but it would be nice to understand this issue more clearly. Also, the inflaton field often takes on values $O(M_P)$. Will a yet-to-be-determined theory of quantum gravity introduce large corrections to our current calculations?

The cosmological constant problem: A very important open question is linked with the cosmological constant problem [26, 27, 28]. The non-zero potential energy of the inflaton during inflation is very similar to a cosmological constant. Why the cosmological constant is extremely close to zero today (at least from a particle physicist's point of view) is perhaps the deepest problem in theoretical physics. One is left wondering whether a resolution of this problem could make the cosmological constant and similar contributions to Einstein's equations identically zero, thus preventing inflation from every occurring. Interestingly, current data is suggesting that there is a non-zero cosmological constant today (see for example Fig. 11 which shows $\Omega_\Lambda = 0$ is strongly excluded). Cosmic acceleration today is a very confusing idea to a theorist, but it actually is helpful to inflation theory in a number of ways: Firstly it shows that the laws of physics do allow a non-zero cosmological constant (or something that behaves in a similar way). Also, it is only thanks to the non-zero Ω_Λ that the current data are consistent with a flat Universe (see Fig. 11).

Wider context and measures: We discussed in Section 3.4 various ideas about how an inflating region might emerge from a chaotic start. This is a very challenging concept to formulate in concrete terms, and not a lot of progress has been made so far. In addition, the fact that so many models are "eternally inflating" makes it challenging to define a unique measure for the tiny fraction of the universe where inflation actually ends. It has even been argued [29, 30, 31] that these measure problems lead to ambiguities in the ultimate predictions from inflation.

In fact it is quite possible that the "chaotic" picture will extend to the actual inflaton as well [32, 33]. Fundamental physics may provide *many* different flat directions that could inflate and subsequently reheat, leading to many different versions of the "Big Bang" emerging from a chaotic start. One then would have to somehow figure out how to extract concrete predictions out of this apparently less focused space of possibilities.

I am actually pretty optimistic that in the long run these measure issues can be resolved [34]. When there are measure ambiguities it is usually time to look carefully at the actual physics questions being posed... that is, what information we are gathering about the universe and how we are gathering it. It is the observations we actually make that ultimately defines a measure for our predictions. Still, we have a long way to go before such optimistic comments can be put to the test. Currently, it is not even very clear just what space we are tying to impose a measure on.

I am a strong opponent of the so-called "anthropic" arguments. All of science is ultimately a process of addressing conditional probability questions. One has the set of all observations and uses a subset of these to determine one's theory and fix its parameters. Then one can check if the theoretical predictions match the rest of the observations. No one really expects we can predict all observations, without "using up" some of them to determine the theory, but the goal of science is use up as few as possible, thus making as many predictions as possible. I believe that this goal must be the only determining factor in deciding *which* observations to sacrifice to determine the theory, and which to try and predict. Perhaps the existence of galaxies will be an effective observation to "use up" in constraining our theories, perhaps it will be the temperature of the CMB. All that matters in the end is that we predict as much as possible.

Thus far, using "conditions for life to exist" has proven an extremely vague and ineffective tool for pinning down cosmology. Some have argued that "there must be at least one galaxy" for life to exist [35, 36, 37], but no one really knows what it takes for life to exist, and certainly if I wanted to try and answer such a question I would not ask a cosmologist. Why even mention life, when one could just as well say "we know at least one galaxy exists" and see what else we can predict? In many cases (including [35, 36, 37]) the actual research can be re-interpreted that way, and my quibble is really only with the authors' choice of wording. It is these sorts of arguments (carefully phrased in terms of concrete observations) that could ultimately help us resolve the measure issues connected with inflation.

7. Conclusions

In theoretical cosmology Cosmic Inflation sits at the interface between the known and unknown. The end of inflation is designed to connect up with the Standard Big Bang model (SBB) which is highly successful at describing the current observations. Before inflation lies a poorly understood chaotic world where apparently anything goes, and even the laws of physics are not well understood.

Given that inflation is straddling these two worlds, it is not surprising that many open questions exist. Despite these open questions, inflation does a strikingly good job at addressing simpler questions that once seemed deep and mysterious when all we had was the SBB to work with. In particular, inflation teaches how it is indeed possible for nature to set up the seemingly highly "unnatural" initial conditions that are required for the SBB. Furthermore, inflation gives us a set of predictive signatures which lie within scope of realistic tests.

Already the confrontation with current observations is strongly favoring inflation. Upcoming large-scale surveys of the Universe promise that the pace of these confrontations will pick up over the next several years. Thus we should know in the foreseeable future whether we should continue to embrace inflation and build upon its successes or return to the drawing boards for another try.

I wish to thank Alex Lewin for comments on this manuscript, and acknowledge support from DOE grant DE-FG03-91ER40674, and UC Davis. Special thanks to the Isaac Newton Institute and the organizers for their hospitality during the this workshop.

References

1. A. Guth, *Phys. Rev. D* **23** 347, (1981)
2. E. Kolb and M. Turner, "The Early Universe" Addison Wesley (1990)
3. A. Linde, "Particle Physics and Inflationary Cosmology" Harwood, Chur, Switzerland (1990)
4. J. Preskill, *Phys. Rev. Lett.* **43** 1387, (1979)
5. A. Guth and E. Weinberg, *Nucl. Phys.* **B212**, 321 (1983)
6. A. Linde, *Phys. Lett.* 108B, 389 (1982)
7. A. Albrecht and P. Steinhardt, *Phys. Rev. Lett* **48**, 1220 (1982)
8. M. Bucher, A. Goldhaber and N. Turok, *Phys. Rev. D* **52**, 3314 (1995)
9. A. Linde and A. Mezhlumian, *Phys. Rev. D* **53** 5538 (1995)
10. J. Hartle and S. Hawking *Phys. Rev. D* **28**, 2960 (1983)
11. A. Liddle and D. Lyth, *Phys. Rept.* **231**, 1 (1993)
12. A. Vilenkin and P. Shellard "Cosmic Strings and other Topological Defects", Cambridge Univeristy Press, (1994)
13. A. Albrecht, D. Coulson, P. Ferreira, and J. Magueijo, *Phys. Rev. Lett.* **76**, 1413 (1996)
14. A. Balbi *et al* astro-ph/0005124
15. P. Steinhardt *Nature* **345**, 41 (1990).
16. D. Lyth and E Stewart *Phys. Lett.* **B252**, 336 (1990)
17. Data compilation provided by L. Knox at http://flight.uchicago.edu/knox/radpack.html
18. J. Weller, R. Battye and A. Albrecht *Phys. Rev D* **60** 103520, 1999
19. A. Albrecht "The status of cosmic defect models of structure formation" Contribution to the XXXVth Rencontres de Moriond "Energy Densities in the Universe" Jan 2000. In preparation.
20. N. Turok *Phys. Rev. Lett.* **77**, 4138 (1996)
21. A. Lewin and A. Albrecht astro-ph/9908061
22. A. Albrecht and J. Magueijo, PRD 59, 43516 (1999)
23. J. Moffat, International Journal of Physics D, Vol. 2, No. 3 (1993) 351-365; Foundations of Physics, Vol. 23 (1993) 411.
24. A. Sarobinsky *JETP Lett* **30**, 682 (1979)
25. R. Battye and P. Shellard *Class. Quant. Grav.* **13**, A239 (1996)
26. S. Carroll astro-ph/0004075
27. S. Weinberg astro-ph/0005265 (Talk given at Dark Matter 2000, February, 2000)
28. E. Witten, hep-ph/0002297
29. A. Linde, D. Linde and A. Mezhlumian *Phys. Rev. D* **49**, 1783 (1994)
30. V. Vanchurin, A. Vilenkin and S. Winitzki **Phys. Rev. D 61**, 083507 (2000)
31. A. Linde, D, Linde, and A. Mezhlumian *Phys. Rev. D* **54**, 2504 (1996)
32. A. Albrecht, In the proceedings of The international workshop on the Birth of the Universe and Fundamental Forces Rome, 1994 F. Occionero ed. (Springer Verlag) 1995
33. A. Vilenkin *Phys. Rev. D* **52** 3365 (1995)
34. See for example A. Vilenkin *Phys. Rev. Lett.* **81** 5501 (1998)
35. S. Weinberg *Phys.Rev. D* **61**, 103505 (2000)
36. M. Tegmark and M. Rees *Astrophys.J.* **499**, 526 (1998)
37. S. Hawking and N. Turok *Phys. Lett.* **B425**, 25 (1998)

BEFORE INFLATION

NEIL TUROK
DAMTP, Centre for Mathematical Sciences,
University of Cambridge Wilberforce Rd, Cambridge CB3 0WA,
United Kingdom.

Abstract.

Observations of the cosmic microwave sky are revealing the primordial non-uniformities from which all structure in the Universe grew. The only known physical mechanism for generating the inhomogeneities we see involves the amplification of quantum fluctuations during a period of inflation. Developing the theory further will require progress in quantum cosmology, connecting inflation to a theory of the initial state of the Universe. I discuss recent work within the framework of the Euclidean no boundary proposal, specifically classical instanton solutions and the computation of fluctuations around them. Within this framework, and for a generic inflationary theory, it appears that an additional anthropic constraint is required to explain the observed Universe. I outline an attempt to impose such a constraint in a precise mathematical manner.

1. Introduction

The maps of the cosmic microwave sky provided by recent experiments [1] signal an important breakthrough in cosmology. The structure revealed provides our most powerful probe of the structure in the Universe at early times, and of its current geometry. The data is consistent the Universe being nearly spatially flat, with a scale invariant spectrum of Gaussian-distributed irregularities in which the overall density fluctuates but the equation of state does not. This simplest form was anticipated on phenomenological grounds as far back as the sixties [2], but later emerged as the prediction of simple models of inflation [3, 4, 5, 6]. The detailed agreement between theory and observations lends impressive support to the view that the Universe is, in some respects at least, remarkably simple.

R.G. Crittenden and N.G. Turok (eds.), Structure Formation in the Universe, 43–61.

In inflationary theories, inhomogeneities of the required form arise a side-effect of the exponential expansion invoked to explain the size and flatness of the Universe. Quantum mechanical vacuum fluctuations in the inflaton field are stretched to large scales during inflation [5, 6], later to re-enter the Hubble radius and seed large scale structure. This magnificent mechanism links quantum mechanics and gravity to the observed structure of the Universe. It provides a direct observational probe of quantum gravity, albeit at leading (Gaussian) order.

The underlying theory is however only provisional, because we still have no compelling model of inflation and because a complete theory of quantum gravity is lacking. Nevertheless the success of the inflationary explanation of the structure in the cosmic microwave sky forces us to take very seriously the idea that quantum mechanics governed the formation of the Universe we see. This moves the topic of quantum cosmology to centre stage. It is a field that *must* be developed if we are attain a deeper level of understanding in cosmology.

The prevalent view is that cosmology will over the next period be observation-driven. It is true that dramatic observational developments are certain, and there will be a lot of phenomenology to do. However if we mean to achieve more than simply measuring the properties of the universe, theoretical developments are also crucial. The great freedom there is in current approaches to phenomenological models of inflation and of the early Universe, combined with the relatively small number of observations (even including the cosmic microwave sky) leads me to conclude that if we are to get anywhere, mathematical requirements of completeness and consistency must play an increasing role. Their power is already convincingly demonstrated by the fact that not one of the existing inflationary models has been made sense of beyond leading (Gaussian) order.

In this talk I will concentrate on one particular incompleteness of inflationary theory. The question is why inflation started in the first place. Vilenkin and others argue that the need for a theory of initial conditions is side-stepped because inflation is self-sustaining to the future (i.e. 'semi-eternal')[7]. Whilst accepting that some theory of the initial conditions is needed at some level, the hope of Guth and others is that the details of that theory are actually irrelevant, since there would typically be an infinite amount of inflation between ourselves and the beginning, during which details of the initial conditions would have been erased. In this view discussions of the initial conditions prior to inflation are fairly unimportant, since 'almost anything' will do. In this talk I criticise the calculational methods which have been used to reach this conclusion, and explain a different point of view according to which inflation is *a priori* very unlikely and in any case only a brief episode in our past.

One can easily imagine an infinite number of possible ansatzes for the initial conditions of the Universe. At the present stage of development of theory, many of these might be perfectly consistent with observation. But among the existing proposals, I think the Euclidean no boundary proposal [8] is appealing because it is based on simple and general ideas which have a rationale beyond cosmology. In a strong sense I think it is 'the most conservative thing you can do'. Of course it may well fail precisely because it is too conservative. Space and time may be emergent rather than fundamental properties. Describing the Universe as a manifold may not be appropriate to its early moments. Nevertheless precisely because the no boundary proposal is *not* just cooked up to make inflation work, its failings and limitations may teach us something deeper about what is in fact required.

In this lecture I focus on one particular approach to the no boundary proposal, using a set of classical instanton solutions of the 'no boundary' form [9, 10]. Our work has focussed on using these to compute the the complete Gaussian correlators [12, 13, 14, 15], in a generic inflationary model (for related work see [25, 26]). These generic instantons are unusual in several respects. In the original 'Einstein frame' they exhibit a curvature singularity which in the Lorentzian spacetime is 'naked'. Nevertheless the Euclidean action for the solutions is finite, suggesting they should contribute to the Euclidean path integral. Second, there is a one parameter family of solutions, with differing action. This is inconsistent with their being true stationary points of the action, but they may nevertheless be legitimate as 'constrained instantons' which are an established tool in other contexts [11]. A second feature is that the quantum fluctuations about these solutions are well defined to leading order, since the Euclidean action selects a unique (Dirichlet) boundary condition at the singularity. This allows one to make unambiguous predictions for the microwave anisotropies in any one of these solutions, which Gratton, Hertog and I have recently computed [12, 13, 15]. (See also [14]).

I then describe recent work by Kirklin, Wiseman and myself [18] showing how the singularity apparent in the original 'Einstein frame' may actually be removed by a change of field variables. Thus the singularity is really only a coordinate singularity on superspace. One still has the problem that the scalar field potential energy diverges at the singularity, and the scalar field equation remains ill defined there. That problem is overcome if one defines the theory as the limit of a family of theories in which $V(\phi)$ takes a particular form as ϕ tends to infinity, but approximates the actual potential at arbitrarily large values of ϕ. The limit is insensitive to the precise form of the regularised potential at large ϕ. The regularised construction allows one to give an improved formulation of the constraint at the singularity, which allows for a detailed computation of the spectrum of homogeneous modes [20]. The latter, we argue, are actually crucial to a proper interpretation of the instantons.

In the regularised theory all quantities appearing are finite. The classical field equations are also satisfied everywhere if one views the instantons as topologically RP^4 rather than S^4. The same construction is applied to instantons possessing two singularities in the Einstein frame. When the singularities are 'blown up' one obtains a regular manifold which is a four dimensional analogue of the Klein bottle.

Finally I discuss the problem of negative modes, complicated by the 'conformal factor' problem of Euclidean general relativity, due to the non-positive character of the Euclidean action. In the regular variables (and with the regularised potential), the problem of negative modes about singular instantons is well defined, and I briefly mention some of our latest results [20] indicating that the most interesting singular instantons (describing Universes with a large amount of inflation) do not possess any negative modes, in contrast to the non-singular instantons of Coleman and De Luccia, and Hawking and Moss. I comment on how this observation might help to resolve the 'empty Universe problem', if one performs a projection onto Universes containing the observer.

2. Inflation and Initial Conditions

Inflation is at heart a simple idea. A cosmological constant inserted into the Friedmann equation leads to exponential expansion of the scale factor (for a Universe which is sufficiently flat and homogeneous). The inflating region rapidly becomes exponentially large and smooth, and all other forms of matter are redshifted away. This basic point seems to have been understood a long time ago. For example an illustration in Peebles' book [21] shows Professor De Sitter blowing up a large balloon. The subtitle states 'What however blows up the ball? What makes the universe expand or swell up? That is done by the Lambda. Another answer cannot be given'.

Guth realised that scalar fields of the type invoked in high energy particle theories could provide a 'temporary Lambda' of just the form required, leading to a period of accelerated expansion *before* the standard hot big bang [3]. He showed how this period of cosmic inflation could solve the riddles of why the observed Universe is so large, so flat and so uniform on large scales. Subsequently Linde, and Albrecht and Steinhardt [4] invented working models involving 'slow-roll' inflation. A scalar field ϕ with potential energy $V(\phi)$ behaves like a ball rolling down a hill. The equations governing the motion of the scalar field ϕ and the scale factor a of the Universe are:

$$\ddot{\phi} + 3\frac{\dot{a}}{a}\dot{\phi} = -V_{,\phi} \tag{1}$$

$$\ddot{a} = \frac{\kappa}{3}a\left(-\dot{\phi}^2 + V\right) \tag{2}$$

$$\left(\frac{\dot{a}}{a}\right)^2 = \frac{\kappa}{3}\left(\frac{1}{2}\dot{\phi}^2 + V\right) - \frac{k}{a^2} \tag{3}$$

where $k = 0, \pm 1$ for flat, closed or open Universes respectively. The line element takes the form $ds^2 = -dt^2 + dr^2/(1-kr^2) + r^2 d\Omega_2^2$ where $d\Omega_2^2$ is the standard line element on S^2. Here $\kappa = 8\pi G$ with G Newtons constant. From the second equation one sees that the condition for accelerated expansion, i.e. inflation, is $V > \dot{\phi}^2$.

If the slope $V_{,\phi}$ is small, the ball takes a long time to roll and the potential V acts in the second equation just like a cosmological constant. But as the field rolls down, V slowly decreases and eventually goes to zero, with the energy in the scalar field being dumped into a bath of excitations of all the fields in the theory. This process is traditionally called re–heating although inflation need not have been preceded by a hot Universe epoch. The decay of the scalar field energy into radiation is responsible for the generation of the plasma of the hot big bang.

An amazing by-product of inflation is that it neatly provides a mechanism for generating the primordial inhomogeneities which later seed structure in the Universe [5, 6]. The downhill roll of the scalar field is subject to continual quantum mechanical fluctuations, which cause it to vary spatially. These fluctuations may be described as due to the Gibbons-Hawking 'temperature' of de Sitter space [22]. The radius of Euclideanised de Sitter space is H^{-1} and this plays the role of a periodicity scale i.e. an inverse temperature $\beta = T^{-1}$ in the Euclidean path integral. The inflaton field acquires fluctuations $\delta\phi \sim H$ on the Hubble radius scale H^{-1} at all times, and these fluctuations become frozen in as a comoving scale leaves the Hubble radius. If the classical field rolls slowly, H is nearly constant and one obtains a nearly scale invariant spectrum of fluctuations in $\delta\phi$. In some parts of the Universe the field fluctuates downhill and in others uphill. The comoving regions where it fluctuates down undergo re–heating earlier, and end up with lower density than those which remain inflating for longer. The fluctuations in the time to re–heating δt lead to density perturbations: $\delta\rho/\rho \sim \delta t H \sim (\delta\phi/\dot{\phi})H \sim H^2/\dot{\phi} \sim H^3/V_{,\phi} \sim (\kappa V)^{\frac{3}{2}}/V_{,\phi}$. For example, for a quadratic potential $m^2\phi^2$, $\delta\rho/\rho$ is proportional to m and fits the observations if $m \sim 10^{-5.5} M_{Pl}$ where $M_{Pl} = (8\pi G)^{-\frac{1}{2}}$ is the reduced Planck mass.

In the simplest models of inflation the fluctuations are adiabatic in character, that is, all particle species are perturbed in fixed ratio. This prediction is not really so much a consequence of inflation but rather of the assumption that no information survives from the inflationary epoch to the hot big bang epoch except the overall density perturbation. In models with more than one scalar field one generically induces perturbations of 'isocurvature' character as well.

So much for the successes of inflation. But let us try to be a bit more critical. A very basic question is

- Why did the scalar field start out up the hill?

In almost all treatments, the answer is simply that the inflationary theorist concerned put it there. In false vacuum inflation, one assumes the field was stuck in the false vacuum. In slow roll inflation, one assumes the field started out large. In some approaches to eternal inflation [7] one considers a theory which has a potential maximum and allows inflating domain walls. But even there one has to assume at least one such domain wall was initially present. At a fundamental level, the question is unanswered.

How far up the hill did the scalar field have to be? The number of inflationary efoldings is given, in the slow roll approximation, by

$$N_e \approx \int_0^{\phi_0} d\phi \frac{V(\phi)}{M_{Pl}^2 V_{,\phi}(\phi)} \tag{4}$$

where the true vacuum is at $\phi = 0$, and inflation starts at $\phi = \phi_0$. For a monomial potential ϕ^n, one has $\phi_0 \approx M_{Pl}\sqrt{2nN_e}$. If we require $N_e > 60$ or so to explain the homogeneity and isotropy of our present Hubble volume, we see that ϕ_0 must be substantially larger then M_{Pl} to obtain the 60 or so efoldings needed to explain the homogeneity and flatness of today's Universe.

Why was the initial value of the scalar field so high? It may be convenient for now to simply say 'why not?', and leave the matter there. But then I think one has to concede that at some level one has simply assumed the desired result.

One attempt to answer to the question of the required initial conditions for inflation is the theory of 'eternal inflation', advocated by Vilenkin, Linde and Guth and others [7]. The idea is that the same quantum fluctuations which produce the density fluctuations have an important backreaction effect upon inflation itself. That is, the scalar field can fluctuate uphill as well as down, and these fluctuations can compete with the classical rolling. Comparing the change in ϕ due to classical rolling in a Hubble time, $\delta\phi \sim H^{-2}V_{,\phi}$, with that due to quantum fluctuations $\delta\phi \sim H$, one sees that for a monomial potential the quantum fluctuations actually dominate at large ϕ. For example, for a quadratic $m^2\phi^2$ potential normalised to COBE this occurs when $\phi > (M_{Pl}/m)^{\frac{1}{2}} M_{Pl} \sim 10^3 M_{Pl}$.

The scenario is that at all times there are some regions of the Universe in which the scalar field takes large values. Classical rolling down from these regions then produces large inflated and reheated regions. However, in the high field regions the field also fluctuates uphill. This process can continuously regenerate the high field regions. People have tried to describe this process using stochastic equations which couple the quantum-driven diffusion to the classical Friedmann equation, but there are many problems with these calculations. First, the averaging scheme used employs a particular time slicing and the stochastic equations derived are not coordinate invariant. (Note that the scalar field itself *cannot* be used as a time coordinate, precisely because the condition for eternal inflation to occur is the same as the condition for the scalar field (classical plus quantum) to cease to be monotonic in time.) Second, the equation ignores spatial gradients and does not incorporate causality properly. Third, the treatment is not quantum mechanical. The subtleties of quantum interference are ignored and effectively it is assumed that the scalar field is 'measured' in each Hubble volume every Hubble time.

The approximate calculations are instructive. However since they in fact violate every known fundamental principle of physics, one should clearly interpret them with caution. It is interesting however that even after the fairly gross approximations made, a 'predictability crisis' emerges which is still unresolved. Simple potentials such as ϕ^2 do not generally lead to a stationary state, since the field is driven to the Planck density where the theory breaks down. A more basic problem is that no way is known to predict the relative probabilities for a discrete set of scalar field vacua. This suggests some profound principle is missing, and I shall suggest below what that principle is.

The spacetime found in the simulations has an infinite number of infinite inflating open 'bubble Universes' (much as in the scenario of open inflation [23]). The source of the 'predictability crisis' is the problem of trying to decide how probable it is for us to be in a region of one type or of the other, when there are infinite volumes of each. In other words,

- Where are we in the infinite 'multiverse'?

This is an infrared (large-scale) catastrophe which is unlikely to go away, and which I think makes it very unlikely that a well defined probability measure will emerge from a 'gods-eye' view in which one attempts to infer probabilities from an inflating spacetime of infinite extent.

It seems to me that one is asking the wrong question in these calculations. The solution may be instead to concentrate on *observable* predictions. Theory should provide a procedure for calculating cosmological correlators like:

$$\langle H_0^m \Omega_0^n (\frac{\delta\rho}{\rho})^p \rangle, \qquad m,n,p \ \epsilon \ Z \tag{5}$$

where we compute the full quantum correlator and then take the classical part to compare with observations. (See e.g. [24] for a discussion of this interpretation of quantum mechanics). We should demand the calculational procedure respects coordinate invariance, causality, and unitarity. Otherwise we shall be inconsistent with general relativity, special relativity or quantum mechanics. As in statistical physics, the role of quantum mechanics is to provide a discrete measure. Causality, I shall argue is equally important since it provides an infrared cutoff.

Causality is built into special and general relativity, and is equally present in quantum field theory. In a fixed background the latter is perfectly causal in the sense that correlators in a given spatial patch are completely determined by the set of correlators on a spatial region crossing the the past light cone of the original patch. This may be seen from the Heisenberg equations of motion. If we only ask questions about what is actually observable on or within our past light cone, we avoid the problem of dealing with the infinite number of infinite open Universes encountered in eternal inflation, just because the bubbles grow at the speed of light, so if you can see a bubble, you must be inside it, and you cannot see other bubbles. Causality indicates it should be possible to define all correlators of interest in a way that never mentions the other bubbles. The question becomes not '*Where are we?*', in some infinite spacetime which is pre-computed for an infinite amount of time, but rather '*What is the probability for the observed Universe to be in a given state?*' To calculate that, in a sum over histories approach we need to sum over the different possible four-Universes which could constitute our past. In the no boundary proposal, crucially, this sum is over *compact* Universes bounded by our past light cone. I believe this framework resolves the infrared problem in approaches to eternal inflation which I mentioned above.

3. The No Boundary Proposal

The no boundary proposal links geometry to complex analysis and to statistical physics. The idea is to contemplate four-geometries in which the real Lorentzian Universe (i.e. with signature $- + ++$) is rounded off on a compact Euclidean four-manifold (i.e. with signature $+ + ++$) with no boundary. This construction can in principle remove the initial singularity in the hot big bang. It can be made quantum mechanical by summing over all geometries and matter histories in a path integral formulation. The 'rounding off' is done in analogy with statistical physics, where one analytically continues to imaginary time. Here too one imagines computing correlators of interest in the Euclidean region, as a function of imaginary time, and then analytically continuing to the Lorentzian time where correlators of observables are needed.

There are many technical obstacles to be overcome in the implementation of this approach. These are not minor problems, and each is potentially fatal. Until they are resolved doubts must remain. Nevertheless I believe progress can be made. What I find most attractive is that the mathematical analogy at the heart of the Euclidean proposal is perhaps the deepest fact we know about quantum field theory, where Euclideanisation is the principal route to describing non-perturbative phenomena and underlies most rigorous results. There is also an analogy with string theory, the theory of two dimensional quantum geometry, where at least at a formal level the the perturbation expansion is defined as a sum over Riemann surfaces, embedded in a Riemannian target space. Scattering amplitudes are given as functions of Euclidean target space momenta, then analytically continued to real Lorentzian values. Euclidean quantum cosmology follows the same philosophy.

Before proceeding let me list a few of the 'technical' difficulties to be faced:

• Einstein gravity is non-renormalisable. This objection relates to the bad 'ultraviolet' properties of the theory, which are certainly important but are not central to the discussion here. Theories such as supergravity with improved ultra-violet properties are not conceptually different as far as the problems we are discussing.

• The Euclidean Einstein action is not positive definite, and therefore the Euclidean path integral is ill defined. This is the 'conformal factor' problem in Euclidean quantum gravity and I shall return to it below. Recent work has shown that at least for some choices of physical variables, and to quadratic order, this problem is overcome[27, 28, 20]. I shall mention this briefly below.

• The sum over topologies in four dimensions is likely to diverge, as happens in string theory. Most likely one will need some formulation in which manifolds of differing topologies are treated together.

These problems are formidable. The main hope, it seems to me, is that our Universe appears to be astonishingly simple, well described by a classical solution of great symmetry with small fluctuations present at a level of a part in a hundred thousand. This suggests that we may be able to accurately describe it using perturbation theory about classical solutions.

4. Generic Instantons

We seek solutions to the Einstein-scalar field equations which describe a background spacetime taking the required Lorentzian-Euclidean form. The main requirement is that there exist a special three-manifold where the normal derivatives of the scalar field and the spatial metric (more precisely the trace of the second fundamental form) vanish. If t is the normal coordinate, then if this condition is fulfilled, the classical solution will be real in both the Euclidean and the Lorentzian regions. Consider the simple case where V is constant and positive so that $V_{\phi} = 0$. Then there is a one parameter family of classical solutions labelled by the value of the scalar field, and in which scalar field is constant. The metric is that for de Sitter space, which may be described globally in closed coordinates: in which form

$$ds^2 = -dt^2 + H^{-2}\cosh^2(Ht)d\Omega_3^2. \tag{6}$$

with $H^2 = \frac{\kappa}{3}V$. This metric possesses an analytic continuation to a Euclidean four sphere, if we set $Ht = -i(\frac{\pi}{2} - \sigma)$. The solution possesses $O(5)$ symmetry in the Euclidean region, $O(4,1)$ in the Lorentzian region. This is too much symmetry to describe our Universe, which only possesses the symmetries of homogeneity and isotropy, a six parameter group.

However, if the potential is sloping, the maximal symmetry of the solution is lower, only $O(4)$ in the Euclidean region. The Euclidean metric is given by

$$ds^2 = d\sigma^2 + b^2(\sigma)d\Omega_3^2 \tag{7}$$

where $d\Omega_3^2$ is the round metric on S^3. The Euclidean Einstein equations are

$$\phi'' + 3\frac{b'}{b}\phi' = -V_{,\phi} \tag{8}$$

$$b'' = \frac{\kappa}{3}b\left(\phi'^2 + V\right) \tag{9}$$

where primes denote derivatives with respect to σ. If ϕ is constant, the second equation has solution $b = H^{-1}\sin(H\sigma)$, describing a round S^4. In general, b is a deformed version of the sine function. If the potential is gently sloping, we might expect a remnant of the one parameter family of solutions to survive, and that is indeed the case. Consider configurations with $O(4)$ as above, and with a regular pole which we shall take to be $\sigma = 0$, where $b \sim \sigma$. If the scalar field is regular there it must obey $\phi = \phi_0 + \frac{1}{2}\ddot{\phi}_0\sigma^2 + ...$, with ϕ_0 an arbitrary constant. If the potential is gently sloping, then $\ddot{\phi}_0$ is small and the scalar field rolls very slowly uphill, so b is nearly sinusoidal. However, once σ is past the maximum of b the ϕ equation exhibits anti-damping, and ϕ rolls off to infinity at some finite value σ_s. It is not hard to show

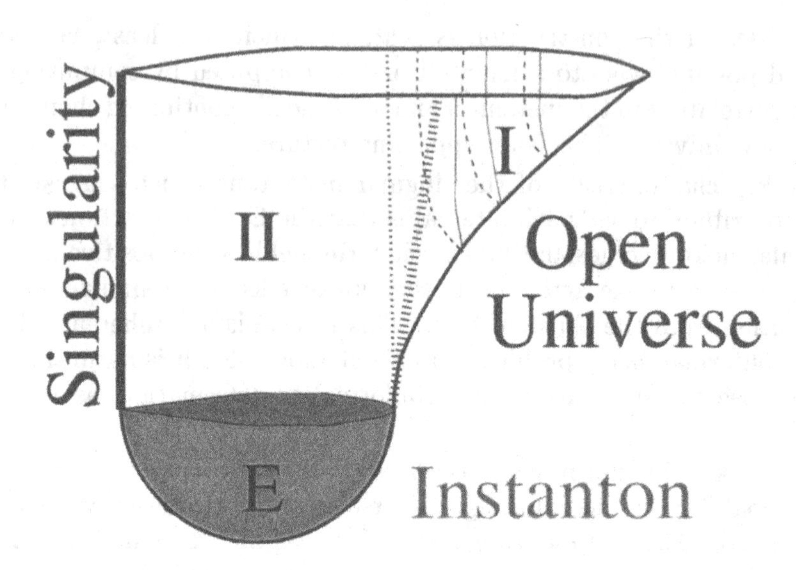

Figure 1. **An infinite open Universe emerges from a 'pea' instanton.**

that ϕ diverges logarithmically in $\sigma_s - \sigma$ and b vanishes as $(\sigma_s - \sigma)^{\frac{1}{3}}$. This behaviour is generic for scalar fields with gentle potentials. It is valid for example for several fields, for a nontrivial Kahler potential (i.e. a kinetic term $K_{ab}(\phi)\partial\phi^a\phi^b$) or for fields with a nontrivial coupling to the Ricci scalar.

As was noted in [9, 10], since all of these solutions have vanishing normal derivatives on 'longitude' three surfaces extending from the regular to the singular pole, they may be may analytically continued to a Lorentzian spacetime, which takes the form shown in Figure 1. Region I is an infinite inflating open Universe, which has a coordinate singularity on the lightcone through which it is connected to Region II, which is an approximately de Sitter region (ϕ is nearly constant) bounded by a time-like singularity. The most surprising result of the construction is that an infinite homogeneous open Universe emerges from a compact Euclidean region, a finite object.

So there is a one parameter family of finite action singular solutions, in which we have a one-to-one relation

$$\phi_0 \leftrightarrow \Omega_0 \tag{10}$$

where Ω_0 is the current density parameter. For the solutions described above, $\Omega_0 < 1$, but if one allows instantons which are singular at both poles, and symmetric about the maximum of b, these analytically continue to closed inflating Universes, and one can obtain any value of $\Omega_0 > 1$.

The idea of the construction is that, in principle at least, *everything* we could possibly wish to calculate could be computed by computing Euclidean correlators in the instanton and analytically continuing them to the Lorentzian Universe. This is an appealing picture.

The key characteristic of the singular instantons which suggests they might contribute to the path integral is that the Euclidean action is finite. The scalar field diverges and its kinetic term yields a the positive logarithmic divergence to the action. But this is cancelled by a similar negative contribution from the Einstein term. This mechanism is inherently linked to the 'conformal factor problem' discussed above, and it is natural to seek to regularise the singularity via a conformal transformation, as I discuss below.

If one takes singular instantons seriously, they allow one to estimate the prior probability for the inflating Universe to begin at a given value of the scalar field ϕ_0. The disappointing result of [9] was that for generic potentials at least the most favoured values of ϕ_0 are rather small and do not lead to much inflation. The most probable Universe is then essentially empty of matter. An attempt was made to rescue the situation with an anthropic argument, but even this led to an unacceptable value for $\Omega \sim 0.01$ [9]. A different argument will be made below, according to which a value of Ω very close to unity is predicted.

Significant criticisms have been made of the use of singular instantons, and of the conclusions drawn from them [17, 16].

First, since the classical field equations break down at the singularity, it is not clear whether they hold there. In fact, the action varies over the one parameter family of solutions, so they cannot all be stationary points of the action. This does not mean they do not contribute to the path integral, nor that they cannot be used to approximate it. But they must be regarded instead as 'constrained instantons', and a suitable constraint must be introduced. This is a relatively well developed procedure both in quantum mechanics and in field theory [11]. One way of imposing a constraint will be described in the next section.

Second, the presence of a singular boundary in the Lorentzian region might lead one to worry that matter or radiation could leak into the spacetime from the singularity. Equivalently, what are the correct boundary conditions at the singularity? Here, at least to quadratic order in the (spatially inhomogeneous) fluctuations I think this worry has been convincingly resolved. As shown in [12, 13], for scalar and tensor perturbations, finiteness of the Euclidean action selects a unique (Dirichlet) boundary condition. For the spatially homogeneous modes, the situation is more involved. The required boundary condition is only provided by a definite regularisation of the singularity. That proposed in [18] and implemented in [20] is described

below.

Third, Vilenkin showed in an interesting paper [17] that analogous asymptotically flat instantons exist, and he argued these would lead to an instability of flat space towards nucleation of singularities. This objection was addressed in [19] where it was shown that if the instantons are constructed as constrained instantons with an appropriate constraint applied on the singular boundary, then they possess no negative mode and therefore do not mediate an instability. The subsequent work of [20] with a better defined constraint confirms this.

Vilenkin then pointed out that constrained instantons might be placed in a 'necklace' of arbitrary length, yielding configurations of arbitrarily negative Euclidean action which would render the Euclidean path integral meaningless.

The last two objections are addressed by the regularisation explained below, in which the classical field equations are satisfied everywhere, and in which a certain constraint is applied yielding a completely non-singular description. In this description, 'necklaces' of the form envisaged by Vilenkin are absent.

5. Resolving the Singularity

We have recently shown that it is possible to regularise the instanton solutions in the following two steps. We first show that the geometry of the instantons may be regularised by 'blowing up' the singularity with a ϕ-dependent conformal transformation i.e. a change of coordinates on field space. Next, we replace the scalar potential $V(\phi)$ by a 'regularised' version, in which we deform $V(\phi)$ at very large ϕ so that it goes to zero at $\phi = \infty$. Then we show that the scalar field may be rewritten in terms of a new 'twisted' field, which is forced to be zero on a special 3-manifold. After all this we show that the 'twisted' scalar field, and the Riemannian metric, satisfy the field equations everywhere. This construction renders the 'singular' instantons 'regular' and makes further analysis of them well defined.

First, we note that if we change from σ to a coordinate $X = \int_{\sigma}^{\sigma_s} d\sigma/b(\sigma)$, the metric becomes $b^2(X)(dX^2 + d\Omega_3^2)$. Near the singularity, one finds $b^2(X) \sim X$, *so the conformal factor has a linear zero*, in these coordinates. This suggests a simple interpretation, which we have explored, which is that the conformal factor $b^2(X)$ is actually a twisted field. We shall implement this below.

We write the Einstein frame metric as

$$g_{\mu\nu}(x) = \Omega^2(x) g^R_{\mu\nu}(x) \tag{11}$$

where $g^R_{\mu\nu}$ is a Riemannian (i.e. positive definite) metric, and we are going to allow $\Omega^2(x)$ to possess zeros on the manifold.

The Einstein-scalar action becomes:

$$\int d^4x\sqrt{g^R}\left(-\frac{1}{2\kappa}\Omega^2 R(g^R) - \frac{3}{\kappa}\left((\nabla\Omega)^2 - \frac{\kappa}{6}\Omega^2(\nabla\phi)^2\right) + \Omega^4 V(\phi)\right) \quad (12)$$

We should also add a surface term $-\frac{1}{\kappa}\int_B \sqrt{h^R}\Omega^2 K^R$ if we wish the action to involve only first derivatives of the metric.

Now the point is that $\Omega(x)$ is to be regarded as a scalar field living on the Riemannian manifold with metric $g^R_{\mu\nu}$. The kinetic terms for the two scalars, $\Omega(x)$ and $\phi(x)$, define the metric on the space of scalar fields, i.e. the superspace metric. The metric clearly has a coordinate singularity at $\Omega = 0$, which we can remove by making the change of coordinates

$$\Omega_1 = \Omega\cosh(\sqrt{\frac{\kappa}{6}}\phi) \qquad \Omega_2 = \Omega\sinh(\sqrt{\frac{\kappa}{6}}\phi). \quad (13)$$

The global structure of superspace is seen to be $1+1$ Minkowski space, and we may change to light cone coordinates

$$\Omega_\pm \equiv \Omega_1 \pm \Omega_2. \quad (14)$$

This change of variables has the effect that the action density is now finite term by term for the 'singular' instantons. To make the whole construction analytic, we need to re-express the scalar field in terms of the light cone variables, which we do via

$$\phi = \sqrt{\frac{3\kappa}{2}}\ln\left(\Omega_+/\Omega_-\right). \quad (15)$$

But now when re-expressed in terms of $\Omega_\pm$ the potential V has a branch cut at $\Omega_- = 0$. We therefore need to modify the potential $V(\phi)$ so that it is analytic as Ω_- tends to zero. Since the term entering the action is $\Omega_+^2\Omega_-^2 V$, the action will be analytic in the vicinity of the conformal zero as long as $V \sim (e^{\sqrt{\frac{2\kappa}{2}}\phi})^n$ at large ϕ, with n an integer ≤ 1.

However there is still a problem. If we continue the solutions through the zero of Ω_-, then we enter a region where the Einstein metric changes from Euclidean to anti-Euclidean. We can avoid this by instead identifying antipodal points on the three-sphere upon which $\Omega_- = 0$. This produces a manifold which is compact, and is topologically RP^4. Since Ω_- vanishes linearly (and is analytic in X) it will solve the field equations across the non-contractible RP^3 if we interpret Ω_- as being 'twisted' on RP^4, an interpretation which is possible since the latter is non-simply connected.

This interpretation requires that the integer n of the previous paragraph be odd, so that the field equations are covariant under the symmetry $\Omega_- \rightarrow -\Omega_-$.

As discussed in [18], in this construction the 'singular' instantons are regularised and solve the appropriate classical field equations everywhere. This is surprising at first sight, since the solutions all have differing action. But this is allowed just because the manifold cannot be covered by a single coordinate system (due to its non-orientability) and the action must be defined by effectively introducing an 'internal boundary', upon which additional data enters. This is quite analogous to the magnetic monopole solution on a two sphere. In that case, the constraint that is used is the magnetic flux over the entire S^2. Here instead it is the volume of the 'internal boundary' upon which the orientation flip occurs, evaluated in the Riemannian frame.

We have had to go to substantial lengths to regularise the singular instantons. But since they are now well defined as constrained objects, one can hope to test whether they provide good approximations to the Euclidean path integral.

6. Fluctuations and Negative Modes

One way to test whether instantons provide sensible approximations to the Euclidean path integral is to examine the fluctuations about them. In fact the two point correlator for scalar and tensor perturbations which have nontrivial dependence on the S^3 coordinates on the instanton and which therefore yield the inhomogeneous density perturbations in the open (or closed) universe, were computed in [12, 13, 15] and translated into predictions for the cosmic microwave anisotropies. (For related work in nonsingular instantons see [25, 26]). As mentioned above, no ambiguities emerged in these calculations, in spite of the presence of the singularity.

The fluctuation modes which are homogeneous on the S^3 slices are much more subtle. A naive treatment instantly encounters the conformal factor problem. Certain gauge invariant fluctuation variables have negative kinetic terms, which means there are an infinite number of negative modes. In recent work [20] (see also [27, 28]), we have shed light on this problem, by showing that with certain choices of physical variables, the kinetic term in the Euclidean action is actually positive definite for the homogenous modes too. For regular instantons we find a finite number of negative modes, always at least one. For singular instantons we find the Euclidean action *does not* uniquely select a boundary condition, and the details of the regularisation therefore matter. Interestingly, we do find that for instantons with large ϕ_0 (therefore giving a nearly flat Universe today), the constraint employed in the RP^4 regularisation explained above actually removes all negative modes. Therefore the constrained Euclidean path integral is actually well defined (only to Gaussian order) for these instantons.

In my view, the presence of physical negative modes is a serious problem for the 'no-boundary' interpretation of cosmological instantons such as Coleman-de Luccia instantons or Hawking-Moss instantons. It means that these instantons can only be regarded as yielding approximate descriptions, appropriate to the decay of an unstable state. They cannot be straightforwardly used as a basis for defining the initial state of the Universe. The situation with 'regularised' singular constrained instantons looks more promising, and deserves greater investigation.

7. A Twisted Universe?

The connection between topology and singular instantons is intriguing. From one point of view, we simply wanted to regularise the instantons, and the topology allowed us to do that. But there may be a more fundamental reason why nontrivial topology is important. Consider a scalar field theory on a circle, with a Z_2 symmetry $\phi \to -\phi$, and with a 'double well' potential. One can now choose the field to be either 'twisted' or 'untwisted'. In the former case the field must aquire a -1 factor as one traverses the circle, and the topology of the configuration space (the set of $\phi(x)$) is that of the Mobius strip. The point is that the lowest energy state in the twisted sector is not one of the two minima of the potential, but is instead a solitonic state where the field has a single zero. There must therefore be a region of large scalar potential energy present.

It is tempting to speculate that this might provide an answer to the question of why the inflaton field started 'up the hill'. That is, in the Euclidean path integral there are naturally distinct topological sectors. If the topology is S^4, the Euclidean ground state is the true vacuum of the theory, the stable minimum describing an empty Universe. However, if the topology is RP^4, there are instead two distinct sectors, in which the conformal factor is respectively twisted and untwisted. In the former, as we have seen, we get a family of singular instantons all yielding inflation and therefore non-empty Universes.

I find it an intriguing idea that the twisted sector of quantum gravity might contain a Universe filled with matter arising naturally, in the appropriate Euclidean 'ground state'.

8. Volume Factors and A-projections

Finally, I want to deal with the question of the value of Ω_0 predicted by the instanton approach [29]. It is clear that for a generic potential, the a

priori most probable Universe does not have much inflation. It could be that the inflaton model is wrong, the Euclidean path integral is wrong, the instanton solutions are wrong, or that all three are wrong! But there is also another possibility, that we are just asking the wrong question. We should not after all be computing correlators of cosmological observables alone, as in Eqn. (5) above. Instead we should insert an operator $\mathcal{P}$ corresponding to *projecting onto the subset of states containing the particular observer who makes the observation.* Of course it would be nice if this insertion had no effect, so that the answer for the most likely Universe did not depend on whether the observer was in it. But we should be open to the possibility that theory will never predict this, and all it will tell us is about the most probable Universe we should see.

This is of course a formulation of the anthropic principle, which seems a step backward to many physicists, from the goal of explaining the Universe from fundamental mathematical principles. Indeed it is, and of course it would be nice to have a theory which explained exactly what we see in the Universe and nothing else. But that seems unlikely since quantum mechanics at best makes statistical predictions. It would still be nice if the correct theory predicted only Universes very like ours. Again that may be too much to ask, and we may be forced to pursue the more limited goal of a theory which allows of all types, but within which those containing ourselves are in agreement with all the observations we can make. Providing the theory satisfies other criteria - simplicity, consistency, I think we would be happy with it. It seems to me that at least within the framework I have discussed i.e. inflation in the context of a generic scalar potential and the Euclidean no boundary approach, this reduced goal is indeed the best we can hope for. There is still a major challenge to be confronted, which is to to properly formulate exactly what the projection $\mathcal{P}$ is. Due to space constraints I can only sketch an approach here. More details will appear in [29].

If we agree to discuss only the observable Universe, then it is clear that the projection $\mathcal{P}$ should be carried out on our past light cone. We shall regard this as the 'observable Universe'. Then we want to impose a condition, that there exists a certain very unlikely field configuration, upon this past light cone. I am really only assuming here that the particular observer concerned is a very rare event - the argument would apply equally for a beetle or even a paper clip!. We do not need to know the details of this configuration. but we shall need to assume is that it resulted in a definite way from a particular configuration of growing mode linear density perturbations. Now these linear density perturbations can be traced back comoving in a simple way, right back to the primordial instanton. We therefore want to ask: How many ways are there of obtaining a Universe containing 'me' from one of the singular instantons. The point now is that there is a 'zero mode' present in the location of the past light cone relative to the perturbations on the instanton. When we perform the path integral we should get a volume factor from integrating this zero mode, and it should be the roughly the volume of the constant Euclidean time slices of the instanton divided by the volume of the comoving 'special region'. If we fix the size of the latter today, to be the spatial resolution with which we want to identify the configuration, then the comoving volume on the instanton is smaller by a factor e^{3N_e} where N_e is the number of efoldings of inflation. Thus one expects the zero mode integration to give a 'volume factor' dependence.

The posterior probability of obtaining a Universe containing the observer, from an instanton solution with scalar field value ϕ_0 at the regular pole is therefore given by

$$\mathrm{Exp}\left(-\frac{S_E(\phi_0)}{\hbar}+3N_e\right)\approx \mathrm{Exp}\left(\frac{24\pi^2 M_{Pl}^4}{\hbar V(\phi_0)}+3\int_0^{\phi_0} d\phi \frac{V(\phi)}{M_{Pl}^2 V_{,\phi}(\phi)}\right) \quad (16)$$

where I have restored Planck's constant $\hbar$ and used the slow-roll formulae for the number of efoldings of inflation given above. For gentle monotonic potentials, the exponent is greatest at very small field values, and at very large field values, where the scalar potential $V(\phi_0)$ becomes of order the Planck density. Interestingly, there is a minimum where the two terms compete, and this is precisely where the quantum fluctuations $\delta\phi \sim H\sqrt{\hbar}$ compete with the classical rolling $H^{-2}V_{,\phi}$.

According to this posterior probability, there is a high probability for us to be in a Universe which started at the Planck density, had the maximum amount of slow-roll classical inflation, and which is extremely flat today. I think this argument holds the prospect of solving the 'predictability crisis' mentioned in Section 2: the 'volume factor' I have alluded to is covariant and slicing-independent. Surprisingly, it is equal in order of magnitude for both closed and open Universes.

The final conclusion is in some respects disappointing. After all there is no hope that a semi-classical calculation will be accurate near the Planck density. And if there has been a very large amount of inflation, then the Universe is extremely flat and there is not much hope of detecting effects coming from the structure of the instanton, since perturbations of those wavelengths are way beyond our Hubble volume today [15]. Nevertheless the overall framework holds some prospect of completeness, and seems to avoid the pitfalls of the 'global view' adopted in eternal inflation. The amount of time that inflation lasted can be estimated, and for example in an $m^2\phi^2$ potential, the rolling time from the Planck density to the bottom is only of order $M_{Pl}m^{-2}$, or about 10^{12} Planck times. Hardly eternal.

I wish to thank many physicists for discussions of these issues, and in particular A. Guth, V. Rubakov and A. Vilenkin. It is a special pleasure to acknowledge my collaborators S. Gratton, S. Hawking, T. Hertog, K. Kirklin and T. Wiseman.

References

1. A. D. Miller *et al.*, Astrophys.J. 524 (1999) L1; P. de Bernardis *et al.*, Nature, 404, 955, (2000); Hanany, S. *et al.*, astro-ph/0005123 (2000).
2. Harrison, E. R., 1970, Phys. Rev., **D1**, 2726; Zeldovich, Ya. B., 1972, M.N.R.A.S., **160**, 1.
3. A. Guth, Phys. Rev **D23,** 347 (1981).
4. A.D. Linde, Phys. Lett. **108B,** 389 (1982); A. Albrecht and P. Steinhardt, Phys. Rev. Lett. **48,** 1220 (1982).
5. S.W. Hawking, Phys. Lett. **115B**, 295 (1982); A.H. Guth and S.-Y. Pi, Phys. Rev. Lett. **49**, 1110 (1982); A.A. Starobinsky, Phys. Lett. **117B**, 175 (1982); J. Bardeen, P. Steinhardt, and M. Turner, Phys. Rev. **D28,** 679 (1983).
6. V.F. Mukhanov, H.A. Feldman and R.H. Brandenberger, Phys. Rep. **215**, 203 (1992).
7. For a recent review and references see V. Vanchurin, A. Vilenkin and S. Winitzki, gr-qc/9905097, Phys. Rev. **D61** 083507 (2000).
8. J.B. Hartle and S.W. Hawking, Phys. Rev. **D28** 2960 (1983).
9. S.W. Hawking and N. Turok, Phys. Lett. **B425** 25 (1998).
10. S.W. Hawking and N. Turok, Phys. Lett. **B432** 271 (1998).
11. I. Affleck, Nucl. Phys. **B191** 429 (1981).
12. S. Gratton and N. Turok, Phys. Rev. **D 60**, 123507 (1999).
13. T. Hertog and N. Turok, Phys. Rev. **D 62** 083514 (2000), astro-ph/9903075.
14. S.W. Hawking, T. Hertog and N. Turok, Phys. Rev. **D62** 063502 (2000).
15. S. Gratton, T. Hertog and N. Turok, Phys. Rev.**D62** 063501 (2000).
16. A. Linde, gr-qc/9802038, Phys. Rev. **D58** (1998) 083514.
17. A. Vilenkin, hep-th/9803084, Phys. Rev. **D57** (1998) 7069; gr-qc/9804051, Phys. Rev. **D58** (1998) 067301; gr-qc/9812027.
18. K. Kirklin, N. Turok and T. Wiseman, hep-th/0005062, submitted to PRD.
19. N. Turok, Phys. Lett. **B458** 202 (1999).
20. S. Gratton and N. Turok, hep-th/0008235 (2000).
21. P.J.E. Peebles, *Principles of Physical Cosmology*, Princeton 1993.
22. G.W. Gibbons and S.W. Hawking, Phys. Rev. **D15** 2738 (1977).
23. M. Bucher, A. Goldhaber and N. Turok, Phys. Rev. **D52**, 3314 (1995).
24. N.D. Mermin, quant-ph/9801057.
25. J. Garriga, X. Montes, M. Sasaki and T. Tanaka, astro-ph/9811257, Nucl. Phys. **B551** (1999) 317; astro-ph/9706229, Nucl. Phys. **B513** (1998) 343.
26. A. Linde, M. Sasaki and T. Tanaka, Phys. Rev. **D59** (1999) 123522.
27. A. Khvedelidze, G. Lavrelashvili, T. Tanaka, gr-qc/0001041.
28. G. Lavrelashvili, Nucl. Phys. Proc. Suppl. **88** (2000) 75.
29. N. Turok, in preparation (2000).

QUANTUM COSMOLOGY

V.A.RUBAKOV
Institute for Nuclear Research
of the Russian Academy of Sciences,
60th October Prospect, 7a, Moscow, 117312, Russia
and
University of Cambridge,
Isaac Newton Institute for Mathematical Sciences,
20 Clarkson Road, Cambridge, CB3 0EH, U.K.

1. Introduction

Most of the evolution of the Universe is likely to have proceeded classically, in the sense that the dominant phenomenon is the classical expansion while quantum fluctuations are small and can be treated as perturbations. Prior to this stage, however, genuinely quantum phenomena almost certainly took place. Although it is not clear whether they have left any observable footprints, it is of interest to try to understand them, as this may shed light on such issues as initial conditions for classical cosmology (are inflationary initial data natural? how long was the inflationary epoch? is open Universe consistent with inflation?), properties of space-time near the cosmological singularity, origin of coupling constants, etc.

To describe the Universe at its quantum phase, one ultimately has to deal with full quantum gravity theory, well beyond the Einstein gravity. It is, however, legitimate to take more modest attitude and consider quantum phenomena below the Planck (or string) energy scale. Then the quantized Einstein gravity (plus quantized matter *fields*) provides an effective "low energy" description which must be tractable, at least in principle, within quantum field theory framework. Processes that should be possible to consider in this way are not necessarily perturbative, as the example of tunnelling in quantum mechanics and field theory shows.

Surprisingly, not so many phenomena are well understood even within this modest approach. Perhaps the most clear process is the decay of a metastable vacuum [1]. It has been clarified recently [2] that at least in

R.G. Crittenden and N.G. Turok (eds.), Structure Formation in the Universe, 63–73.

the range of parameters where the treatment of space-time in terms of background de Sitter metrics is reliable, the Coleman–De Luccia instanton indeed describes the false vacuum decay, provided the quantum fluctuations above the classical false vacuum are in de Sitter-invariant (conformal) vacuum state. This is in full accord with the results of Refs. [1, 3]. It is likely that this conclusion holds also when the quantum properties of metrics are taken into account. Furthermore, the Hawking–Moss instanton [4] can be interpreted [2] as a limiting case of constrained instantons that describe the false vacuum decay in an appropriate region of parameter space, again in agreement with previous analyses [5, 6]. Hence, there emerges a coherent picture of the false vacuum decay with gravity effects included.

One may try to apply laws of quantum mechanics to the Universe as a whole, and consider the wave function of the Universe. Although research in this direction began more than 30 years ago [7], the situation here is still intriguing and controversial. The main purpose of this contribution is to make a few comments on this subject. Namely, we will discuss which analogies to ordinary quantum mechanics are likely to work in quantum cosmology, and which are rather misleading. We begin with quantum mechanics, and only then turn to the wave function of the Universe.

2. Wave function in quantum mechanics

To set the stage, let us consider a quantum mechanical system with two dynamical coordinates, x and y. Let the Hamiltonian be

$$\hat{H} = \hat{H}_0 + \hat{H}_y$$

where

$$\hat{H}_0 = \frac{1}{2}\hat{p}_x^2 + V_0(x)$$

$$\hat{H}_y = \frac{1}{2}\hat{p}_y^2 + \frac{1}{2}\omega^2(x)y^2 + \frac{1}{4}\lambda(x)y^4 + \dots$$

Let us assume that the potential $V_0(x)$ is such that the motion along the coordinate x is semiclassical, while the dynamics along the coordinate y can be treated in perturbation theory about the semiclassical motion along x. This approach is close in spirit to the Born–Oppenheimer approximation. We will consider solutions to the *stationary* Schrödinger equation with fixed energy E.

Let us first discuss the dynamics in the classically allowed region of x, where $E > V_0(x)$. In this region, there are two sets of solutions with the semiclassical parts of the wave functions equal to

$$\Psi \propto \mathrm{e}^{+iS(x)} \tag{1}$$

and

$$\Psi \propto e^{-iS(x)}$$

where

$$S(x) = \int^x dx' \sqrt{2(E - V_0(x'))}$$

These two sets of solutions correspond to motion right and left, respectively. Note that this interpretation is based on the fact that there exists *extrinsic time* t inherent in the problem: the complete, time-dependent wave functions are $\exp(-iEt + iS(x))$ and $\exp(-iEt - iS(x))$; the wave packets constructed out of the wave functions of these two types indeed move right and left, respectively, as t increases.

Let us now consider the dynamics along the coordinate y, still using the time-independent Schrödinger equation in the allowed region of x. This is done for, say, right-moving system by writing, instead of eq.(1),

$$\Psi(x, y) = \frac{1}{\sqrt{p_x(x)}} \tilde{\Psi}(x, y) e^{iS(x)}$$

where $p_x = \partial S / \partial x$. To the first order in $\hbar$ one obtains that the time-independent Schrödinger equation reduces to

$$i \frac{\partial \tilde{\Psi}}{\partial x} \frac{\partial S}{\partial x} = \hat{H}_y \tilde{\Psi} \tag{2}$$

This can be cast into the form of *time-dependent* Schrödinger equation by changing variables from x to τ related by $x = x_c(\tau)$, where $x_c(\tau)$ is the solution of the classical equation of motion for x in "time" τ, which has energy E and obeys

$$\frac{\partial S}{\partial x}(x = x_c) = \frac{\partial x_c}{\partial \tau} \tag{3}$$

After this change of variables, $\tilde{\Psi}$ becomes a function of y and τ and obeys the following equation,

$$i \frac{\partial \tilde{\Psi}(y; \tau)}{\partial \tau} = \hat{H}_y(\hat{y}, \hat{p}_y; \tau) \tilde{\Psi}(y; \tau) \tag{4}$$

where the explicit dependence of $\hat{H}_y$ on τ comes from $x_c(\tau)$. We see that there have emerged intrinsic time τ which parameterizes the classical trajectory $x_c(\tau)$ and also the y-dependent part of the wave function. We note again that in quantum mechanics, the arrow of intrinsic time, which is set by the sign convention in eq.(3), is determined by the arrow of extrinsic time t.

Note also that one is free to choose any representation for operators $\hat{y}$ and $\hat{p}_y$ and write, instead of eq.(4),

$$i\frac{\partial|\tilde{\Psi}\rangle}{\partial\tau} = \hat{H}_y(\tau)|\tilde{\Psi}\rangle \qquad (5)$$

To solve this equation, one may find convenient to switch to the Heisenberg representation, as usual.

We now turn to the discussion of the region of x where the classical motion is forbidden and the system has to tunnel. To simplify formulas, we set $E = 0$ in what follows. If the system tunnels from left to right, the dominant semiclassical wave function is

$$\Psi \propto \mathrm{e}^{-S(x)}$$

where $S(x) = \int^x dx' \sqrt{2V_0(x')}$ and obeys the following equation,

$$-\frac{1}{2}\left(\frac{\partial S}{\partial x}\right)^2 + V_0(x) = 0$$

This equation may be formally considered as the classical Hamilton–Jacobi equation in Euclidean ("imaginary") time. The zero energy classical trajectory $x_c(\tau)$ in Euclidean time τ obeys

$$\frac{d^2x_c}{d\tau^2} = +\frac{\partial V_0}{\partial x}(x = x_c)$$

and hence

$$\frac{dx_c}{d\tau} = \frac{\partial S}{\partial x}(x = x_c)$$

Then $S(x)$ can be calculated as the value of the Euclidean action along this trajectory.

To find the equation governing the dynamics along y-direction in the classically forbidden region of x, we again write

$$\Psi(x, y) = \frac{1}{\sqrt{p_x}}\tilde{\Psi}(x, y)\mathrm{e}^{-S(x)}$$

and obtain, changing variables from x to τ, $x = x_c(\tau)$, that $\tilde{\Psi}$ obeys the time-dependent Schrödinger equation, now in Euclidean time,

$$\frac{\partial\tilde{\Psi}(y;\tau)}{\partial\tau} = -\hat{H}_y(\hat{y}, \hat{p}_y; \tau)\tilde{\Psi}(y;\tau) \qquad (6)$$

The minus sign on the right hand side of this equation *is crucial* for the stability of the approximation we use. Indeed, the system described by eq.(6)

tends to de-excite, rather than excite, as "time" τ increases, so that the part $\tilde{\Psi}$ of the wave function remains always subdominant as compared to the leading semiclassical exponential. The physics behind this property is quite clear: we consider tunneling at fixed energy, so the de-excitation of fluctuations along y means the transfer of energy to the tunneling subsystem, which makes tunneling (exponentially) more probable. Inversely, if fluctuations along y get excited, the kinetic energy along x decreases, and tunneling gets suppressed stronger.

3. Wave function of the Universe

To discuss specific aspects of quantum cosmology, let us consider the closed Friedmann–Robertson–Walker Universe with the scale factor a. Let us introduce the cosmological constant Λ, minimal scalar field $\phi(x)$ with a scalar potential $V(\phi)$ and also massless conformal scalar field. We are going to treat the dynamics of the scale factor in a semiclassical manner; in this respect a is analogous to the variable x of the previous section. The minimal scalar field (as well as gravitons) will be considered within perturbation theory, so each of the modes $\phi_{\mathbf{k}}$ will be analogous to the variable y of the previous section.

The basic equation in quantum cosmology is the Wheeler–De Witt equation, which in our case reads

$$\left[-\frac{1}{2}\hat{p}_a^2 - \frac{1}{2}a^2 + \Lambda a^4 + \hat{H}_\phi\right]\Psi = -\epsilon\Psi \tag{7}$$

where we have set $3M_{Pl}^2/16\pi = 1$ and ignored the operator ordering problems which are irrelevant for our discussion. Here

$$\hat{H}_\phi = \int \frac{d^3x}{2\pi^2}\left[\frac{1}{2a^2}\hat{p}_\phi^2 + \frac{a^2}{2}(\partial_i\hat{\phi})^2 + a^4 V(\hat{\phi})\right]$$

is the term due to the minimal scalar field; at the classical level $\hat{H}_\phi$ is the energy of matter defined with respect to conformal time. The non-negative constant ϵ on the right hand side of eq.(7) is the contribution of the conformal scalar field; the only purpose of introducing the latter field is to allow for non-zero ϵ. We do not consider gravitons in what follows, as they are similar to the quanta of the minimal scalar field ϕ.

In the spirit of the Born–Oppenheimer approximation, let us first neglect the conformal energy of the field ϕ, i.e., omit the term $\hat{H}_\phi$ in eq.(7). Then the Wheeler–De Witt equation takes the form of the time-independent Schrödinger equation in quantum mechanics of one generalized coordinate a with energy ϵ and potential

$$U(a) = \frac{1}{2}a^2 - \Lambda a^4$$

At $16\Lambda^2\epsilon < 1$, there are two classically allowed regions: at small a ($0 < a^2 < [1-\sqrt{1-16\Lambda^2\epsilon}]/4\Lambda$) and at large a ($\infty > a^2 > [1+\sqrt{1-16\Lambda^2\epsilon}]/4\Lambda$). At the classical level, the former region corresponds to an expanding and recollapsing Friedmann-like closed Universe, while the latter corresponds to the de Sitter-like behavior. As $\epsilon \to 0$, the first classically allowed region disappears, while the second becomes exactly de Sitter.

In between these two regions, classical evolution is impossible (if one neglects $\hat{H}_\phi$), and one has to consider classically forbidden "motion". Let us discuss classically allowed and classically forbidden regions separately.

3.1. CLASSICALLY ALLOWED REGION: ISSUE OF ARROW OF TIME

To be specific, let us consider classically allowed de Sitter-like region where the scale factor a is large. In the leading order, there are again two types of semiclassical wave functions,

$$\Psi \propto \mathrm{e}^{-iS(a)} \tag{8}$$

and

$$\Psi \propto \mathrm{e}^{+iS(a)} \tag{9}$$

where

$$S(a) = \int^a da' \sqrt{2(\epsilon - U(a'))}$$

Classically, the momentum is related to the derivative of the conformal factor with respect to conformal time,

$$\frac{da}{d\eta} = -p_a$$

For the two semiclassical wave functions one has

$$\hat{p}_a \Psi = \left(\mp \frac{\partial S}{\partial a}\right) \Psi$$

where upper and lower signs refer to eq.(8) and eq.(9), respectively. Hence, one is tempted to interpret the wave functions (8) and (9) as describing expanding and contracting Universes, respectively. Indeed, the Hartle–Hawking wave function [8] that in the allowed region is a superposition

$$\Psi_{HH} \propto \mathrm{e}^{-iS(a)} + \mathrm{e}^{+iS(a)} \tag{10}$$

is often interpreted as describing a collapsing and re-expanding de Sitter-like Universe. Similar interpretation is often given to the Linde wave function [9]. On the other hand, the tunneling wave functions [10, 11, 12] which contain one wave in the allowed region,

$$\Psi_{tun} \propto \mathrm{e}^{-iS(a)}$$

are often assumed to be the only ones that correspond to an expanding, but not contracting, Universe; this is, at least partially, the basis for the tunneling interpretation.

An important difference with conventional quantum mechanics is, however, the absence of extrinsic time in quantum cosmology. Hence, the arrow of intrinsic time has yet to be determined. In other words, there is no *a priori* reason to interpret the wave functions (8) and (9) as describing expanding and contracting Universes, respectively. The sign of the semiclassical exponent does not by itself determine the arrow of time.

Were the scale factor the only dynamical variable, it would be impossible to decide whether, say, the wave function (8) corresponds to expanding or contracting Universe. If the matter fields (and/or gravitons) are included, this should be possible. Before discussing this point, let us derive the equation for the wave function describing matter [13, 14, 15], again in the spirit of the Born–Oppenheimer approximation.

Let us extend the wave functions (8) and (9) to contain the dependence on the matter variables,

$$|\Psi(a)\rangle = \frac{1}{\sqrt{p_a}} e^{\mp iS(a)} |\tilde{\Psi}(a)\rangle \tag{11}$$

where at given a both $|\Psi(a)\rangle$ and $|\tilde{\Psi}(a)\rangle$ belong to the Hilbert space in which $\hat{\phi}(\mathbf{x})$ and $\hat{p}_\phi(\mathbf{x})$ act. As an example, one may (but does not have to) choose the generalized coordinate representation; then $|\Psi(a)\rangle$ becomes a function $\Psi(\{\phi_\mathbf{k}\}; a)$ of the Fourier components of ϕ. In the first order in $\hbar$ one obtains from eq.(7)

$$\pm i\sqrt{\epsilon - U(a)} \frac{\partial|\tilde{\Psi}(a)\rangle}{\partial a} = \hat{H}_\phi |\tilde{\Psi}(a)\rangle \tag{12}$$

in complete analogy to eq.(2).

The arrow of time is determined now by where (at what a) and which initial conditions are imposed on $|\tilde{\Psi}(a)\rangle$. As an example, let us assume that the initial conditions for the evolution in real intrinsic time are imposed at small a (at the turning point $a^2 = [1 + \sqrt{1 - 16\Lambda^2\epsilon}]/4\Lambda$), and that at that point $|\tilde{\Psi}\rangle$ describes smooth distribution of the scalar field. This type of initial data are characteristic, in particular, to the Hartle–Hawking no-boundary wave function. As a increases, the system will become more and more disordered, independently of the sign in eq.(12). With thermodynamical arrow of time, *both* wave functions (11) will describe expanding Universe.

If, with these initial conditions, one changes variables from a to η using

$$\frac{\partial a}{\partial \eta} = \sqrt{\epsilon - U(a)}$$

then η increases with a, so that η is the conformal intrinsic time, independently of the choice of sign in eq.(11). In the case of positive sign, eq.(12) becomes the conventional Schrödinger equation for quantized matter in the expanding Universe,

$$i\frac{\partial|\tilde{\Psi}\rangle}{\partial\eta} = \hat{H}_\phi(\eta)|\tilde{\Psi}\rangle \tag{13}$$

where the matter Hamiltonian depends on η through $a(\eta)$. On the other hand, in the case of negative sign one obtains "wrong sign" Schrödinger equation,

$$i\frac{\partial|\tilde{\Psi}\rangle}{\partial\eta} = -\hat{H}_\phi(\eta)|\tilde{\Psi}\rangle$$

This little problem is easily cured by considering, instead of $|\tilde{\Psi}\rangle$, its T-conjugate, $|\tilde{\Psi}^{(T)}\rangle$; if the generalized coordinate representation is chosen for $|\tilde{\Psi}\rangle$, then T-conjugation is merely complex conjugation, $\tilde{\Psi}^{(T)}(\phi_{\mathbf{k}};\eta) = \tilde{\Psi}^*(\phi_{\mathbf{k}};\eta)$. The T-conjugate wave function obeys conventional Schrödinger equation, but with CP-transformed Hamiltonian. Hence, the interpretation of *both* wave functions (11) as describing the expanding Universe is self-consistent; the only peculiarity is that the wave function $\mathrm{e}^{+iS}|\tilde{\Psi}\rangle$ corresponds to the Universe in which matter is CP-conjugate. In particular, we argue that both components of the Hartle–Hawking wave function (10) correspond to expanding Universes.

In more generic cases (in particular, when the matter degrees of freedom cannot be treated perturbatively, see, e.g., Refs. [16, 17] and references therein), the situation may be much more complicated. Still, the arrow of time is generally expected to be one of the key issues in the interpretation of the wave function of the Universe.

3.2. CLASSICALLY FORBIDDEN REGION: ISSUE OF STABILITY OF THE BORN-OPPENHEIMER APPROXIMATION

We now consider the region of the scale factor that is classically forbidden in the absence of $\hat{H}_\phi$, i.e., $a_1 < a < a_2$, where

$$a_{1,2}^2 = \frac{1 \mp \sqrt{1 - 16\Lambda^2\epsilon}}{4\Lambda}$$

If $\hat{H}_\phi$ is switched off, there are two semiclassical solutions to the Wheeler–De Witt equation,

$$\Psi \propto \mathrm{e}^{-S(a)} \tag{14}$$

and

$$\Psi \propto \mathrm{e}^{+S(a)} \tag{15}$$

where

$$S(a) = \int_{a_1}^{a} da' \sqrt{2(U(a') - \epsilon)}$$

is defined in such a way that it always increases at large a. The wave function (14) decays as a increases, so it may be interpreted as describing tunneling from classically allowed Friedmann region to de Sitter-like one. It is convenient to introduce Euclidean conformal time parameter τ and consider Euclidean trajectory $a_c(\tau)$ obeying

$$\frac{da_c}{d\tau} = \frac{\partial S}{\partial a}(a = a_c)$$

At $\epsilon = 0$ the Euclidean four-geometry corresponding to this trajectory is a four-sphere, the standard de Sitter instanton.

. Let us now turn on the scalar field Hamiltonian $\hat{H}_\phi$, and try to apply the procedure of the Born–Oppenheimer type. We write, instead of eq.(14), for the wave function decaying at large a,

$$|\Psi(a)\rangle = \frac{1}{\sqrt{p_a}} \mathrm{e}^{-S(a)} |\tilde{\Psi}(a)\rangle$$

and obtain in the first order in $\hbar$ that $|\tilde{\Psi}(a)\rangle$ obeys the "wrong sign" Euclidean Schrödinger equation [12]

$$\frac{\partial|\tilde{\Psi}(\tau)\rangle}{\partial\tau} = +\hat{H}_\phi(\tau)|\tilde{\Psi}(\tau)\rangle \tag{16}$$

where the change of variables from a to τ with $a = a_c(\tau)$ has been performed. The sign on the right hand side of eq.(16) is opposite to that appearing in the usual quantum mechanics, eq.(6), and is directly related to the sign of $\hat{p}_a^2$-term in the Wheeler–De Witt equation (7).

The "wrong" sign in eq.(16) implies that the approximation we use is in fact unstable, if generic "initial" conditions are imposed at small a, say, at $a = a_1$. Note that imposing initial conditions in this way is natural if one interprets the wave function decaying at large a as describing tunneling from small to large a. The formal reason for the instability of the approximation is that the degrees of freedom of the scalar field get excited as a increases in the forbidden region. The rate at which this excitation occurs is generically high [12], and the approximation breaks down well before a gets close to the second turning point a_2.

In the path integral framework, breaking of the Born–Oppenheimer-type approximation for the wave function decaying at large a is also manifest [18]. This wave function corresponds to the Euclidean path integral with "wrong" sign of the action,

$$\int Dg \, D\phi \, \mathrm{e}^{+S[g,\phi]}$$

The instanton action then gives the factor $e^{-S_{inst}}$, but the integral over ϕ (and gravitons) diverges.

The physics behind this instability is that tunneling of a Universe filled with matter is exponentially more probable as compared to empty Universe. Hence, the matter degrees of freedom tend to get excited in the forbidden region, thus making tunneling easier. Note that this property is peculiar to quantum cosmology: in quantum mechanics the situation is opposite, as we discussed in the previous section.

There are exceptional cases in which matter degrees of freedom do not get excited in the forbidden region, e.g., because of symmetry. In our model this would be the case if $\epsilon = 0$ and the scalar field ϕ was in the de Sitter-invariant state, cf. Ref. [19]. Such cases do not seem generic, however.

Breaking of the Born–Oppenheimer approximation does not necessarily mean that tunneling-like transitions from small a to large a with generic state of matter at small a do not make sense. Rather, it is the semiclassical expansion that does not work in this case, so the state of the Universe after the transition may be quite unusual. Presently, neither the properties of this state, nor the properties of the wave function in the forbidden region are understood (except for special cases mentioned above).

The situation is different for the wave functions increasing towards large a, eq.(15). In that case the matter wave function obeys the usual Euclidean Schrödinger equation, $\partial|\tilde{\Psi}(\tau)\rangle/\partial\tau = -\hat{H}_\phi(\tau)|\tilde{\Psi}(\tau)\rangle$, where τ is still assumed to increase with a. Hence, it is possible to impose fairly general initial conditions at small a, and the approximation will not break down. In particular, the Hartle–Hawking wave function is a legitimate approximate solution to the Wheeler–De Witt equation in the forbidden region. This is in accord with the path-integral treatment: the increasing wave function (15) corresponds to the standard sign of the Euclidean action in the path integral.

The non-semiclassical behavior of the tunneling wave functions, signalled by the instability of the Born–Oppenheimer-type approximation, is a special, and potentially interesting, feature of quantum cosmology. It is a challenging technical problem to develop techniques adequate to this situation. It is not excluded also that the properties of the tunneling wave functions are rich and complex, and that understanding them may shed light on the beginning of our Universe.

The author is indebted to A. Albrecht, J. Goldstone, N. Turok, W. Unruh and A. Vilenkin for helpful discussions.

References

1. Coleman, S. and De Luccia, F. (1980) *Phys. Rev.*, **Vol. D21**, p. 3305

2. Rubakov, V.A and Sibiryakov, S.M. (1999) False vacuum decay in de Sitter space-time, gr-qc/9905093
3. Guth, A.H., and Weinberg, E. (1983) *Nucl. Phys.*, **Vol. B212**, p. 321
4. Hawking, S.W. and Moss, I.G. (1983) *Nuc. Phys.*, **Vol. B224**, p. 180
5. Starobinsky, A.A. (1984) In: Fundamental Interactions. Moscow, MGPI, p. 55
6. Goncharov, A.S and Linde, A.D (1986) *Elem. Chast. At. Yad.*, **Vol. 17**, p. 837
7. De Witt, B.S. (1967) *Phys. Rev.*, **Vol. 160**, p. 1113
8. Hartle, J.B. and Hawking, S.W. (1983) *Phys. Rev.*, **Vol. D28**, p. 2960
9. Linde, A.D. (1984) *Lett. Nuovo Cimento*, **Vol. 39**, p. 401
10. Vilenkin, A. (1984) *Phys. Rev.*, **Vol. D30**, p. 509;
    Vilenkin, A. (1986) *Phys. Rev.*, **Vol. D33**, p. 3560
11. Zeldovich, Ya. B. and Starobinsky, A.A. (1984) *Sov. Astron. Lett.*, **Vol. 10**, p. 135
12. Rubakov, V.A. (1984) *Phys. Lett.*, **Vol. 148B**, p. 280
13. Lapchinsky, V.G. and Rubakov, V.A. (1979) *Acta Phys. Polonica*, **Vol. B10**, p. 1041
14. Halliwell, J. and Hawking, S.W. (1985) *Phys. Rev.*, **Vol. D31**, p. 1777
15. Banks, T., Fischler, W. and Susskind, L. (1985) *Nucl. Phys.*, **Vol. B262**, p. 159
16. Khalatnikov, I.M and Kamenshchik, A.Yu. (1998) *Usp. Fiz. Nauk*, **Vol. 168**, p. 593
17. Unruh, W.G and Jheeta, M. (1998) Complex paths and the Hartle Hawking wave function for slow roll cosmologies, gr-qc/9812017
18. Hawking, S.W. and Turok, N.G. (1998) Comment on 'Quantum creation of an open Universe', by Andrei Linde, gr-qc/9802062
19. Vachaspati, T. and Vilenkin, A. (1988) *Phys. Rev.*, **Vol. D37**, p. 898

THE FUTURE OF QUANTUM COSMOLOGY

S. W. HAWKING
DAMTP, Centre for Mathematical Sciences
University of Cambridge, Wilberforce Rd, Cambridge CB3 0WA

1. Introduction

In this lecture, I will describe what I see as the framework for quantum cosmology, on the basis of M-theory. I shall adopt the no boundary proposal and shall argue that the anthropic principle is essential if one is to pick out a solution to represent our universe from the whole zoo of solutions allowed by M-theory.

Cosmology used to be regarded as a pseudo science, an area where wild speculation was unconstrained by any reliable observations. We now have lots and lots of observational data, and a generally agreed picture of how the universe is evolving.

But cosmology is still not a proper science in the sense that, as usually practiced, it has no predictive power. Our observations tell us the present state of the universe, and we can run the equations backward to calculate what the universe was like at earlier times. But all that tells us is that the universe is as it is now because it was as it was then. To go further, and be a real science, cosmology would have to predict how the universe should be. We could then test its predictions against observation, like in any other science.

The task of making predictions in cosmology is made more difficult by the singularity theorems that Roger Penrose and I proved:

The Universe must have had a beginning if

1. Einstein's General Theory of Relativity is correct
2. The energy density is positive
3. The universe contains the amount of matter we observe

These showed that if General Relativity were correct, the universe would have begun with a singularity. Of course, we would expect classical General Relativity to break down near a singularity, when quantum gravitational

R.G. Crittenden and N.G. Turok (eds.), Structure Formation in the Universe, 75–89.

effects have to be taken into account. So what the singularity theorems are really telling us is that the universe had a quantum origin, and that we need a theory of quantum cosmology if we are to predict the present state of the universe.

A theory of quantum cosmology has three aspects:

1. Local theory - M Theory
2. Boundary conditions - No boundary proposal
3. Anthropic principle

The first is the local theory that the fields in spacetime obey. The second is the boundary conditions for the fields. I shall argue that the anthropic principle is an essential third element.

1.1. P-BRANE DEMOCRACY

We hold these truths as self evident: All p-branes are created equal.

As far as the local theory is concerned the best, and indeed the only, consistent way we know to describe gravitational forces is curved spacetime. The theory has to incorporate supersymmetry, because otherwise the uncancelled vacuum energies of all the modes would curl spacetime into a tiny ball. These two requirements seemed to point to supergravity theories, at least until 1985. But then the fashion changed suddenly. People declared that supergravity was only a low energy effective theory, because the higher loops probably diverged, though no one was brave (or fool-hardy) enough to calculate an eight loop diagram. Instead, the fundamental theory was claimed to be superstrings, which were thought to be finite to all loops. But it was discovered that strings were just one member of a wider class of extended objects, called p-branes. It seems natural to adopt the principle of p-brane democracy.

Yet for $p > 1$, the quantum theory of p-branes diverges for higher loops. I think we should interpret these loop divergences not as a break down of the supergravity theories, but as a break down of naive perturbation theory. In gauge theories, we know that perturbation theory breaks down at strong coupling. In quantum gravity, the role of the gauge coupling is played by the energy of a particle. In a quantum loop, one integrates over all energies, so one would expect perturbation theory to break down.

In gauge theories, one can often use duality to relate a strongly coupled theory, where perturbation theory is bad, to a weakly coupled one, in which it is good. The situation seems to be similar in gravity, with the relation between ultra-violet and infra-red cut-offs in the AdS-CFT correspondence. I shall therefore not worry about the higher loop divergences and use eleven dimensional supergravity as the local description of the universe. This also goes under the name of M-theory, for those that rubbished supergravity in the 80s and don't want to admit it was basically correct. In fact, as I shall show, it seems the origin of the universe is in a regime in which first order perturbation theory is a good approximation.

1.2. BOUNDARY CONDITIONS FOR QUANTUM COSMOLOGY

The second pillar of quantum cosmology is boundary conditions for the local theory. There are three candidates: the pre big bang scenario, the tunnelling hypothesis, and the no boundary proposal.

The pre big bang scenario claims that the boundary condition is some vacuum state in the infinite past. But if this vacuum state develops into the universe we have now, it must be unstable. And if it is unstable, it wouldn't be a vacuum state, and it wouldn't have lasted an infinite time before becoming unstable.

The quantum tunnelling hypothesis is not actually a boundary condition on the spacetime fields, but on the Wheeler-Dewitt equation. However, the Wheeler-Dewitt equation acts on the infinite dimensional space of all fields on a hyper-surface and is not well defined. Also, the $3+1$, or $10+1$, split is putting apart that which God, or Einstein, has joined together. In my opinion, therefore, neither the pre bang scenario, nor quantum tunnelling hypothesis, are viable.

To determine what happens in the universe, we need to specify the boundary conditions on the field configurations that are summed over in the path integral. One natural choice would be metrics that are asymptotically Euclidean, or asymptotically Anti-de Sitter. These would be the relevant boundary conditions for scattering calculations, where one sends particles in from infinity and measures what comes back out.

However, they are not the appropriate boundary conditions for cosmology. We have no reason to believe the universe is asymptotically Euclidean or Anti-de Sitter. Even if it were, we are not concerned about measurements at infinity, but in a finite region in the interior. For such measurements, there will be a contribution from metrics that are compact, without boundary. The action of a compact metric is given by integrating the Lagrangian. Thus, its contribution to the path integral is well defined.

By contrast, the action of a non-compact, or singular, metric involves a surface term at infinity or at the singularity. One can add an arbitrary quantity to this surface term. It therefore seems more natural to adopt what Jim Hartle and I called, the 'no boundary proposal'. The quantum state of the universe is defined by a Euclidean path integral over compact metrics. In other words, the boundary condition of the universe is that it has no boundary.

1.3. THE ANTHROPIC PRINCIPLE

There are compact Ricci flat metrics of any dimension, many with high dimensional moduli spaces. Thus eleven dimensional supergravity, or M-theory, admits a very large number of solutions and compactifications. There may be some principle that we haven't yet thought of that restricts the possible models to a small subclass, but it seems unlikely. Thus I believe that we have to invoke the anthropic principle. Many physicists dislike the anthropic principle. They feel it is messy and vague, that it can be used to explain almost anything, and that it has little predictive power. I sympathize with these feelings, but the anthropic principle seems essential in quantum cosmology. Otherwise, why should we live in a four dimensional world and not eleven, or some other number of dimensions? The anthropic answer is that two spatial dimensions are not enough for complicated structures like intelligent beings.

On the other hand, four or more spatial dimensions would mean that gravitational and electric forces would fall off faster than the inverse square law. In this situation, planets would not have stable orbits around their star, nor would electrons have stable orbits around the nucleus of an atom. Thus intelligent life, at least as we know it, could exist only in four dimensions. I very much doubt we will find a non-anthropic explanation.

The anthropic principle is usually said to have weak and strong versions. According to the strong anthropic principle, there are millions of different universes, each with different values of the physical constants. Only those universes with suitable physical constants will contain intelligent life. With the weak anthropic principle, there is only a single universe. But the effective couplings are supposed to vary with position, and intelligent life occurs only in those regions in which the couplings have the right values. Even those who reject the strong anthropic principle, would accept some weak anthropic arguments. For instance, the reason stars are roughly half way through their evolution is that life could not have developed before stars, or have continued when they burnt out.

When one goes to quantum cosmology, however, and uses the no boundary proposal, the distinction between the weak and strong anthropic principles disappears. The different physical constants are just different moduli of the internal space, in the compactification of M-theory, or eleven dimensional supergravity. All possible moduli will occur in the path integral over compact metrics. By contrast, if the path integral was over non compact metrics, one would have to specify the values of the moduli at infinity. Each set of moduli at infinity would define a different super-selection sector of the theory, and there would be no summation over sectors. It would then be just an accident that the moduli at infinity have those particular values, like

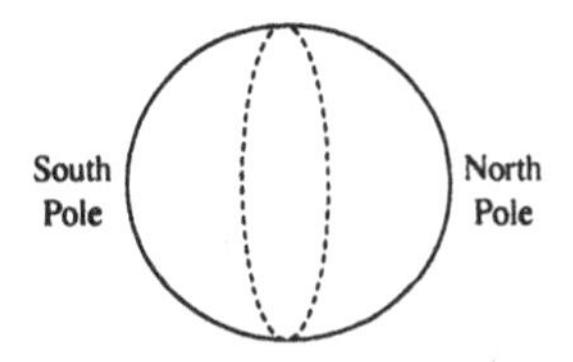

Figure 1. Euclidean four-sphere

four uncompactified dimensions, that allow intelligent life. Thus it seems that the anthropic principle really requires the no boundary proposal, and vice versa.

One can make the anthropic principle precise by using Bayesian statistics,

$$P(\Omega_{matter}, \Omega_\Lambda \mid Galaxy) \propto P(Galaxy \mid \Omega_{matter}, \Omega_\Lambda) \times P(\Omega_{matter}, \Omega_\Lambda). \quad (1)$$

One takes the a-priori probability of a class of histories, to be the e to the minus the Euclidean action, given by the no boundary proposal. One then weights this a-priori probability, with the probability that the class of histories contain intelligent life. As physicists, we don't want to be drawn into to the fine details of chemistry and biology, but we can reckon certain features as essential prerequisites of life as we know it. Among these are the existence of galaxies and stars, and physical constants near what we observe. There may be some other region of moduli space that allows some different form of intelligent life, but it is likely to be an isolated island. I shall therefore ignore this possibility and just weight the a-priori probability with the probability to contain galaxies.

2. Instanton Solutions

The simplest compact metric that could represent a four dimensional universe would be the product of a four-sphere, with a compact internal space:

$$ds^2 = d\sigma^2 + \frac{1}{H^2}\sin^2 H\sigma(d\chi^2 + \sin^2\chi d\Omega^2). \quad (2)$$

(See Figure 1.) But the world we live in has a metric with Lorentzian signature, rather than a positive definite Euclidean one. So one has to analytically continue the four sphere metric to complex values of the coordinates.

There are several ways of doing this:

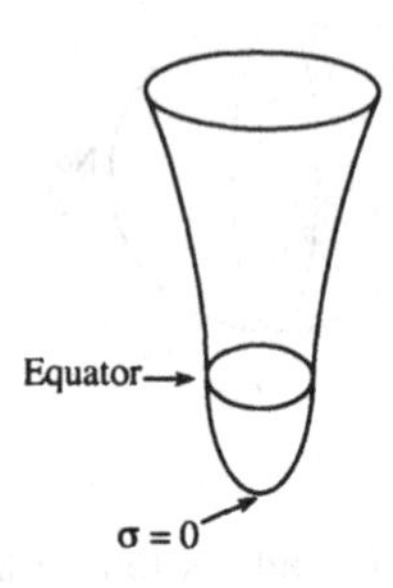

Figure 2. Analytical continuation to a closed universe: analytically continue $\sigma = \sigma_{equator} + it$.

One can analytically continue the coordinate, σ, as $\sigma_{equator} + it$. One obtains a Lorentzian metric, which is a closed Friedmann solution, with a scale factor that goes like $\cosh(Ht)$,

$$ds^2 = -dt^2 + \frac{1}{H^2}\cosh^2 Ht(d\chi^2 + \sin^2\chi d\Omega^2). \tag{3}$$

So this is a closed universe that starts at the Euclidean instanton and expands exponentially.

However, one can analytically continue the four sphere in another way. Define $t = i\sigma$, and $\chi = i\psi$. This gives an open Friedmann universe, with a scale factor like $\sinh(Ht)$,

$$ds^2 = -dt^2 + (\frac{1}{H}\sinh Ht)^2(d\psi^2 + \sinh^2\psi d\Omega^2). \tag{4}$$

Thus one can get an apparently spatially infinite universe from the no boundary proposal. The reason is that one is using as a time coordinate the hyperboloids of constant distance inside the light cone of a point in de Sitter space. The point itself, and its light cone, are the big bang of the Friedmann model, where the scale factor goes to zero. But they are not singular. Instead, the spacetime continues through the light cone to a region beyond. It is this region that deserves the name the 'Pre Big Bang Scenario', rather than the misguided model that commonly bears that title.

If the Euclidean four-sphere were perfectly round, both the closed and open analytical continuations would inflate for ever. This would mean they would never form galaxies. A perfectly round four-sphere has a lower action, and hence a higher a-priori probability than any other four-metric of the same volume. However, one has to weight this probability with the

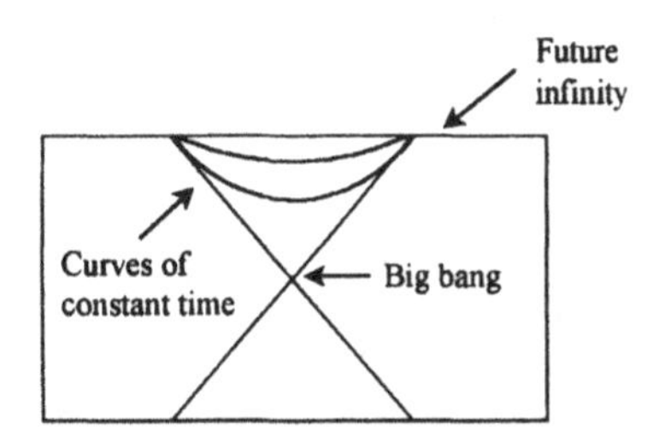

Figure 3. Penrose diagram of an open analytical continuation

probability of intelligent life, which is zero. Thus we can forget about round four-spheres.

On the other hand, if the four-sphere is not perfectly round, the analytical continuation will start out expanding exponentially, but it can change over later to radiation or matter dominated and can become very large and flat.

This means there are equal opportunities for dimensions. All dimensions, in the compact Euclidean geometry, start out with curvatures of the same order. But in the Lorentzian analytical continuation, some dimensions can remain small, while others inflate and become large. However, equal opportunities for dimensions might allow more than four to inflate. So, we will still need the anthropic principle to explain why the world is four dimensional.

In the semi-classical approximation, which turns out to be very good, the dominant contribution comes from metrics near instantons. These are solutions of the Euclidean field equations. So we need to study deformed four-spheres in the effective theory obtained by dimensional reduction of eleven dimensional supergravity, to four dimensions. These Kaluza-Klein theories contain various scalar fields that come from the three-index field and the moduli of the internal space. For simplicity, I will describe only the single scalar field case.

The scalar field, ϕ, will have a potential, $V(\phi)$. The energy-momentum tensor is given by,

$$T_{\mu\nu} = \phi_{,\mu}\phi_{,\nu} - \frac{1}{2}g_{\mu\nu}[\phi_{,\lambda}\phi^{,\lambda} + V(\phi)] \tag{5}$$

In regions where the gradients of ϕ are small, the energy momentum tensor will act like a cosmological constant, $\Lambda = 8\pi GV$, where G is Newton's

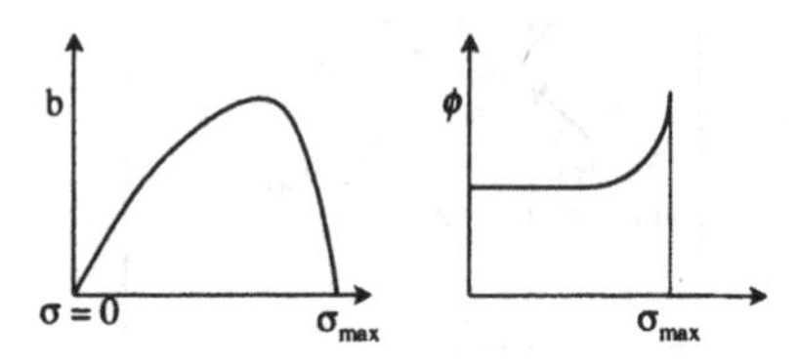

Figure 4. O(4) Instantons

constant in four dimensions. Thus it will curve the Euclidean metric like a four-sphere.

However, if the field ϕ is not at a stationary point of V, it can not have zero gradient everywhere. This means that the solution can not have O(5) symmetry, like the round four-sphere. The most it can have is O(4) symmetry. In other words, the solution is a deformed four-sphere.

One can write the metric of an O(4) instanton in terms of a function, $b(\sigma)$,

$$ds^2 = d\sigma^2 + b^2(\sigma)(d\chi^2 + \sin^2 \chi d\Omega^2). \tag{6}$$

Here b is the radius of a three sphere of constant distance, σ, from the north pole of the instanton. If the instanton were a perfectly round four-sphere, b would be a sine function of σ. It would have one zero at the north pole, and a second at the south pole, which would also be a regular point of the geometry. However, if the scalar field at the north pole is not at a stationary point of the potential, it will vary over the four-sphere. If the potential is carefully adjusted, and has a false vacuum local minimum, it is possible to obtain a solution that is non-singular over the whole four sphere. This is known as the Coleman-De Lucia instanton.

However, for general potentials without a false vacuum, the behavior is different. The scalar field will be almost constant over most of the four-sphere, but will diverge near the south pole. This behavior is independent of the precise shape of the potential and holds for any polynomial potential and for any exponential potential with an exponent, a, less than 2. The scale factor, b, will go to zero at the south pole, like distance to the third. This means the south pole is actually a singularity of the four dimensional geometry. However, it is a very mild singularity, with a finite value of the trace K surface term on a boundary around the singularity at the south

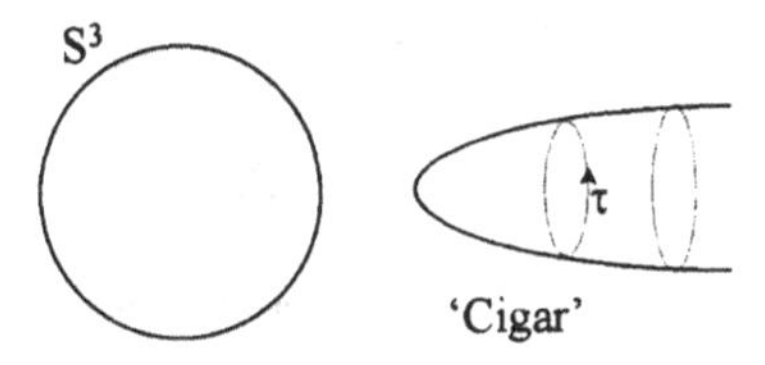

Figure 5. Five dimensional view of the singularity.

pole. This means the actions of perturbations of the four dimensional geometry are well defined, despite the singularity. One can therefore calculate the fluctuations in the microwave background, as I shall describe later.

The deep reason behind this good behavior of the singularity was first seen by Garriga. He dimensionally reduced five dimensional Euclidean Schwarzschild along the τ direction (see Figure 5) to get a four dimensional geometry and a scalar field. These were singular at the horizon, in the same manner as at the south pole of the instanton. In other words, the singularity at the south pole can be just an artefact of dimensional reduction and the higher dimensional space can be non-singular. This is true quite generally. The scale factor, b, will go like distance to the third when the internal space collapses to zero size in one direction.

When one analytically continues the deformed sphere to a Lorentzian metric, one obtains an open universe, which is inflating initially. One can think of this as a bubble in a closed, de Sitter-like universe. In this way, it is similar to the single bubble inflationary universes that one obtains from Coleman-De Lucia instantons. The difference is the Coleman-De Lucia instantons required carefully adjusted potentials, with false vacuum local minima. But the singular Hawking-Turok instanton will work for any reasonable potential. The price one pays for a general potential is a singularity at the south pole. In the analytically continued Lorentzian spacetime, this singularity would be time-like and naked. One might think that anything could come out of this naked singularity and propagate through the big bang light cone into the open inflating region. Thus one would not be able to predict what would happen. However, as I already said, the singularity at the south pole of the four-sphere is so mild that the actions of the instanton, and of perturbations around it, are well defined.

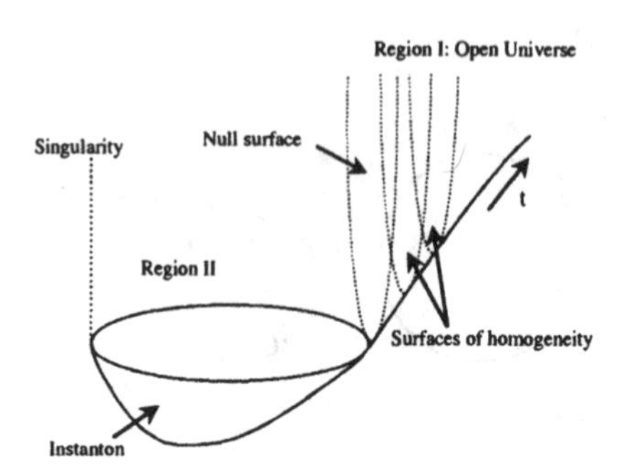

Figure 6. The Hawking-Turok instanton

This behavior of the singularity means one can determine the relative probabilities of the instanton and of perturbations around it. The action of the instanton itself is negative, but the effect of perturbations around the instanton is to increase the action. That is, to make the action less negative. According to the no boundary proposal, the probability of a field configuration is e to minus its action. Thus perturbations around the instanton have a lower probability than the unperturbed background. This means that the more quantum fluctuations are suppressed, the bigger the fluctuation, as one would hope. This is not the case with some versions of the tunnelling boundary condition.

3. The observed universe

How well do these singular instantons account for the universe we live in? The hot big bang model seems to describe the universe very well, but it leaves unexplained a number of features.

Problems of a Hot Big Bang

1. Isotropy
2. Amplitude of fluctuations
3. Density of matter
4. Vacuum energy

There is the overall isotropy of the universe and the origin and spectrum of small departures from isotropy. Then there is the fact that the density was sufficiently low to let the universe expand to its present size, but not so low that the universe is empty now. And the fact that despite symmetry breaking, the energy of the vacuum is either exactly zero or at least very small.

Inflation was supposed to solve the problems of the hot big bang model. It does a good job with the first problem, the isotropy of the universe. If the inflation continues for long enough, the universe would now be spatially flat, which would imply that the sum of the matter and vacuum energies had the critical value.

But inflation, by itself, places no limits on the other linear combination of matter and vacuum energies and does not give an answer to problem two, the amplitude of the fluctuations. These have to be fed in as fine tunings of the scalar potential, V. Also, without a theory of initial conditions, it is not clear why the universe should start out inflating in the first place.

The instantons I have described predict that the universe starts out in an inflating, de Sitter-like state. Thus they solve the first problem, the fact that the universe is isotropic. However, there are difficulties with the other three problems. According to the no boundary proposal, the a-priori probability of an instanton is e to the minus the Euclidean action. But if the Ricci scalar is positive, as is likely for a compact instanton with an isometry group, the Euclidean action will be negative.

The larger the instanton, the more negative will be the action, and so the higher the a-priori probability. Thus the no boundary proposal favours large instantons. In a way, this is a good thing, because it means that the instantons are likely to be in the regime where the semi-classical approximation is good. However, a larger instanton means starting at the north pole with a lower value of the scalar potential, V. If the form of V is given, this in turn means a shorter period of inflation. Thus the universe may not achieve the number of e-foldings needed to ensure $\Omega_{matter} + \Omega_{\Lambda}$ is near to one now.

In the case of the open Lorentzian analytical continuation considered here, the no boundary a-priori probabilities would be heavily weighted towards $\Omega_{matter} + \Omega_{\Lambda} = 0$. Obviously, in such an empty universe, galaxies would not form and intelligent life would not develop. So one has to invoke the anthropic principle.

If one is going to have to appeal to the anthropic principle, one may as well use it also for the other fine tuning problems of the hot big bang. These are the amplitude of the fluctuations and the fact that the vacuum energy now is incredibly near zero. The amplitude of the scalar perturbations depends on both the potential and its derivative. In most potentials the scalar perturbations are of the same form as the tensor perturbations, but are larger by a factor of about ten. For simplicity, I shall consider just the tensor perturbations. They arise from quantum fluctuations of the metric, which freeze in amplitude when their co-moving wavelength leaves the horizon during inflation.

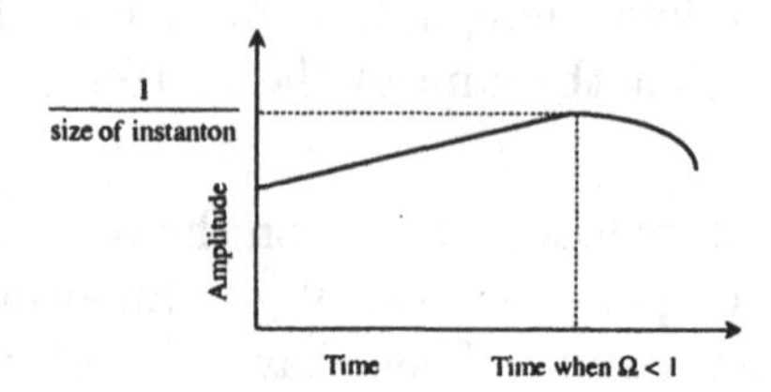

Figure 7. Amplitude of perturbations when they come into the visible universe

Thus, the amplitude of the tensor perturbation will be roughly one over the horizon size, in Planck units. Longer co-moving wavelengths will leave the horizon earlier during inflation. Thus the spectrum of the tensor perturbations at the time they re-enter the horizon will slowly increase with wave length, up to a maximum of one over the size of the instanton.

The time at which the maximum amplitude re-enters the horizon is also the time at which Ω begins to drop below one. There are two competing effects. One is the a-priori probability from the no boundary proposal, which wants to make the instantons large. The other is the probability of the formation of galaxies. This requires sufficient inflation to keep Ω near to one, and a sufficient amplitude of the fluctuations. Both these favour small instanton sizes. Where the balance occurs depends on whether we weight with the density of galaxies per unit proper volume, or by the total number of galaxies. If we weight with the present proper density of galaxies, the probability distribution for Ω would be sharply peaked at about $\Omega = 10^{-3}$.

This is the minimum value that would give one galaxy in the observable universe and clearly does not agree with observation. On the other hand, one might think that one should weight with a factor proportional to the total number of galaxies in the universe. In this case, one would multiply the probability by a factor e^{3n}, where n is the number of e-foldings during inflation. This would lead to the prediction that $\Omega = 1$, which seems to be consistent with observation, as I shall discuss.

So far I haven't taken into account the anthropic requirement that the cosmological constant is very small now. Eleven dimensional supergravity contains a three-form gauge field, with a four form field strength. When reduced to four dimensions, this acts as a cosmological constant. For real components in the Lorentzian four dimensional space, this cosmological

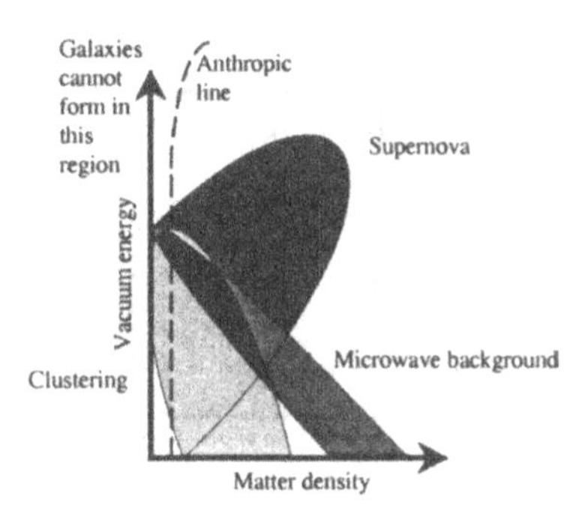

Figure 8. Comparison of supernova, microwave background and clustering regions

constant is negative. Thus it can cancel the positive cosmological constant that arises from super symmetry breaking. Supersymmetry breaking is an anthropic requirement. One could not build intelligent beings from mass less particles. They would fly apart.

Unless the positive contribution from symmetry breaking cancels almost exactly with the negative four-form, galaxies wouldn't form, and again, intelligent life wouldn't develop. I very much doubt we will find a non-anthropic explanation for the cosmological constant.

In the eleven dimensional geometry, the integral of the four-form over any four cycle, or its dual over any seven cycle, have to be integers. This means that the four-form is quantized and can not be adjusted to cancel the symmetry breaking exactly. In fact, for reasonable sizes of the internal dimensions, the quantum steps in the cosmological constant would be much larger than the observational limits. At first, I thought this was a set back for the idea there was an anthropically controlled cancellation of the cosmological constant. But then I realized that it was positively in favour. The fact that we exist shows that there must be a solution to the anthropic constraints.

But the fact that the quantum steps in the cosmological constant are so large means that this solution is probably unique. This helps with the problems of low Ω, or Ω exactly one, I described earlier. If there were a continuous family of solutions, the strong dependence of the Euclidean action and the amount of inflation on the size of the instanton would bias the probability either to the lowest Ω, or $\Omega = 1$. This would give either a single galaxy in an otherwise empty universe or a universe with Ω exactly one.

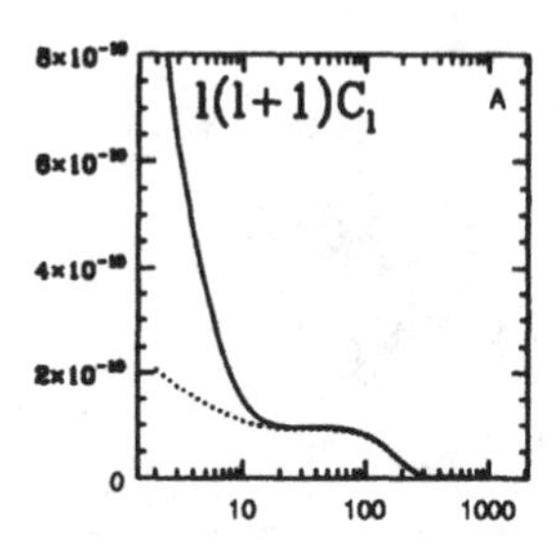

Figure 9. Predictions for the microwave background spectrum resulting from gravitational waves with the Hawking-Turok instanton (dotted) and with the Coleman-De Luccia instanton (solid).

But if there is only one instanton in the anthropically allowed range, the biasing towards large instantons has no effect. Thus Ω_{matter} and Ω_Λ could be somewhere in the anthropically allowed region, though it would be below the $\Omega_{matter} + \Omega_\Lambda = 1$ line if the universe is one of these open analytical continuations. This is consistent with the observations.

Figure 8 shows the present state of cosmological observations of Ω_{matter} and Ω_Λ. The red elliptic region is the three sigma limits of the supernova observations. The blue region is from clustering observations and the purple is from the Doppler peak in the microwave. They seem to have a common intersection, on or below the $\Omega_{total} = 1$ line.

Assuming that one can find a model that predicts a reasonable Ω, how can we test it by observation? The best way is by observing the spectrum of fluctuations in the microwave background. This is a very clean measurement of the quantum fluctuations about the initial instanton. However, there is an important difference between the non-singular Coleman-De Lucia instantons and the singular instantons I have described.

As I said, quantum fluctuations around the instanton are well defined, despite the singularity. Perturbations of the Euclidean instanton have finite action if and only they obey a Dirichlet boundary condition at the singularity. Perturbation modes that do not obey this boundary condition will have infinite action and will be suppressed. The Dirichlet boundary condition also arises if the singularity is resolved in higher dimensions.

When one analytically continues to Lorentzian spacetime, the Dirichlet boundary condition implies that perturbations reflect at the time like singularity. This has an effect on the two point correlation function of the perturbations. It is very small for the density perturbations, but calculations by Gratton, Hertog and Turok indicate a significant difference for gravitational waves if Ω is less than one.

The present observations of the microwave fluctuations are certainly not sensitive enough to detect this effect. But it may be possible with the new observations that will be coming in from the MAP satellite in 2001 and the Planck satellite in 2006. Thus the no boundary proposal and the singular instanton are real science. They can be falsified by observation.

I will finish on that note.

Part II
THE BACKGROUND UNIVERSE

THE GLOBAL COSMOLOGICAL PARAMETERS

MASATAKA FUKUGITA
University of Tokyo, Institute for Cosmic Ray Research
Tanashi, Tokyo 188, Japan, and
Institute for Advanced Study, Princeton, NJ 08540, U. S. A.

1. Introduction

In these lectures I shall discuss the status of the determination of the three cosmological parameters which enter the Einstein equation and govern geometry and evolution of space-time of the Universe: the Hubble constant H_0, the mass density parameter Ω and the cosmological constant λ.

Among the three parameters, the Hubble constant is the dimensionfull quantity which sets the basic size and age of the Universe. The perennial effort to determine H_0 dates back to Hubble (1925) and has a long history of disconcordance. Recent progress has done much to resolve the long-standing discrepancy concerning the extragalactic distance scale, but there are some newly revealed uncertainties in the distance scale within the Milky Way. The emphasis in this lecture is on discussion of these uncertainties.

The mass density parameter directly determines the formation of cosmic structure. So, as our understanding of the cosmic structure formation is tightened, we should have a convergence of the Ω parameter. An important test is to examine whether the Ω parameter extracted from cosmic structure formation agrees with the value estimated in more direct ways. This gives an essential verification for the theory of structure formation.

The third important parameter in the Friedmann universe is the cosmological constant Λ. We now have some evidence for a non-zero Λ which, if confirmed, would have most profound implications for fundamental physics. This lecture will focus on the strength of this 'evidence'.

We take the normalisation

$$\Omega + \lambda = 1 \tag{1}$$

R.G. Crittenden and N.G. Turok (eds.), Structure Formation in the Universe, 93–130.

for the flat curvature, where $\lambda = \Lambda/3H_0^2$ with Λ the constant entering in the Einstein equation. The case with $\Omega = 1$ and $\lambda = 0$ is referred to as the Einstein-de Sitter (EdS) universe. We often use distance modulus

$$m - M = 5\log(d_L/10\text{pc}) \tag{2}$$

instead of the distance d_L. For conciseness, we shall omit the units for the Hubble constant, (km $\text{s}^{-1}\text{Mpc}^{-1}$).

After the Summer Institute there appeared several important papers on the distance scale. I try to incorporate these results in this article.

2. The Hubble Constant

2.1. HISTORICAL NOTE

The global value of H_0 has long been uncertain by a factor of two. Before 1980 the dispute was basically between two schools: Sandage and collaborators had insisted on $H_0 = 50$ (Sandage & Tammann 1982); de Vaucouleurs and collaborators preferred a high value $H_0 = 90 - 100$ (de Vaucouleurs 1981). Conspicuous progress was brought by the discovery of an empirical but tight relationship between galaxy's luminosity and rotation velocity, known as the Tully-Fisher relation (Tully & Fisher 1977). The use of the Tully-Fisher relation has largely reduced subjective elements in the distance work, and $H_0 = 80 - 90$ has been derived from a straightforward reading of the Tully-Fisher relation. Representative of this work are the papers of Aaronson et al. (1986) and Pierce & Tully (1988). A doubt was whether the result was marred with the Malmquist bias — whether the sample selects preferentially bright galaxies, and hence the result was biased towards a shorter distance (Kraan-Korteweg, Cameron & Tammann 1988; Sandage 1993a). A related dispute was over the distance to the Virgo cluster, whether it is 16 Mpc or 22 Mpc: the different results depending on which sample one used.

The next momentous advancement was seen in 1989–1990 when a few qualified distance indicators were discovered. One of them is a technique using planetary nebula luminosity function (PNLF), the shape of which looked universal (Jacoby et al. 1990a). Another important technique is the use of surface brightness fluctuations (SBF), utilizing the fact that the images of distant galaxies show a smoother light distribution; while surface brightness does not depend on the distance, pixel-to-pixel fluctuations in a CCD camera decreases as d_L^{-1} (Tonry & Schneider 1988). They proposed that this smoothness can be a distance indicator if the stellar population is uniform. What was important is that the two completely independent methods predicted distances to individual galaxies in excellent agreement with each other (Ciardullo, Jacoby & Tonry 1993). The PNLF/SBF distance also agreed with the value from the Tully-Fisher relation, with a somewhat larger scatter. These new techniques, when calibrated with the distance to M31, yielded a value around $H_0 = 80$ and the Virgo distance of 15 Mpc (For a review of the methods, see Jacoby et al. 1992).

Around the same time the use of Type Ia supernovae (SNeIa) became popular (Tammann & Leibundgut 1990; Leibundgut & Pinto 1992; Branch & Miller 1993). The principle is that the maximum brightness of SNIa is nearly constant, which can be used as an absolute standard candle. Arnett, Branch and Wheeler proposed that the maximum brightness is reliably calculable using models which are constrained from observations of released kinetic energy (Arnett, Branch & Wheeler 1985; Branch 1992). This led to $H_0 = 50 - 55$, in agreement with the calibration based on the first Cepheid measurement of the nearest SNIa host galaxy using the pre-refurbished *Hubble Space Telescope* (HST) (Sandage et al. 1992). In the early nineties the discrepancy was dichotomous as whether $H_0 = 80$ *or* 50. (see Fukugita, Hogan & Peebles 1993 for the status at that time; see also van den Bergh 1989, 1994).

The next major advancement was brought with the refurbishment mission of HST, which enabled one to resolve Cepheids in galaxies as distant as 20 Mpc (1994). This secured the distance to the Virgo cluster and tightened the calibrations of the extragalactic distance indicators, resulting in $H_0 = (70 - 75) \pm 10$, 10% lower than the 'high value'. Another important contribution was the discovery that the maximum brightness of SNeIa varies from supernova to supernova, and that it correlates with the decline rate of brightness (Pskovskiĭ 1984; Phillips 1993; Riess, Press & Kirshner 1995; Hamuy et al. 1996a). This correction, combined with the direct calibration of the maximum brightness of several SNeIa with HST Cepheid observations, raised the 'low value' of H_0 to $65\,^{+5}_{-10}$, appreciably higher than 55. This seemed to resolve the long-standing controversy.

All methods mentioned above use distance ladders and take the distance to Large Magellanic Clouds (LMC) to be 50 kpc ($m - M = 18.5$) as the zero point. Before 1997 few doubts were cast on the distance to LMC (TABLE 1 shows a summary of the distance to LMC known as of 1997). With the exception of RR Lyr, the distance converged to $m - M = 18.5 \pm 0.1$, i.e., within 5% error, and the discrepancy of the RR Lyr distance was blamed on its larger calibration error. It had been believed that the Hipparcos mission (ESA 1997) would secure the distance within MW and tighten the distance to LMC. To our surprise, the work using the Hipparcos catalogue revealed the contrary; the distance to LMC was more uncertain than we had thought, introducing new difficulties into the determination of H_0. In this connection, the age of the Universe turned out to be more uncertain than it was believed.

During the nineties, efforts have also been conducted to determine the Hubble constant without resorting to astronomical ladders. They are called 'physical methods'. The advantage of the ladder is that the error of each ladder can be documented relatively easily, while the disadvantage is that these errors accumulate. Physical methods are free from the accumulation of errors, but on the other hand it is not easy to document the systematic errors. Therefore, the central problem is how to minimise the model dependence and document realistic systematic errors. Nearly ten years of effort has brought results that can be compared with the distances from ladders. The physical methods include the expansion photosphere model (EPM) for type II SNe (Schmidt, Kirshner & Eastman 1992) and gravitational lensing time delay (Refsdal 1964). Use of SNeIa maximum brightness was once taken to be a physical method (Branch 1992), but then 'degraded' to be a ladder, which however significantly enhanced its accuracy.

TABLE 1. Distance to LMC as of 1997

Method	Ref	Distance moduli
Cepheid optical PL	Feast & Walker 1987	18.47±0.15
Cepheid optical PL	Madore & Freedman 1991	18.50±0.10
Cepheid IR PL	Laney & Stobie 1994	18.53±0.04
Mira PL	Feast & Walker 1987	18.48±(0.06)
SN1987A ring echo	Panagia et al. 1991	18.50±0.13
SN1987A EPM	Schmidt et al. 1992	18.45±0.13
RR Lyrae	van den Bergh 1995	18.23±0.04

2.2. EXTRAGALACTIC DISTANCE SCALE

The measurement of cosmological distances traditionally employs distance ladders (see Weinberg 1972). The most traditional ladders are shown in TABLE 2. The listings written in italic indicate new methods which circumvent intermediate rungs. The most important milestone of the ladder is LMC at 50kpc ($m-M = 18.5$). A distance indicator of particular historical importance (Hubble 1925) is the Cepheid period-luminosity (PL) relation, which is given a great confidence, but we note that it requires a few rungs of ladders to calibrate its zero point.

Prior to the HST work there were only 4–5 galaxies with Cepheid distances which could be used to calibrate secondary indicators. The reach of the ground-based Cepheid measurement is about 3 Mpc, which means that

TABLE 2. Traditional distance ladders

Method	Distance range	typical targets
Population I stars		
trigonometric or kinematic methods (ground)	<50 pc	Hyades, nearby dwarfs
main sequence fitting (FG stars) Pop. I	<200pc	Pleiades
trigonometric method (Hipparcos)	<500pc	nearby open clusters
main sequence fitting (B stars)	40pc−10kpc	open clusters
Cepheids [Population I] (ground)	1kpc−3Mpc	LMC, M31, M81
Cepheids [Population I](HST)	<30Mpc	Virgo included
secondary (extragalactic) indicators	700kpc−100Mpc	
Population II stars		
trigonometric method (Hipparcos)	<500pc	nearby subdwarfs
subdwarf main sequence fitting	100pc−10kpc	global clusters
cluster RR Lyr	5kpc−100kpc	LMC, age determinations

one cannot increase the number of calibrating galaxies from the ground. Pierce et al. (1994) could finally measure Cepheids in NGC 4571 in the Virgo cluster at 15 Mpc, but only with the best seeing conditions and difficult observations. The refurbishment of HST achieved a sufficient power to resolve Cepheids at the Virgo cluster (Freedman et al. 1994). Now 28 nearby spiral galaxies within 25 Mpc are given distances measured using the Cepheid PL relation (Ferrarese et al. 1999b). A typical random error is 4-5% (0.08-0.10 mag), and the systematic error (from photometry) is 5% (0.1 mag) excluding the uncertainty of the LMC distance, to which the HST-Key Project(KP) group assigns 6.5% error (0.13 mag). The prime use of these galaxies is to calibrate secondary distance indicators which penetrate into a sufficient depth that perturbations in the Hubble flow are small enough compared with the flow itself.

Cepheids are Population I stars, so reside only in spiral galaxies. The calibration is therefore direct for TF and some SNeIa. For early type galaxies (fundamental plane or $D_n - \sigma$, and SBF) the calibration is not very tight; one must either use some groups where both early and late galaxies coexist, or regard the bulges of spiral galaxies as belonging to the same class as early galaxies and avoid contaminations from discs. Additional observations have been made for the galaxies that host SNeIa (Saha et al. 1999). The results are summarised in TABLE 3. We include a few earlier SNIa results which employ a partial list of Cepheid calibrators.

We accentuate the results with the two methods, SBF and SNeIa, in

TABLE 3. Hubble constant

Secondary indicators	Refs	Hubble constant
Tully-Fisher	HST-KP (Sakai et al. 1999)	$71 \pm 4 \pm 7$
Fundamental Plane	HST-KP (Kelson et al. 1999)	$78 \pm 8 \pm 10$
SBF	HST-KP (Ferrarese et al. 1999a)	$69 \pm 4 \pm 6$
SBF	Tonry et al. (1999)	$\underline{77 \pm 4 \pm 7}$
SNeIa	Riess et al. (1995)	67 ± 7
SNeIa	Hamuy et al. (1996b)	$63 \pm 3 \pm 3$
SNeIa	Jha et al. (1999)	$64.4^{+5.6}_{-5.1}$
SNeIa	Suntzeff et al. (1999)	65.6 ± 1.8
SNeIa	HST-KP (Gibson et al. 1999)	$\underline{68 \pm 2 \pm 5}$
SNeIa	Saha et al. (1999)	60 ± 2
Summary (see text)		$(64 - 78) \pm 7$

particular to those we underlined in the table. A cross correlation analysis showed that the relative distances agree well between SBF and others, including the Cepheid (Tonry et al. 1997; Freedman et al. 1997), and that it is probably the best secondary indicator presently available together with SNeIa; Also important is that there are now 300 galaxies measured with SBF, which are essential to make corrections for peculiar velocity flows for their ≤ 4000 km s^{-1} sample (Tonry et al. 1999). (PNLF is an indicator of comparable quality, but it requires more expensive observations so that applications are rather limited; see Jacoby et al. 1996 for the recent work.) The final value of Tonry et al. from their I band survey is $H_0 = 77 \pm 8$, in which ± 4 is allotted to uncertainties in the flow model and another ± 4 to SBF calibration procedure in addition to the error of the Cepheid distance ± 6 (a quadrature sum is taken). There are a several other pieces of the SBF work to determine H_0, which generally result in $H_0 = 70 - 90$ (e.g., Thomsen et al. 1997 using HST; Jensen et al. 1999 with K band; see a review by Blakeslee et al. 1998). The new calibration made by the HST-KP group (Ferrarese et al. 1999a) would decrease H_0 only by 2%. The difference in the final H_0 between Tonry et al. and Ferrarese et al. comes from using different targets (the latter authors use only 4 clusters) and flow models.

It is impressive that analyses of SNeIa Hubble diagram give virtually the same answer, even though the samples are all derived primarily from the Calán-Tololo sample of Hamuy et al. (1996b). A smaller H_0 of Saha et al. (1997) basically reflects the absence of the the luminosity-decline rate correction, which pushes up H_0 by 10%. The other notable difference is a slightly higher value of HST-KP (Gibson et al. 1999), who made a reanalysis for all Cepheid observations performed by other groups and showed that their distances (to SN host galaxies) are all farther than would be derived from the HST-KP procedure. The average offset is as large as 0.16 mag (8%). This correction applies to all results other than HST-KP should we keep uniformity of the Cepheid data reduction. This is important especially when one compares the SN results with those from other secondary distance indicators, since the calibrations for the latter exclusively use HST-KP photometry. Taking the luminosity-decline rate correlation to be real and adopting Cepheid distance from the HST-KP data reduction, I adopt $H_0 =$ 68 from SNeIa.

We present two plots in Figure 1, (a) the estimates of maximum brightness of different authors and (b) the decline rate Δm_{15}, the amount of the decrease in brightness over 15 days following maximum light, both as a function of metallicity [O/H]. The second plot shows how metallicity effects are absorbed by the $M_V^{\rm max} - \Delta m_{15}$ relation and the first proves that there is little metallicity dependence in the corrected maximum brightness, though some scatter is seen among authors.

Leaving out the uncertainty of the Cepheid distance, H_0 from Tonry et al.'s SBF is 77±6, and that from SNeIa (HST-KP) is 68±4. The difference is 13%, and the two values overlap at $H_0 = 71$. Allowing for individual two sigma errors, the overlap is in a range of $H_0 = 65 - 76$. An additional uncertainty is 6% error ($\delta H_0 = \pm 4.5$) from the Cepheid distance which is common to both, still excluding the uncertainty of the LMC distance. We may summarise $H_0 = 71 \pm 7$ or 64–78 as our current standard, provided that LMC is at 50 kpc. All numbers in the table are within this range, except for the central value of Saha et al (1997).

In passing, let us note that $H_0 = 75 \pm 15$ (Freedman et al. 1997) obtained directly from the Cepheid galaxy sample agrees with the global value, implying that peculiar velocities are not so large even in a 10–20 Mpc region.

This convergence is a great achievement, but keep in mind that the SNeIa results are still lower than those from other secondary indicators[1] by 10%. There are additional problems. First, all these analyses are based on a LMC distance modulus of $m - M = 18.50$ (Feast & Walker 1987; Madore & Freedman 1991), which has recently been cast into doubt. In addition,

[1] A remark is given to the TF distance. While Sakai et al. (HST-KP) derived $H_0 = 71 \pm$ 8 using Giovanelli et al.'s (1997) cluster sample, Tammann and collaborators (Tammann 1999; Sandage & Tammann 1997) insist on a low value $H_0 = 53 - 56$. Their cluster result (Federspiel et al. 1998) neglects the depth effect of the Virgo cluster: contrary to ellipticals, spiral galaxies are distributed elongating along the line of sight (Yasuda et al. 1996). Hence identifying the centre of gravity of the spiral galaxy distribution with the true core leads to an offset. In fact the presence of substructure behind the Virgo core is confirmed with the Cepheid for NGC4649. Tammann et al.'s field result comes from the allocation of an unusually large dispersion to the TF relation, which largely amplifies the Malmquist bias. Tully (1999) obtained $H_0 = 82 \pm 16$ (Tully et al. 1998).

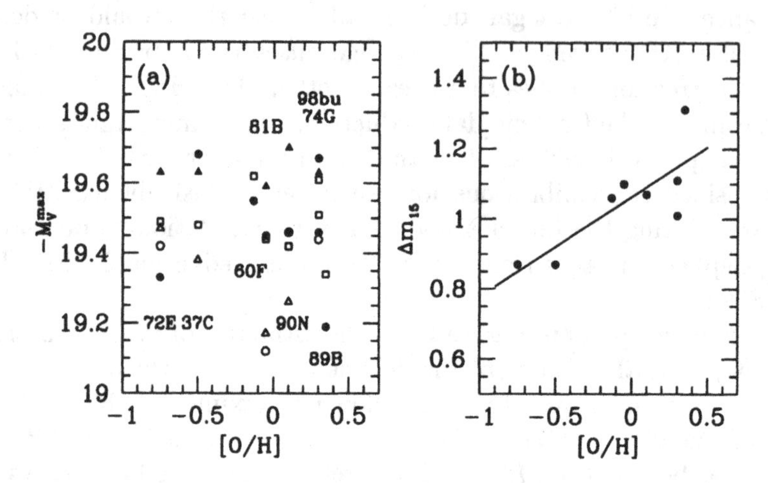

Figure 1. (a) Maximum brightness of SNeIa (in the V band) adopted by different authors (see TABLE 3) as a function of [O/H] of host galaxies: solid circles, Gibson et al. (1999); solid triangles, Suntzeff et al. 1999; open circles, Jha et al. (1999); open triangles, Hamuy et al. (1996b); open square, Saha et al. (1999). Note that Saha et al.'s calibration is not necessarily brighter, which is mainly due to a different treatment of extinction. (b) The decline rate Δm_{15} measured in the B band (Phillips et al. 1999) as a function of [O/H] (Gibson et al. 1999). The slope of the curve is $\partial\Delta m_{15}/\partial[\mathrm{O/H}] \simeq 0.28$.

metallicity effects could lead to systematic errors. Finally, should we derive the Hubble constant with the error of 10%, the problem of dust extinction could be an issue, and it is potentially coupled with the metallicity. We now consider these issues in greater detail.

2.3. DISTANCE TO LMC

The present status of the LMC distance is given in TABLE 4. The most traditional paths to the LMC distance follow the ladder shown in the upper half of TABLE 2. The Hipparcos satellite can measure a parallax down to 2 milli arcsec (mas), corresponding to a distance of 500 pc (ESA 1997). It was a reasonable expectation that one could obtain the geometric distance to the Pleiades cluster, circumventing the main sequence fitting from nearby parallax stars to the Pleiades and thus securing the Galactic distance scale. Hipparcos observations have also opened a number of novel methods that can be used to estimate the distance to LMC. This and related activities, however, have actually brought confusions, rather than securing the distance scale within the MW. We discuss several issues in order.

TABLE 4. Distance to LMC: Year 1997/1998

Method	Ref	Distance moduli
Cepheid PL	Feast & Catchpole 1997	18.70 ± 0.10
	Paturel et al. 1997	18.7
	Madore & Freedman 1998	18.57 ± 0.11
	Luri et al. 1998	18.29 ± 0.17
	Luri et al. 1998	18.21 ± 0.20
	(traditional) w/ new Pleiades	18.26
RR Lyrae (stat. para)	Fernley et al. 1998	18.31 ± 0.10
	Luri et al. 1998	18.37 ± 0.23
	Udalski 1998/Gould et al. 1998	18.09 ± 0.16
RR Lyrae (subdwarf)	Reid 1997	18.65
	Gratton et al. 1997	18.60 ± 0.07
Mira	van Leeuwen et al. 1997	18.54 ± 0.18
	Whitelock et al. 1997	18.60 ± 0.18
Red clump	Udalski et al. 1998a	18.08 ± 0.15
	Stanek et al. 1998	18.07 ± 0.04
	Cole 1998	18.36 ± 0.17
Eclipsing binaries	Guinan et al. 1998	18.30 ± 0.07
	(Udalski et al. 1998b)	18.19 ± 0.13(?)
SN 1987A Ring echo	Gould & Uza 1998	<18.37 ± 0.04
	Sonneborn et al. 1997	18.43 ± 0.10
	Panagia et al. 1997	18.58 ± 0.03
	Lundqvist & Sonneborn1997	18.67 ± 0.08
Cepheid PL	Sekiguchi & Fukugita 1998	18.10-18.60
	Sandage et al. 1999	18.57±0.05
Cepheid PL (BW method)	Gieren et al. 1997	18.49 ± 0.05

2.3.1. *"The Pleiades problem"*

The Pleiades cluster at 130 pc has been taken to be the first milestone of the distance work, since it has nearly solar abundance of heavy elements. This cluster is already too far to obtain a reliable parallax with the ground based observations, and its distance is estimated by tying it with nearby stars with solar metallicity employing main sequence fitting of FGK dwarfs (e.g., van Leeuwen 1983). The distance obtained this way agrees with an estimate via the Hyades, the nearest cluster to which geometric distance is available from the ground (Hanson 1980; van Altena et al. 1997), after a correction for large metallicity of the Hyades (VandenBerg & Bridges 1984). It was then a natural exercise to confirm these estimates with a parallax measured by the Hipparcos. The result showed that the Pleiades distance is shorter by 0.25 mag (12%) (van Leeuwen & Hansen-Ruiz 1997, Mermilliod et al 1997)! This is summarised in TABLE 5.

Mermilliod et al.'s (1997) (see also de Zeeuw et al. 1997) have shown that such a disagreement is seen not only for the Pleiades but also for other open clusters to some degree. A noteworthy example is that the locus of the Praesepe ([Fe/H]=+0.095) agrees with that of the Coma Ber ([Fe/H]=−0.065) *without* metallicity corrections, while we anticipate the former to be 0.25 mag brighter due to higher metallicity.

This is a serious problem, since the disagreement means that either our understanding of FGK dwarfs, for which we have the best knowledge for stellar evolution, is incomplete, or the Hipparcos parallax contains systematic errors (Pinsonneault et al. 1998; Narayanan & Gould 1999). The origin is not understood yet.

TABLE 5. Pleiades distance summary

Author/Method		Distance modulus
van Leeuwen (1983)		5.57±0.08
Lingå (1987)		5.61
Hyades (Perryman et al. 1998)	3.33±0.01	
Pleiades−Hyades	2.52±0.05	
metallicity correction	−0.22±0.03	
		5.63±0.06
van Leeuwen & Hansen-Ruiz 1997		5.32±0.05
Mermilliod et al. 1997		5.33±0.06
van Leeuwen 1999		5.37±0.07

2.3.2. *Metallicity effects in the LMC Cepheid calibration*

The Cepheid distance to LMC is based on the calibration using open cluster Cepheids, the distances to which are estimated by B star main sequence fitting that ties to the Pleiades (Sandage & Tammann 1968, Caldwell, 1983, Feast & Walker 1987, Laney & Stobie 1994). Metallicity has been measured for some of these calibrator Cepheids (Fry & Carney 1997). The residual of the PL fit shows a strong metallicity (Z) dependence. This means *either* the Cepheid PL relation suffers from a large Z effect, or the distances to open clusters contain significant Z-dependent errors (Sekiguchi & Fukugita 1998). A correction for this effect changes the distance to LMC in either way, depending upon which interpretation is correct.

This metallicity dependence problem can be avoided if parallaxes are used to find the distances to calibrator Cepheids. Attempts were made (Feast & Catchpole 1997; Luri et al. 1998; Madore & Freedman 1998) using field Cepheids in the Hipparcos catalogue. Unfortunately, Cepheid parallax data are so noisy (only 6 have errors less than 30%) that they do not allow calibrations tighter than ladders. Another skepticism is that 2/3 of Cepheids in the nearby sample (e.g., 14/26 in the Feast-Catchpole sample) are known to have companion stars, which would disturb the parallax (Szabados 1997).

2.3.3. *Red clump*

The OGLE group revived the use of the red clump (He burning stage of Population I stars) as a distance indicator. Paczyński & Stanek (1998) showed that the I band luminosity of the red clumps depends little on metallicity (see, Cole 1998, however), and gave a calibration using the Hipparcos parallax for nearby He burning stars. Udalski et al. (1998a) and Stanek et al. (1998) applied this to LMC, and obtained a distance modulus 18.1±0.1, much shorter than those from other methods. This is a modern version of an analysis of Mateo & Hodge (1986), who reported 18.1±0.3. We should also recall that earlier analyses using MS fitting of OB stars resulted in a short distance of 18.2–18.3 (Schommer et al. 1984; Conti et al. 1986), though somewhat dismissed in the modern literature.

2.3.4. *Detached eclipsing binaries*

Detached double-spectroscopic eclipsing binaries provide us with a unique chance to obtain the distance in a semi-geometric way out to LMC or even farther. From the information given by the light curve and velocity curve, one can solve for the orbital parameters and stellar radii (Andersen 1991, Paczyński 1997; Bell et al 1993 for an earlier application to LMC HV2226; Torres et al. 1997 for an application to the Hyades). If surface brightness of the two stars is known from colour or spectrum, one can obtain the distance as $d = (F/f)^{1/2}R_i$ where F and f are fluxes at the source and the observer and R_i is stellar radius. Guinan et al. (1998) applied this method to HV2274 in LMC and derived $m - M = 18.30 \pm 0.07$ with the aid of Kurucz' model atmosphere to estimate surface brightness from the spectrum. Udalski et al. (1998b) claimed that the extinction used is too small by an amount of $\Delta E(B-V) = 0.037$ mag based on OGLE multicolour photometry. If we accept this correction the distance becomes 0.11 mag shorter, i.e., $m - M = 18.19$.

2.3.5. *RR Lyr problems*
In the first approximation the luminosity of RR Lyr is constant, but in reality it depends on metallicity. The dependence is usually expressed as

$$\langle M_V(\text{RR Lyr})\rangle = a[\text{Fe/H}] + b\,. \tag{3}$$

Much effort has been invested to determine a and b. The problem is again how to estimate the distance to RR Lyr. Unlike the case with Cepheids, there are no unique ladders for the calibration, and a variety of methods have been used, of which the best known is the Baade-Wesselink method. The calibration from the ground may be summarised as

$$\langle M_V(\text{RR Lyr})\rangle = 0.2[\text{Fe/H}] + 1.04. \tag{4}$$

With this calibration we are led to the LMC distance of $m - M \simeq 18.3$, as we saw in TABLE 1 above.

The Hipparcos catalogue contains a number of field subdwarfs with parallax. This makes a ladder available to calibrate RR Lyr in globular clusters. Gratton et al. (1997) and Reid (1997) carried out this subdwarf fitting. Gratton et al. gave

$$\langle M_V(\text{RR Lyr})\rangle = (0.22 \pm 0.09)[\text{Fe/H}] + 0.76. \tag{5}$$

Their data are plotted in Figure 2, together with (4) and (5). Reid's result is also consistent. This zero point, being brighter by 0.3 mag (at [Fe/H]=–1.8) compared to (4), would bring the LMC distance to $m - M = 18.5 - 18.6$.

There are a few analyses using the statistical parallax for field RR Lyr in the Hipparcos catalogue. Fernley et al. (1998) reported that their halo RR Lyr lie almost exactly on the curve of (4), rather than (5), and concluded a confirmation of the ground-based calibration. This is also endorsed by an analysis of Gould & Popowski (1998).

The distance to eponymous RR Lyr was measured by Hipparcos. We see (Fig. 2) that RR Lyr itself does not fall on (5), but almost exactly on (4), although the error is fairly large. The uncertainties by 0.3 mag in the RR Lyr calibration translate to the LMC distance modulus 18.25–18.55.

2.3.6. *Conclusions on the LMC distance*
The distance to LMC is uncertain as much as 0.4 mag (20% in distance), ranging from 18.20 to 18.60. The results are rather bimodal around the two values close to the edges. A geometric method with SN1987A ring echo initiated by Panagia et al. 1991 does not differentiate between these two values: the data are too noisy and the result depends on the model of the light curve and emission lines that is adopted (Gould & Uza 1997; Sonneborn et al. 1997; see Fig. 8 of the latter literature for the data quality).

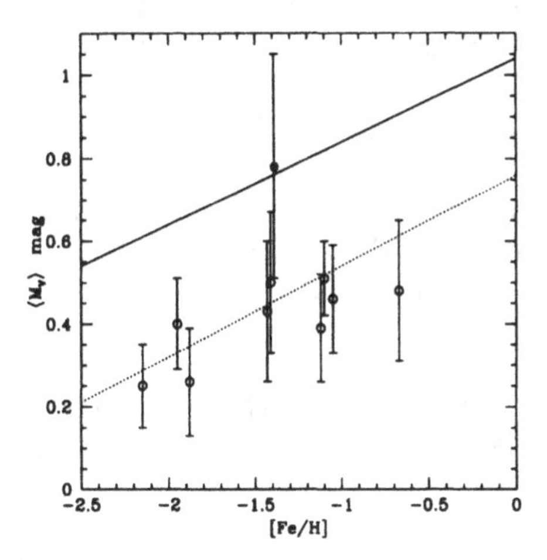

Figure 2. Calibrations of RR Lyr. The open points are taken from Gratton et al. (1997) with the dotted line indicating (5). The solid line is the ground-based calibration (4). The solid point denotes the eponymous RR Lyr measured by the Hipparcos satellite.

As we have seen in this section, recent observations with new techniques seem to tip the list to the lower value. This is clearly a systematic effect, so that we cannot simply take an 'average of all observations'. Rather, we should leave both possibilities open.

2.3.7. *Age of the globular clusters*

The RR Lyr calibration is also crucial in the estimation of the age of globular clusters, since the stellar age is proportional to the inverse of luminosity, i.e., inverse square of the distance. The modern evolution tracks of the main sequence agree reasonably well among authors. There are some disagreements in colours around the turn-off point, largely depending on the treatment of convection, but the luminosity is little affected (e.g., Renzini 1991; Vandenberg et al. 1996, especially their Fig. 1). Absolute magnitude at the turn-off point M_V^{TO} of the main sequence is hence a good indicator of the age, as (Renzini 1991),

$$\log t_9 = -0.41 + 0.37 M_V^{TO} - 0.43Y - 0.13[\mathrm{Fe/H}], \tag{6}$$

in units of Gyr, or

$$\log t_9 = -0.41 + (0.37a - 0.13)[\mathrm{Fe/H}] + 0.37[(M_V^{TO} - M_V^{\mathrm{RR}}) + b]] - 0.43Y, \tag{7}$$

if (3) is included. The difference of the magnitudes between the turn-off point and RR Lyr ($M_V^{TO} - M_V^{\mathrm{RR}}$) varies little among clusters and is measured to be 3.5±0.1 mag (Buonanno et al. 1989; see Chaboyer et al. 1996 for a compilation). The metallicity dependence of the cluster age disappears if $a = 0.35$, i.e., the globular cluster formation is coeval (Sandage 1993b). Both (4) and (5), however, give $a \simeq 0.2$,[2] which indicates that metal-poor clusters appear older.

The dichotomous calibrations of RR Lyr obviously affect the age of globular clusters. Another large uncertainty is whether the age-metallicity correlation is real, indicating metal-poor clusters formed earlier, or is merely due to a systematic error, with the formation of globular cluster being coeval. The possibilities are four-fold:

b	$(m-M)_{\rm LMC}$	t_0(noncoeval)	t_0(coeval)
1.05	18.25	18Gyr	15Gyr
0.75	18.55	14Gyr	12Gyr

In addition there are ±10% errors from various sources (Renzini 1991; Bolte & Hogan 1995; VandenBerg et al. 1996; Chaboyer et al. 1996). Figure 3 shows the age of various clusters from Gratton et al. (1997) and Chaboyer et al. (1998) both using the calibration close to (5). The [Fe/H] dependence is apparent.

The claims of Gratton et al. (1997), Reid (1997) and Chaboyer et al. (1998) for young universe (11-12 Gyr) assume a coeval-formation interpretation together with the long RR Lyr calibration and take a mean of globular cluster ages. Three other possibilities, however, are not excluded.

2.4. METALLICITY PROBLEMS WITH CEPHEIDS

In most applications of the Cepheid PL relation, metallicity effects are neglected, motivated by theoretical arguments that they will be very small. This results from double cancellations of the metallicity dependences between core luminosity and atmosphere, as well as between the effects of the helium abundance and of heavier elements. The expected effect is (Stothers 1988, Iben& Tuggle 1975; Chiosi et al. 1993)

$$\gamma_\lambda \equiv \partial M_\lambda / \partial [\mathrm{Fe/H}] \simeq \pm 0.05 \text{ dex mag}^{-1} \tag{8}$$

for the $\lambda = V, I$ pass bands. A new calculation of Sandage et al. (1999) gives $|\gamma_\lambda| < 0.1$ for $\lambda = B, V, I$.

When one is concerned with a 10% systematic error in the cosmic distance scale, the metallicity effect must be scrutinised. If it were as large as -0.5, say, the true Cepheid distance to normal spiral galaxies would be

[2] A remark is made on a recent analysis of Kovács & Jurcsik (1996), who obtained $a < 0.19$ from a model-independent approach using the Fourier coefficients of the light curves that correlate with the metal abundance.

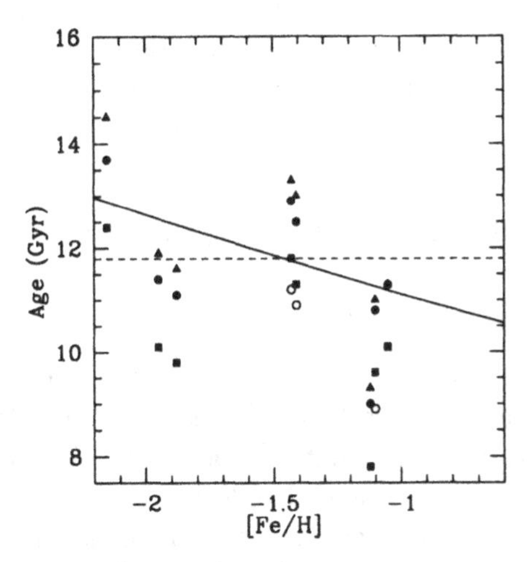

Figure 3. Age of globular clusters as a function of [Fe/H]. The solid points are from Gratton et al. (1997) for three different stellar evolution models. The open points are from Chaboyer et al. (1998). The solid line shows (7) but offset by −0.06. The dashed line is 11.8 Gyr of Gratton et al.

longer by 10% relative to LMC ([O/H]=−0.4). The calibrator SNe used in earlier papers (SN1937C, SN1972E, 1981B and 1990N) all reside in low Z galaxies, but recent additions include SNe in high Z galaxies (1989B, and notably 1998bu), thus the sample spans a wider metallicity range (see Figure 1 above). There is now no relative difference in metallicity effects any more between the SBF and SNIa calibrator samples (the offset is Δ[O/H] < 0.1). Therefore, we cannot ascribe the difference in H_0 to the metallicity effect of the Cepheid PL relation: the effect slightly reduces H_0 from both methods if the sign of γ is negative.

However, it is important to know the magnitude of γ. Observationally, Freedman & Madore (1990: FM) showed with the M31 data that the metallicity dependence is small ($\gamma_{BVRI} = -0.32 \pm 0.21$). Gould (1994), however, reanalysed the same data and concluded it to be as large as $\gamma = -0.88 \pm 0.16$. The EROS collaboration derived $\gamma_{VI} = -0.44$ from a comparison between LMC and SMC (Beaulieu et al. 1997). Kochanek (1997) suggested $\gamma_{VI} = -0.14 \pm 0.14$ from a global fit of galaxies with Cepheid observations. The metallicity dependence for Galactic Cepheids discussed in section 2.3.2 corresponds to $\gamma_{VJHK} \approx -2$. Kennicutt et al. (1998) pointed out that the metallicity gradient of M31 used by Freedman & Madore is a factor of three too large and argued that the above values should be $\gamma_{BVRI} = -0.94 \pm 0.78$ (FM) and -2.1 ± 1.1 (Gould).

Kennicutt et al. (1998) derived from HST observations of two fields in M101 that $\gamma_{VI} = -0.24 \pm 0.16$, which is the value currently adopted in metallicity dependence analyses of the HST-KP group. If this is the true value, the effect on the distance scale is of the order of 5±3% (H_0 gets smaller). I would emphasize, however, that independent confirmations are necessary for this γ value, since the M101 analysis is based only on V and I bands, and the effect of extinction might not be completely disentangled.

2.5. CROSS-CHECK OF THE CEPHEID DISTANCES

2.5.1. *Tests with geometric methods*

NGC4258 (M106) is a Seyfert 2 galaxy with H_2O maser emission from clouds orbiting around a black hole of mass $4 \times 10^7 M_\odot$ located at the centre. Precise VLBA measurements of Doppler velocities show that the motion of the clouds is very close to Keplerian and is perturbed very little (Miyoshi et al. 1995). A complete determination is made for the orbital parameters including centripetal acceleration and a bulk proper motion of the emission system. This yields a geometric distance to NGC4258 to be 7.2 ± 0.3 Mpc (Herrnstein et al. 1999).

Maoz et al. (1999) measured the distance to NGC4258 using the conventional Cepheid PL relation, and gave 8.1±0.4 Mpc with $(m-M)_{\rm LMC} = 18.5$. This distance is 13% longer than that from the maser measurement. The short LMC distance would bring the Cepheid distance in a perfect agreement with the geometric distance. This is, however, only one example, and it can merely be a statistical effect: the deviation is only twice the error, so it may happen with a chance probability of 5%.

2.5.2. *Further checks for M31*

A number of distance estimates are available for the nearest giant spiral M31, and they are shown in TABLE 6 (the underlined numbers are the zero point). Stanek & Garnavich (1998) applied the red clump method to M31, and obtained $m - M = 24.47 \pm 0.06$, which agrees with the M31 modulus 24.44 ± 0.10 from the Cepheid based on the $(m - M)_{\rm LMC} = 18.5$ calibration, whereas the same method gives 18.1 for the LMC distance. Namely, M31−LMC largely disagrees between the two. This discrepancy might be ascribed to a metallicity effect of either Cepheids or red clumps, or to the systematic error of either indicator. Mochejska et al. (1999) ascribed it to the error of the Cepheid distance from a crowded stellar population. On the other hand, tip of giant branch (TRGB) gives $(m - M)_{\rm M31-LMC}$ in agreement with the value derived from the Cepheid. The difference from PNLF is also consistent. The value from RR Lyr, however, is consistent with that from red clumps (the numbers in the Table is derived using (4) for the zero point). The results are dichotomous again.

TABLE 6. Relative distance of M31 to LMC

method	M31	LMC	M31−LMC	refs.
Cepheid	24.44±0.10	18.5	5.94±0.10	Ferrarese et al. 1999b
red clump	24.47±0.06	18.07±0.04	6.40±0.07	Stanek & Garnavich 1998
TRGB	24.41±(0.19)	18.5	5.91±(0.19)	Ferrarese et al. 1999b
RR Lyr (B)	24.50±(0.15)	18.30	6.20±	Pritchet & vd Bergh 1989
PNLF	24.44	18.56±0.18	5.82±0.18	Jacoby et al. 1990b

2.6. PHYSICAL METHODS

2.6.1. *Expansion photosphere model (EPM) for Type II supernovae*

This is a variant of the Baade-Wesselink method. If a supernova is a black body emitter one can calculate source brightness from temperature; the distance can then be estimated by comparing source brightness with the observed flux. In SNeII atmosphere the flux is diluted due to electron scattering opacity. If this greyness is calculated source brightness can be inferred. Schmidt, Kirshner & Eastman (1992) developed this approach and obtained the distances to SNeII in agreement with those from the ladder. The point is that EPM gives absolute distance without zero point calibrations. The Hubble constant they obtained is 73±9 (Schmidt et al. 1994).

A possible source of systematic errors is in the estimation of the temperature from the spectrum or colour. The SNeII physics also might not be uniform, as we see occasional large scatters in a cross-correlation analysis.

2.6.2. *Gravitational lensing time delay*

When quasar image is split into two or more by gravitational lensing, we expect the time delay among images, arising from different path lengths and gravitational potentials among image positions. The time delay between images A and B takes the form

$$\Delta t = \frac{1+z_L}{H_0}\left(\frac{D_{OL}D_{OS}}{D_{LS}}H_0\right)\left[\frac{1}{2}(|\theta_A|^2 - |\theta_B|^2 - \Delta\phi|_{A-B})\right] \qquad (9)$$

where θ is the angular difference between the source and image, $\Delta\phi$ is the difference in the potential and D_{IJ} is the angular diameter distances. The time delay is observable if the source is variable, and can be used to infer H_0 (Refsdal 1964). Crucial in this argument is a proper modelling of the mass distribution of the deflector. The DD/D factor depends on Ω only weakly; its λ dependence is even weaker.

The first case where H_0 is derived is with the 0957+561 lens system. The deflector is complicated by the fact that a giant elliptical galaxy is embedded into a cluster. Falco, Gorenstein & Shapiro (1991) noted an ambiguity associated with a galaxy mass – cluster mass separation, which does not change any observed lens properties but affects the derived Hubble constant. One way to resolve this degeneracy is to use the velocity dispersion of the central galaxy (Falco et al. 1991; Grogin & Narayan 1996). Kundić et al. (1997b), having resolved a long-standing uncertainty about the time delay, obtained $H_0 = 64 \pm 13$ employing the Grogin-Narayan model. Tonry & Franx (1998) revised it to 71 ± 7 with their new velocity dispersion measurement near the central galaxy. More recently, Bernstein & Fischer (1999) searched a wider variety of models, also using weak lensing information to constrain the mass surface density of the cluster component, and concluded $H_0 = 77^{+29}_{-24}$, the large error representing uncertainties associated with the choice of models.

The second example, PG1115+080, is again an unfortunate case. The deflector is elliptical galaxy embedded in a Hickson-type compact group of galaxies (Kundić et al. 1997a). Keeton & Kochanek (1997) and Courbin et al. (1997) derived $(51-53) \pm 15$ from the time delay measured by Schechter et al. (1997). Impey et al. (1998) examined the dependence of the derived H_0 on the assumption for the dark matter distribution, and found it to vary from 44 ± 4 (corresponding to M/L linearly increasing with the distance) to 65 ± 5 (when M/L is constant over a large scale). The latter situation may sound strange, but it seems not too unusual for elliptical galaxies, a typical example being seen in NGC5128 (Peng et al. 1998).

Recently, time delays have been measured for three more lenses, B0218+357, B1608+656 and PKS1830-211. B0218+357 is a rather clean, isolated spiral galaxy lens, and Biggs (1999) derived $H_0 = 69^{+13}_{-19}$ (the central value will be 74 if $\Omega = 0.3$) with a simple galaxy model of a singular isothermal ellipsoid. For B1608+656, Koopmans & Fassnacht (1999) obtained 64 ± 7 for $\Omega = 0.3$ (59 ± 7 for EdS). For PKS1830-211, they gave 75^{+18}_{-10} for EdS and 85^{+20}_{-11} for $\Omega = 0.3$ from the time delay measured by Lovell et al. (1998). More work is clearly needed to exhaust the class of models, but these three lens systems seem considerably simpler than the first two examples. Koopmans & Fassnacht concluded 74 ± 8 for low density cosmologies (69 ± 7 for EdS) from four (excluding the second) lensing systems using the simplest model of deflectors. It is encouraging to find a good agreement with the values from the ladder argument, though the current results from lenses are still less accurate than the ladder value. It would be important to ask whether $H_0 < 60$ or > 80 is possible within a reasonable class of deflector models.

2.6.3. *Zeldovich-Sunyaev effect*

The observation of the Zeldovich-Sunyaev (ZS) effect for clusters tells us about the cluster depth (times electron density), which, when combined with angular diameter (times electron density square) from X ray observations, gives us the distance to the cluster provided that cluster is spherical (Birkinshaw et al. 1991, Myers et al. 1997). This is often taken as a physical method to measure H_0. I give little weight to this method in these lectures, since it is difficult to estimate the systematic errors. The currently available results wildly vary from a cluster to a cluster. The most important is a bias towards elongation. None of the known clusters are quite spherical, and selection effects bias towards clusters elongated along the line of sight because of higher surface brightness. This may happen even if one uses a large sample. Additional systematics arise from the sensitivity of the ZS effect to the cluster envelopes; one must resort to a model to correct for this effect.

2.6.4. *Physical methods: summary*

Physical methods now yield the Hubble constant which can be compared with that from ladders. TABLE 7 presents a summary of H_0 from the physical methods. However, effort is still needed to determine systematic errors associated with the use of specific methods.

TABLE 7. Hubble constant from ladders and physical methods

method	H_0	reference
ladders	**71±7(×0.95-1.15)**	
physical: EPM	73±9	Schmidt et al. 1994
physical: lensing (low Ω)	74±8	Koopmans & Fassnacht 1999
(EdS)	69±7	
physical: ZS	(54±14)	Myers et al 1997

2.7. CONCLUSIONS ON H_0

The progress in determining the extragalactic distance scale has been dramatic. The ladders yield values convergent within 10%, which is compared to a factor 1.6 disagreement in the early nineties. A new uncertainty, however, becomes manifest in the Galactic distance scale: there is a 15–20% uncertainty in the distance to LMC. Therefore, we may summarise

$$H_0 = (71 \pm 7) \times \begin{matrix} 1.15 \\ 0.95 \end{matrix} \tag{10}$$

as a currently acceptable value of the Hubble constant. This agrees with that from HST-KP (Mould et al. 1999) up to the uncertainty from the LMC distance, though we followed a different path of argument. This allows $H_0 = 90$ at the high end (if Tonry et al's SBF is weighted) and 60 at the low end (if the SNeIa results are weighted). Note that H_0 from both EPM and gravitational lensing are consistent with the ladder value for $(m-M)_{\rm LMC} = 18.5$. With the shorter LMC distance the overlap is marginal.

The short LMC distance will also cause trouble for the H_0−age consistency. The LMC distance modulus of $m - M = 18.25$ would raise the lower limit of H_0 to 72, and increase the lower limit of age from ≈11.5 Gyr to ≈14.5 Gyr at the same time. There is then no solution for a $\lambda = 0$ universe. With a non-zero λ, a unique solution is $H_0 \simeq 72$, $\Omega \simeq 0.25$, $\lambda \simeq 0.75$ with coeval globular cluster formation (see Figure 6 below).

In the future it is likely that more effort will be expended for geometric methods. The great advantage is that it is free from errors arising from the chemical composition. In the surface brightness method, the chemical composition may still enter into the game, but its effect is tolerable and can even be reduced to a negligible level by using near infrared observations.

Ultimately, gravitational wave observations could provide us with a novel method. For instance, for coalescing binary neutron stars the distance can be calculated as $d \sim \nu^{-2}\varepsilon^{-1}\tau^{-1}$, where ε is metric perturbations, ν is the frequency and $\tau = \nu/\dot{\nu}$ is a characteristic time of the collapse (Schutz 1986). The position of the object may be difficult to infer, but there might be a gamma ray burst associated with the coalescence.

3. The density parameter

3.1. MODEL-INDEPENDENT DETERMINATIONS

3.1.1. *Luminosity density* × $\langle M/L \rangle$

The mass density can be obtained by multiplying the luminosity density with galaxy's average mass to light ratio $\langle M/L \rangle$. The local luminosity density, evaluated by integrating the luminosity function, is reasonably well converged to $\mathcal{L}_B = (2.0 \pm 0.4) \times 10^8 h L_\odot$ Mpc^{-3} from many observations. The M/L_B of galaxies generally increases with the scale. When the mass is integrated to ≈ 100 kpc, a typical M/L_B is about $(100-200)h$ in solar units, and it may still increase outward (e.g., Faber & Gallagher 1979; Little & Tremaine 1987; Kochanek 1996; Bahcall et al. 1995; Zaritsky et al. 1997). The virial radius in a spherical collapse model is $r = 0.13$ Mpc $\Omega^{-0.15}[M/10^{12}M_\odot]^{1/2}_{<100\rm kpc}$. If the dark matter distribution is isothermal within the virial radius, the value of M/L_B inside the virial radius is $(150-400)h$ for L^* galaxies. This is about the value of M/L_B for groups and clusters, $(150-500)h$. Multiplying the two values we get

$\Omega = 0.20 \times 2^{\pm 1}$. See also Fukugita, Hogan & Peebles (1998) for variants of this argument.

Carlberg et al. (1996, 1997a) tried to make the argument more quantitative using their cluster sample and a built-in field galaxy sample. They estimated $M/L_r \simeq (210 \pm 60)h$ for field galaxies from the cluster value $(289 \pm 50)h$. Their luminosity density of field galaxies is $\mathcal{L}_r = (1.7 \pm 0.2) \times 10^8 hL_\odot$ Mpc^{-3}, and therefore $\Omega_0 = 0.19 \pm 0.06$. Note that $M/L_B \simeq 1.4 \times M/L_r$ in solar units for the respective pass bands.

The important assumption for these calculations is the absence of copious matter outside the clusters. This is a question difficult to answer, but the observation of weak lensing around the clusters indicate that the distributions of dark mass and galaxies are similar at least in the vicinity of clusters (Tyson & Fischer 1995; Squires et al. 1996).

Some attempts have also been made to estimate the mass on a supercluster scale. Small et al. (1998) inferred $M/L_B \simeq 560h$ for the Corona Borearis supercluster, by applying the virial theorem (inspired by an N body simulation). On the other hand, Kaiser et al. (1998) estimated $M/L_B \simeq 250$ from a measurement of the gravitational shear of weak lensing caused by a supercluster MS0302+17 [3]; the result is not well convergent, but it seems unlikely that Ω is larger than 0.5.

3.1.2. *H_0 versus cosmic age*

For $H_0 \geq 60$, the age is 10.9 Gyr for the EdS universe. This is too short. Ω must be smaller than unity. If we take $t_0 > 11.5$ Gyr $\Omega < 0.7$. The limit is weak, but the significance is that EdS universe is nearly excluded.

3.1.3. *Type Ia supernova Hubble diagram*

The type Ia supernova Hubble diagram now reaches $z \simeq 0.4 - 0.8$. It can be used to infer the mass density parameter and the cosmological constant. As we discuss later (section 4.1) the observation favours a low Ω and a positive λ. If we take their formal errors, $\Omega < 0.1$ is allowed only at three sigma for a zero λ universe (Riess et al. 1998; Perlmutter et al. 1999). A zero λ open universe may not be excluded yet if some allowance is taken for systematic effects, but EdS geometry is far away from the observation. The best favoured value is approximately,

$$\Omega \approx 0.8\lambda - 0.4 \ . \tag{11}$$

[3] They suggest $\Omega \simeq 0.04$ on the basis that only early-type galaxy population traces the mass distribution and the luminosity density is multiplied by the fraction of early-type galaxies (20%). It seems possible that late type galaxies reside in low density regions, causing only a small shear, which is buried in noise, and escaped from the measurement.

3.1.4. *Baryon fractions*

A cluster is a virialised object with the cooling time scale longer than the dynamical time scale, and hence the physics is governed only by gravity (except for cooling flows in high density regions). The gas in clusters is shock heated to the virial temperature $T \simeq 7 \times 10^7 (\sigma/1000\text{km s}^{-1})^2$ K, and thus emits X rays by thermal bremsstrahlung. From the luminosity and temperature of X rays one can infer the mass of the X ray emitting gas. It has been known that the gas amounts to a substantial fraction of the dynamical mass, which means that baryons reside more in the gas than in stars by an order of magnitude (Forman & Jones 1982). The argument was then elaborated by White et al. (1993b) based on ROSAT observations. From 19 clusters White & Fabian (1995) obtained $M_{\rm gas}/M_{\rm grav} = 0.056h^{-2/3}$, where $M_{\rm grav}$ is the dynamical mass. By requiring that the cluster baryon fraction agrees with Ω_B/Ω in the field, we have $\Omega = 0.066h^{-1/2}\eta_{10} = 0.39(\eta_{10}/5)$, where η_{10} is the baryon to photon ratio in units of 10^{-10} and the last number assumes $h = 0.7$.

An independent estimate is made from the Zeldovich-Sunyaev effect observed in clusters (Myers et al. 1997; Grego et al. 1999): $M_{\rm gas}/M_{\rm grav} = 0.082h^{-1}$ is translated to $\Omega = 0.044h^{-1}\eta_{10} = 0.31(\eta_{10}/5)$.

If we insert a probable value of the baryon to photon ratio from primordial nucleosynthesis calculations, $\eta_{10} = 3 - 5$, we have $\Omega = 0.2 - 0.4$.

3.1.5. *Peculiar velocity - density relation*

This is one of the most traditional methods to estimate the cosmic mass density. The principles are spelled out by Peebles (1980). There are two basic tools depending on the scale. For small scales ($r < 1$ Mpc) the perturbations developed into a non-linear regime, and the statistical equilibrium argument is invoked for ensemble averages that the peculiar acceleration induced by a pair of galaxies is balanced by relative motions (cosmic virial theorem). For a large scale ($r > 10$ Mpc), where perturbations are still in a linear regime, the basic equation is

$$\nabla \cdot \vec{v} + H_0 \Omega^{0.6} \delta = 0 \tag{12}$$

with δ the density contrast. The contribution from a cosmological constant is negligible. The problem inherent in all arguments involving velocity is the uncertainty regarding the extent to which galaxies trace the mass distribution (biasing), or how much mass is present far away from galaxies.

Small-scale velocity fields: The status is summarised in Peebles (1999), where he has concluded $\Omega(10\text{kpc} \lesssim r \lesssim 1\text{Mpc}) = 0.15 \pm 0.10$ from the pair wise velocity dispersion (with samples excluding clusters) and the three point correlation function of galaxies via a statistical stability argument.

Bartlett & Blanchard (1996) argued that it is possible to reconcile the observed velocity dispersion with $\Omega \sim 1$ if one assumes galactic halo extended beyond > 300kpc. As Peebles (1999) argued, however, the halo is unlikely to be extended that much as indicated by the agreement of MW's mass at 100-200kpc and the mass estimate for MW+M31 in the Local Group.

Beyond a 10 Mpc scale, linear perturbation theory applies. An integral form of (12) for a spherical symmetric case ($v/H_0 r = \Omega^{0.6}\langle\delta\rangle/3$) applied to the Virgocentric flow gives $\Omega \simeq 0.2$ for $v \simeq 200 - 400$km s^{-1} and $\langle\delta\rangle \sim 2$, assuming no biasing (Davis and Peebles 1983). Recently, Tonry et al. (1999) argued that the peculiar velocity ascribed to Virgo cluster is only 140 km s^{-1}, while the rest of the peculiar velocity flow is attributed to the Hyd-Cen supercluster and the quadrupole field. For this case $\Omega \simeq 0.06$. We may have $\Omega \sim 1$ only when half the mass is well outside the galaxies.

Peebles (1995) argued that the configuration and kinematics of galaxies are grown following the least action principle from the nearly homogeneous primeval mass distribution. Applying this formalism to Local Group galaxies, he inferred $\Omega = 0.15 \pm 0.15$. On the other hand, Branchini & Carlberg (1994) and Dunn & Laflamme (1995) argued that this conclusion is not tenable if mass is distributed smoothly outside galaxies as in $\Omega = 1$ CDM models. This seems, however, not very likely unless mass distribution is extended over 10 Mpc scale (Peebles 1999).

Large scale velocity fields: There are a few methods to analyse the large-scale velocity fields based on (12). The direct use of (12) is a comparison of the density field derived from redshift surveys with measured peculiar velocities. Alternatively, one may use the density field reconstructed from observed velocity field for comparison with the actual density field, as in the `POTENT` programme (Dekel et al. 1990). A variant of the first method is to observe the anisotropy in redshift space (redshift distortion) (Kaiser 1987). As linear theory applies, Ω always appears in the combination $\beta = \Omega^{0.6}/b$ where b is a linear biasing factor of galaxies against the mass distribution and can be inferred through non-linear effects. Much effort has been invested in such analyses (see e.g., Strauss & Willick 1995; Dekel et al. 1997; Hamilton 1998), but the results are still controversial. The value of $\Omega^{0.6}/b$ derived from many analyses varies from 0.3 to 1.1, though we see a general trend to favour a high value. Notably, the most recent `POTENT` analysis using the Mark III compilation of velocities (Willick et al. 1997) indicates a high density universe $\Omega = 0.5 - 0.7$, and $\Omega > 0.3$ only at a 99% confidence level (Dekel et al. 1999).

The difficulty is that one needs accurate information for velocity fields, for which an accurate estimate of the distances is crucial. Random errors of the distance indicators introduce large noise in the velocity field. This seems particularly serious in the POTENT algorithm, in which the derivative $\nabla \cdot \vec{v}/\Omega^{0.6}$ and its square are numerically computed; this procedure enhances noise, especially for a small Ω. The difficulty of inferring large scale velocity field may also be represented by the 'great attractor problem'. Lynden-Bell et al. (1988) found a large-scale velocity field towards the Hyd-Cen supercluster, but also argued that this supercluster is also moving towards the same direction attracted by a 'great (giant) attractor'. With Tonry et al.'s (1999) new estimate of the distance using SBF, this velocity field is modest, and Hyd-Cen itself serves as the great attractor that pulls the Virgo cluster, with the conclusion that Ω is small.

3.2. MODEL-DEPENDENT DETERMINATIONS

The following derivations of the mass density parameter are based on the hierarchical clustering model of cosmic structure formation assuming the cold dark matter dominance. The extraction of Ω is, therefore, indirect, but on the other hand, it is reasonable to appeal to such models since Ω is the parameter that predominantly controls structure formation. Note that CDM model is the only model known today that successfully predicts widely different observations, yet there are no observations strong enough to refute its validity. We do not discuss results from cosmological models where physical processes other than gravity play a major role.

3.2.1. *Shape parameter of the transfer function*

The initial perturbations of the density fluctuation $P(k) = |\delta_k|^2 \sim k^n$ receive a modification as $P(k) = |\delta_k|^2 \sim k^n T(k)$ as they grow, where $T(k)$ is called the transfer function. Fluctuations of a small scale that enter the horizon in the radiation dominant epoch do not grow for a while, till the universe becomes matter dominated. The transfer function $T(k)$ thus damps for small scales as $\sim k^{-4}$, whereas it stays close to unity for long-wave lengths. The transition region is controlled by a parameter $k \sim 2\pi/ct_{\rm eq}$, ct_{eq} being the horizon size at the time of matter-radiation equality, i.e., a characteristic length of $6.5(\Omega h)^{-1}h^{-1}$ Mpc. The parameter $\Gamma = \Omega h$ determines the behaviour of the transfer function and is called the shape parameter. To give a sufficient power to several tens of Mpc, Γ must be as small as 0.2 (Efstathiou et al. 1990). This small value ($\Gamma = 0.15 - 0.25$) is supported by later analyses (e.g., Peacock & Dodds 1994; Eke et al. 1998).

3.2.2. *Evolution of the rich cluster abundance*

The cluster abundance at $z \approx 0$ requires the rms mass fluctuation $\sigma_8 = \langle(\delta M/M)\rangle^{1/2}|_{r=8h^{-1}{\rm Mpc}}$ to satisfy (White et al. 1993a; Eke et al. 1996; Pen 1998; Viana & Liddle 1999; see also Henry & Arnaud 1991)

$$\sigma_8 \approx 0.6\Omega^{-0.5} . \tag{13}$$

The evolution of the cluster abundance is sensitive to σ_8 in early epochs of growth for a given mass; it is $z \gtrsim 0.3$ for rich clusters. The rich cluster abundance at $z \sim 0.3 - 1$, when compared with that at a low z, determines both σ_8 and Ω (Oukbir & Blanchard 1992). Carlberg et al. (1997b) derived $\Omega = 0.4 \pm 0.2$, and Bahcall & Fan (1998) obtained $\Omega = 0.2^{+0.3}_{-0.1}$ corresponding to a slow growth of the abundance. On the other hand, Blanchard & Bartlett (1998) obtained $\Omega \simeq 1$ from a more rapid growth. A high value is also claimed by Reichart et al. (1999), while Eke et al. (1998) reported $\Omega = 0.43 \pm 0.25$ for an open, and $\Omega = 0.36 \pm 0.25$ for a flat universe.

The controversy among authors arises from different estimates of the cluster mass at high z. This is a subtle effect, since the mass varies little over the range of relevant redshift, while the cluster number density evolution is sufficiently rapid at fixed mass (Pen 1998). At low z we have an established mass temperature relation, and the cluster mass is securely estimated (Henry & Arnaud 1991). At high z, however, such direct information is not available. Blanchard & Bartlett and Eke et al. used mass temperature relations as a function of z derived from hydrodynamic simulations. Reichart et al. used an extrapolated mass X-ray luminosity relation. Bahcall and Fan used more direct estimates of the cluster mass at higher z for three clusters. A change of a factor of two in the mass estimate would modify the conclusion.

3.2.3. *Cluster abundance versus the COBE normalisation*

There are a number of ways to infer σ_8 from galaxy clustering and peculiar velocity fields. The problem with the information from galaxy clustering is that it involves an unknown biasing factor, which hinders us from determining an accurate σ_8. The velocity data are susceptible to noise from the distance indicators. Therefore, the cluster abundance discussed above seems to give us a unique method to derive an accurate estimate of σ_8 for a low z universe. Another place we can extract an accurate σ_8 is the fluctuation power imprinted on cosmic microwave background radiation (CBR) anisotropies. Currently only the COBE observation (Bennett et al. 1996) gives sufficiently accurate $\sigma_8 = \sigma_8(H_0, \Omega, \lambda, \Omega_B, ...)$. Assuming the model transformation function, the matching of COBE σ_8 with that from the cluster abundance gives a significant constraint on cosmological parameters $\Omega = \Omega(H_0, \lambda)$ (Efstathiou et al. 1992; Eke et al. 1996)Figure 4 shows allowed regions for two cases, open and flat universes, assuming a flat perturbation spectrum $n = 1$ and ignoring possible tensor perturbations.

The transfer function is modified if $n \neq 1$. The possible presence of the tensor perturbations in CBR anisotropies causes another uncertainty. The

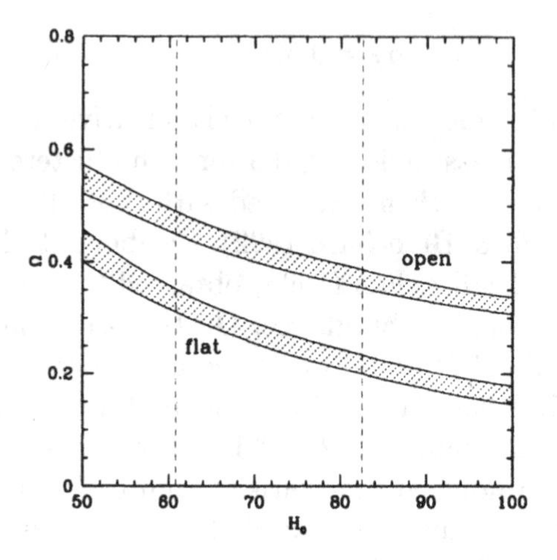

Figure 4. Parameter regions allowed by matching the rms fluctuations from COBE with those from the cluster abundance. A flat spectrum ($n = 1$) is assumed and the tensor perturbations are neglected. The lower band is for a flat universe, and the upper one for a universe with $\Lambda = 0$.

COBE data alone say n being between 0.9 and 1.5 (Bennett et al. 1996), but the allowed range is narrowed to $n = 0.9 - 1.2$ if supplemented by smaller angular-scale CBR anisotropy data (Hancock et al. 1998; Lineweaver 1998; Efstathiou et al. 1998; Tegmark 1999). The presence of the tensor mode would make the range of n more uncertain as well as it reduces the value of σ_8. The limit of n when the tensor mode is maximally allowed is about < 1.3[4]. Notwithstanding these uncertainties, $\Omega > 0.5$ is difficult to reconcile with the matching condition. On the other hand, a too small Ω ($\lesssim 0.15$) is not consistent with the cluster abundance.

3.2.4. *Power spectrum in nonlinear galaxy clustering*

Peacock (1997) argued that the power spectrum in a small scale region ($k^{-1} < 3h^{-1}$ Mpc), where nonlinear effect is dominant, shows more power than is expected in $\Omega = 1$ cosmological models. He showed that the excess power is understood if the mass density is $\Omega \approx 0.3$.

3.2.5. *CBR anisotropy harmonics*

The ℓ distribution of the CBR harmonics C_ℓ depends on many cosmological parameters. Precise measurements of the harmonics will allow an accurate determination of the cosmological parameters up to geometrical degeneracy (Zaldarriaga et al. 1997; Efstathiou & Bond 1999; Eisenstein et al. 1999). At

[4]In Tegmark's analysis $n < 1.5$ is quoted as an upper bound, but this is obtained by making Ω_B (and H_0) a free parameter. If one would fix the baryon abundance, the allowed range is narrower, $n \lesssim 1.3$.

present the data do not give any constraint on Ω, but on some combination of Ω and λ; so we defer the discussion to the next section.

4. Cosmological constant

Currently three tests yield useful results on the problem as to the existence of the cosmological constant: (i) the Hubble diagram for distant type Ia supernovae; (ii) gravitational lensing frequencies for quasars; (iii) position of the acoustic peak in the harmonics of CBR anisotropies.

4.1. TYPE IA SUPERNOVA HUBBLE DIAGRAM

The luminosity distance receives a cosmology dependent correction as z increases; in a way Ω pulls down d_L and λ pushes it up. (In the first order of z the correction enters in the combination of $q_0 = \Omega/2 - \lambda$, so this is often referred to as a q_0 test.) The discovery of two groups that distant supernovae are fainter than are expected from the local sample, even fainter than are expected for $q_0 = 0$, points to the presence of $\lambda > 0$ (Riess et al. 1998; Schmidt et al. 1998; Perlmutter et al. 1999).

The general difficulty with such a Hubble diagram analysis is that one has to differentiate among a few interesting cosmologies with small differences of brightness. For instance, at $z = 0.4$ where many supernovae are observed, the difference is $\Delta m = 0.12$ mag between $(\Omega, \lambda) = (0.3, 0.7)$ and $(0, 0)$, and $\Delta m = 0.22$ from $(0, 0)$ to $(1.0, 0)$. Therefore, an accuracy of ($\lesssim 5\%$) must be attained including systematics to conclude the presence of Λ. On the other hand, there are a number of potential sources of errors:
(i) K corrections evaluated by integrating spectrophotometric data that are dominated by many strong features;
(ii) relative fluxes at the zero point (zero mag) across the colour bands;
(iii) dust obscuration in a host galaxy;
(iv) subtraction of light from host galaxies;
(v) identification of the maximum brightness epoch, and estimates of the maximum brightness including a Δm_{15} correction;
(vi) selection effects (for high z SNe);
(vii) evolution effects.

Except for (vii), for which we cannot guess much[5], the most important seems to be combined effects of (i), (ii) and (iii). It is not easy a task to reproduce a broad band flux by integrating over spectrophotometric data convoluted with filter response functions, especially when spectrum

[5]Riess et al. (1999) showed that the rise time is different between low z and high z samples, indicating some evolution of the SNIa population. The effect on the cosmological parameter is not clear.

TABLE 8. Estimates of maximum brightness on SNe: 1997 vs. 1999 from Perlmutter et al. (1997; 1999).

SN	1997 value	1999 value	difference
SN1992bi	(23.26±0.24)	23.11±0.46	(0.15)
SN1994H	22.08±0.11	21.72±0.22	0.36
SN1994al	22.79±0.27	22.55±0.25	0.24
SN1994F	(21.80±0.69)	22.26±0.33	(−0.58)
SN1994am	22.02±0.14	22.26±0.20	−0.24
SN1994G	22.36±0.35	22.13±0.49	0.23
SN1994an	22.01±0.33	22.58±0.37	−0.57

Note: The numbers in the parentheses are not used in the final result of the 1997 paper.

contains strong features. (Even for the spectrophotometric standard stars, the synthetic magnitude contains an error of 0.02–0.05 mag, especially when the colour band involves the Balmer or Paschen regions.) Whereas Perlmutter et al. assigns 0.02 mag to the error of (i) [and (ii)], a comparison of the two values of estimated maximum brightness in their 1997 paper (Perlmutter et al. 1997, where they claimed evidence for a high Ω universe) and the 1999 paper (TABLE 8) shows a general difficulty in the evaluation of the K correction (the difference dominantly comes from different K corrections). Schmidt et al. claim that their K correction errors are 0.03% mag. Dust obscuration (iii) is also an important source of errors, since the error of (i)+(ii) propagates to $E(B-V)$ and is then amplified with the R factor. So a 0.02 mag error in colour results in a 0.06 mag error in A_V.

We note that each SN datum contains ±0.2 mag (20%) error. The issue is whether this error is almost purely of random nature and systematics are controlled to a level of $\lesssim$0.05.

4.2. GRAVITATIONAL LENSING FREQUENCIES FOR QUASARS

The gravitational lensing optical depth is given by

$$d\tau = \mu F H_0^3 (1+z_L)^3 \left[\frac{D_{OL} D_{LS}}{D_{OS}}\right]^2 \frac{dt}{dz} dz \tag{14}$$

where $F = \langle 16\pi^3 n_g \sigma_g^4 H_0^{-3} \rangle$, and μ is a magnification factor. The cosmological factor in (14) is very sensitive to the cosmological constant, when it dominates (Fukugita & Turner 1991). F is the astrophysical factor that depends on the galaxy number density n_g and the mass distribution of galaxies, which is usually assumed to be a singular isothermal sphere with velocity dispersion σ_g. Figure 5 shows a typical calculation for the expected

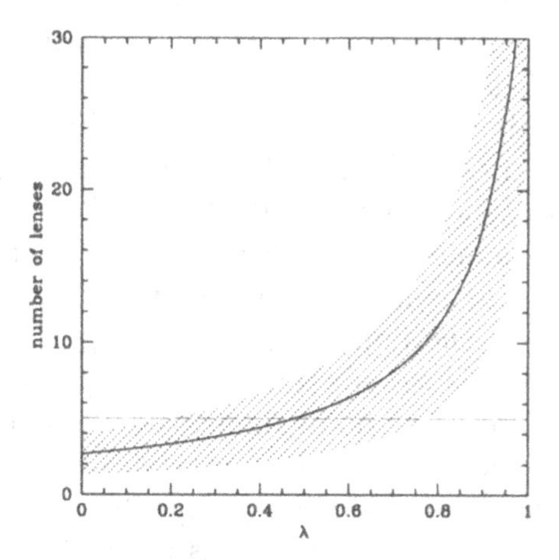

Figure 5. Gravitational lensing frequencies as a function of Λ in a flat universe. The expected number is given for 504 quasars of the HST Snapshot Survey sample. The shade means the region within a ±50% uncertainty. The observed number is 5 (dashed line).

number of strong lenses for 504 quasars of the HST Snapshot Survey (Maoz et al. 1993) sample: the observed number is 5 (4 if 0957+561 is excluded). The curve shows a high sensitivity to λ for $\lambda > 0.7$, but in contrast a nearly flat dependence for a lower λ. It is likely that $\lambda > 0.8$ is excluded. On the other hand, a more stringent limit is liable to be elusive. Fifty percent uncertainty in the F factor, say, would change largely a limit on, or a likely value of, λ.

In order to acquire information for a smaller λ, an accurate estimate is essential for the F factor, which receives the following uncertainties in: (1) the luminosity density and the fraction of early-type galaxies (the lensing power of E and S0 galaxies is much higher than that of spirals, and F is roughly proportional to the luminosity density of early-type galaxies); (2) σ_g-luminosity relation (Faber-Jackson relation); (3) the relation between σ(dark matter) and σ(star); (4) the model profile of dark haloes, specifically the validity of the singular isothermal sphere approximation (note that dark matter distributions seem more complicated in elliptical galaxies than in spiral galaxies, see Fukugita & Peebles 1999); (5) the core radius which leads to a substantial reduction in $d\tau$; (6) selection effects of the observations; (7) dust obscuration; (8) evolution of early-type galaxies.

There are continuous efforts for nearly a decade that have brought substantial improvement in reducing these uncertainties (Maoz & Rix 1993; Kochanek 1996; Falco et al. 1998). Nevertheless, the issue (1) still remains as a cause of a large uncertainty. While the total luminosity density is known to an uncertainty of 20% or so, the fraction of early type galaxies is more uncertain. It varies from 0.20 to 0.41 depending on the literature.

Including other items, it is likely that an estimate of F has a 50% uncertainty. For the curve in Figure 5 a change of F by $\pm 50\%$ brings the most likely value of λ to 0.75 or 0.2.

Kochanek and collaborators have made detailed considerations on the above uncertainties, and carried out elaborate statistical analyses. In their latest publication they concluded $\lambda < 0.62$ at 95% confidence level from an optical sample (Kochanek 1996). They took the fraction of early-type galaxies to be 0.44 and assigned a rather small 1σ error. (The predicted frequency comes close to the upper envelope of Fig. 5, and the observed number of lenses in the HST sample is taken to be 4). If one would adopt a smaller early-type fraction, the limit is immediately loosened by a substantial amount. Since the uncertainty is dominated by systematics rather than statistical, it seems dangerous to give significance to statistics. Statistical significance depends on artificial elements as to what are assumed in the input. A similar comment also applies to the recent work claiming for a positive λ (Chiba & Yoshii 1997; Cheng & Krauss 1998). I would conclude a conservative limit being $\lambda < 0.8$.

4.3. HARMONICS OF CBR ANISOTROPIES

This is a topic discussed repeatedly in this Summer Institute, so I will only briefly mention it for completeness. The positions of the acoustic peaks are particularly sensitive to Ω and λ, and even low accuracy data available at present lead to a meaningful constraint on a combination of Ω and λ.

The first acoustic peak appears at $\ell = \pi$(the distance to the last-scattering surface)/(the sound horizon) (Hu & Sugiyama 1995). Its position ℓ_1 is approximated as

$$\ell_1 \simeq 220 \left(\frac{1-\lambda}{\Omega}\right)^{1/2} , \tag{15}$$

for the parameter range that concerns us. This means that the position of the acoustic peak is about $\ell \simeq 220$ if $\Omega + \lambda = 1$, but it shifts to a high ℓ as $\Omega^{-1/2}$ if $\lambda = 0$. On the other hand, there is little power to determine Ω separately from λ, unless full information of C_ℓ is used. The harmonics C_ℓ measured at small angles revealed the acoustic peak (Scott et al. 1996), and its position favours a universe not far from flat (Hancock et al. 1998). More exhaustive analyses of Lineweaver (1998) and Efstathiou et al. (1999) show a limit $\Omega + \lambda/2 > 0.52$ (1σ). (The contours of the confidence level fall approximately on the curve given by (15) with ℓ_1=constant.) This means that a zero Λ universe is already marginal, when combined with Ω from other arguments. If a flat universe is chosen from CBR, a non-zero Λ will be compelling.

TABLE 9. Summary of Ω and λ.

method	Ω_0	Λ?	model used?
H_0 vs t_0	< 0.7		
luminosity density +M/L	0.1-0.4		
cluster baryon fraction	0.15-0.35		
SNeIa Hubble diagram	≤ 0.3	$\lambda \approx 0.7$	
small-scale velocity field (summary)	0.2 ± 0.15		
(pairwise velocity)	0.15 ± 0.1		
(Local Group kinematics)	0.15 ± 0.15		
(Virgocentric flow)	0.2 ± 0.2		
large-scale vel field	$0.2-1$		
cluster evolution (low Ω sol'n)	$0.2^{+0.3}_{-0.1}$		yes
(high Ω sol'n)	$\sim$1		yes
COBE-cluster matching	0.35-0.45 (if $\lambda = 0$)		yes
	0.20-0.40 (if $\lambda \neq 0$)		yes
shape parameter Γ	$0.2 - 0.4$		yes
CBR acoustic peak	free (if flat)	$\gtrsim 1 - 2\Omega$	yes
	> 0.5 (if open)		yes
gravitational lensing		$\lambda < 0.8$	
summary	0.15–0.45 (if open)		
	0.2–0.4 (if flat)		
		0.6–0.7(?)	

5. Conclusions

The status of Ω and λ is summarised in TABLE 9. We have a reasonable convergence of the Ω parameter towards a low value $\Omega = 0.15 - 0.4$. The convergence of Ω is significantly better with the presence of the cosmological constant that makes the universe flat. Particularly encouraging is that the Ω parameters derived with the aid of structure formation models agree with each other. This is taken to be an important test for the cosmological model, just as in particle physics when many different phenomena are reduced to a few convergent parameters to test the model. There are yet a still highly discrepant results on Ω, but it is not too difficult to speculate their origins. On the other hand, the current 'low Ω' means the values that vary almost by a factor of three and effort is needed to make these converge.

The cosmological constant has been an anathema over many years because of our ignorance of any mechanism that could give rise to a very small vacuum energy of $(3\ \text{meV})^4$, and neither can we understand a zero cosmo-

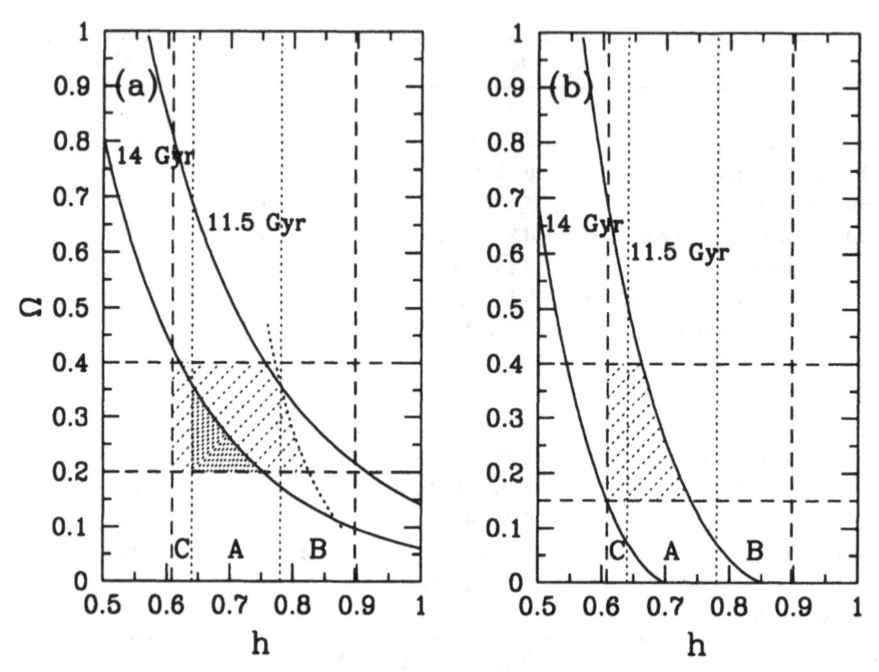

Figure 6. Consistent parameter ranges in the $H_0 - \Omega$ space for (a) a flat universe and (b) an open universe. A is the range of the Hubble constant when $(m-M)_{\rm LMC} = 18.5$. B or C is allowed only when the LMC distance is shorter by 0.3 mag, or longer by 0.1 mag. Note in panel (a) that most of the range of B is forbidden by the compatibility of age and H_0 that are simultaneously driven by the RR Lyr calibration (section 2.7). Also note that the age range between ≈11.5 Gyr and ≈14 Gyr is possible only with the interpretation that globular cluster formation is coeval (section 2.4). The most naturally-looking parameter region is given a thick shade.

logical constant. In mid-nineties the atmosphere was changing in favour for a non-zero Λ. The prime motivation was the Hubble constant−age problem, but the introduction of a non-zero Λ was helpful in many respects. One theoretical motivation was to satisfy flatness which is expected in inflationary scenarios (Peebles 1984). Ostriker & Steinhardt (1995) proclaimed a 'cosmic concordance' with a flat universe mildly dominated by Λ. By 1997, only one observation contradicted with the presence of a moderate value of Λ; this was the SNeIa Hubble diagram presented by the *Supernova Cosmology Project* (Perlmutter et al. 1997); see Fukugita 1997. In the next two years the situation changed. Two groups analysing SNeIa Hubble diagram, including the *Supernova Cosmology Project*, now claim a low Ω and a positive Λ. On the other hand, the Hubble constant−age problem became less severe due to our cognition of larger uncertainties, especially in the age estimate. The indications from SNeIa Hubble diagram are very interesting and important, but the conclusions are susceptible to small systematic effects. They should be taken with caution. We should perhaps wait for small-scale CBR anisotropy observations to confirm a nearly flat universe before concluding the presence of Λ.

In these lectures we have not considered classical tests, number counts, angular-size redshift relations, and magnitude-redshift relations of galaxies (Sandage 1961; 1988), in those testing for Ω and Λ. Unlike clusters or large scale structure, where no physics other than gravity plays a role, the evolution of galaxies is compounded by rich physics. Unless we understand their astrophysics, these objects cannot be used as testing candles. It has been known that galaxy number counts is understood more naturally with a low matter density universe under the assumption that the number of galaxies are conserved, but it is possible to predict the correct counts with an $\Omega = 1$ model where galaxies form through hierarchical merging, by tuning parameters that control physics (Cole et al. 1994; Kauffmann et al. 1994). It is important to work out whether the model works for any cosmological parameters or it works only for a restricted parameter range. This does not help much to extract the cosmological parameters, but it can falsify the model itself.

We have seen impressive progress in the determination of the Hubble constant. The old discrepancy is basically solved. On the other hand, a new uncertainty emerged in more local distance scales. The most pressing issue is to settle the value of the distance to LMC. There are also a few issues to be worked out should one try to determine H_0 to an accuracy of a 10% error or less. They include understanding of metallicity effects and interstellar extinction. The future effort will give more weight to geometric or semi-geometric methods. From the view point of observations the work will go to infra-red colour bands to minimise these problems.

In conclusion, I present in Figure 6 allowed ranges of H_0 and Ω (and λ) for the case of (a) flat and (b) open universes. With the flat case we cut the lower limit of Ω at 0.2 due to a strong constraint from lensing. An ample amount of parameter space is allowed for a flat universe. A high value of $H_0 > 82$, which would be driven only by a short LMC distance, is excluded by consistency with the age of globular clusters as noted earlier. Therefore, we are led to the range $H_0 \simeq 60 - 82$ from the consistency conditions. For an open universe the coeval-formation interpretation is compelling for globular clusters, or else no region is allowed. The allowed H_0 is limited to $60 - 70$. No solution is available if LMC takes a short distance.

I would like to thank Rob Crittenden for his careful reading and many useful suggestions on this manuscript. This work is supported in part by Grant in Aid of the Ministry of Education in Tokyo and Raymond and

Beverly Sackler Fellowship in Princeton.

References

1. Aaronson, M. et al. 1986, ApJ, 302, 536
2. van Altena,, B. et al. 1997, Baltic Astron., 6(1), 27
3. Andersen, J. 1991, A& AR, 3, 91
4. Arnett, W. D., Branch, D. & Wheeler, J. C. 1985, Nature, 314, 337
5. Bahcall, N. A. & Fan, X. 1998, ApJ, 504, 1
6. Bahcall, N. A., Lubin, L. M. & Dorman, V. 1995, ApJ, 447, L81
7. Bartlett, J. G. & Blanchard, A. 1996, A& A, 307, 1
8. Beaulieu et al. 1997, A& A, 318, L47
9. Bell, S. A., Hill, G., Hilditch, R. W., Clausen, J. V. & Reynolds, A. P. 1993, MNRAS, 265, 1047
10. Bennett, C. L. et al. 1996, ApJ. 464, L1
11. van den Bergh, S. 1989, A& AR, 1, 111
12. van den Bergh, S. 1994, PASP, 106, 1113
13. van den Bergh, S. 1995, ApJ, 446, 39
14. Bernstein, G. & Fischer, P. 1999, astro-ph/9903274
15. Biggs, A. D. et al. 1999, MNRAS, 304, 349
16. Birkinshaw, M., Hughes, J. P. & Arnaud, K. A. 1991, ApJ, 379, 466
17. Blakeslee, J. P., Ajhar, E. A. & Tonry, J. L. 1998, in *Post Hipparcos Cosmic Candles*, eds. A. Heck & F. Caputo (Kluwer, Boston), p.181
18. Blanchard, A. & Bartlett, J. G. 1998, A& A, 332, L49
19. Bolte, M. & Hogan, C. J. 1995, Nature, 376, 399
20. Branch, D. 1992, ApJ, 392, 35
21. Branch, D. & Miller, D. L. 1993, ApJ, 405, L5
22. Branchini, E. & Carlberg, R. G. 1994, ApJ, 434, 37
23. Buonanno, R., Corsi, C. E. & Fusi Pecci, F. 1989, A& A, 216, 80
24. Caldwell, J. A. R. 1983, Observatory, 103, 244
25. Carlberg, R. G., Yee, H. K. C. & Ellingson, E. 1997a, ApJ, 478, 462
26. Carlberg, R. G., Morris, S. L., Yee, H. K. C. & Ellingson, E. 1997b, ApJ, 479, L19
27. Carlberg, R. G. et al. 1996, ApJ, 462, 32
28. Chaboyer, B., Demarque, P. & Sarajedini, A. 1996, ApJ, 459, 558
29. Chaboyer, B., Demarque, P. Kernan, P. J. & Krauss, L. M. 1998, ApJ, 494, 96
30. Cheng, Y.-C. N. & Krauss, L. M. 1998, astro-ph/9810393
31. Chiba, M. & Yoshii, Y 1997, ApJ, 490, L73
32. Chiosi, C., Wood, P. R. & Capitanio, N. 1993, ApJS, 86, 541
33. Ciardullo, R., Jacoby, G. H. & Tonry, J. L. 1993, ApJ, 419, 479
34. Cole, A. A. 1998, ApJ 500, L137
35. Cole, S. et al. 1994, MNRAS, 271, 781
36. Conti, P. S., Garmany, C. D. & Massey, P. 1986, AJ, 92, 48
37. Courbin, F. et al. 1997, A& A, 324, L1
38. Davis, M. & Peebles, P. J. E. 1983, ARA& A, 21, 109
39. Dekel, A., Bertschinger, E & Faber, S. M. 1990, ApJ, 364, 349
40. Dekel, A., Burstein, D. & White, S. D. M. 1997, in *Critical Dialogues in Cosmology*, ed. N. Turok (World Scientific, Singapore), p. 175
41. Dekel, A. et al. 1999, ApJ, 522, 1
42. Dunn, A. M. & Laflamme, R. 1995, ApJ, 443, L1
43. Efstathiou, G. & Bond, J. R. 1999, MNRAS, 304, 75
44. Efstathiou, G., Bond, J. R. & White, S. D. M. 1992, MNRAS, 258, 1p
45. Efstathiou, G., Sutherland, W. J. & Maddox, S. J. 1990, Nature, 348, 705
46. Efstathiou, G., Bridle, S. L., Lasenby, A. N., Hobson, M. P. & Ellis, R. S. 1999, MNRAS, 303, 47

47. Eisenstein, D. J., Hu, W. & Tegmark, M. 1999, ApJ. 518, 2
48. Eke, V. R., Cole, S., & Frenk, C. S. 1996, MNRAS, 282, 263
49. Eke, V. R., Cole, S., Frenk, C. S. & Henry, J. P. 1998, MNRAS, 298, 1145
50. ESA 1997, *The Hipparcos and Tycho Catalogues* SP-1200 (ESA, Noordwijk)
51. Faber, S. M. & Gallagher, J. S 1979, ARA& A, 17, 135
52. Falco, E. E., Gorenstein, M. V.& Shapiro, I. I. 1991, ApJ, 372, 364
53. Falco, E. E., Kochanek, C. S. & Muñoz, J. A. 1998, ApJ, 494, 47
54. Feast, M. W. & Catchpole, R. M. 1997 MNRAS, 286, L1
55. Feast, M. W. & Walker, A. R. 1987, ARA& A, 25, 345
56. Federspiel, M., Tammann, G. A. & Sandage, A, 1998, ApJ 495, 115
57. Fernley, J. et al. 1998, A&A, 330, 515
58. Ferrarese, L et al. 1999a, astro-ph/9908192
59. Ferrarese, L et al. 1999b, astro-ph/9910501
60. Forman, W. & Jones, C. 1982, ARA& A, 20, 547
61. Freedman, W. L. & Madore, B. F. 1990, ApJ, 365, 186
62. Freedman, W. L., Madore, B. F. & Kennicutt, R. C. 1997, in *The Extragalactic distance scale*, eds. M. Livio, M. Donahue, & N. Panagia (Cambridge University Press, Cambridge), p.171
63. Freedman, W. L. et al. 1994, Nature, 371, 757
64. Fry, A. M. & Carney, B. W. 1997, AJ, 113, 1073
65. Fukugita, M. 1997, in *Critical Dialogues in Cosmology*, ed. N. Turok (World Scientific, Singapore, 1997), p.204
66. Fukugita, M. & Peebles, P. J. E. 1999, ApJ. 524, L31
67. Fukugita, M. & Turner, E. L. 1991, MNRAS, 253, 99
68. Fukugita, M., Hogan, C. J. & Peebles, P. J. E. 1993, Nature, 366, 309
69. Fukugita, M., Hogan, C. J. & Peebles, P. J. E. 1998, ApJ, 503, 518
70. Gibson, B. K. et al. 1999, astro-ph/9908149, ApJ, in press
71. Gieren, W. P., Fouqué, P. & Gómez, M. 1998, ApJ, 496, 17
72. Giovanelli, R. et al. 1997, AJ, 113, 22
73. Gould, A 1994, ApJ, 426, 542
74. Gould, A & Popowski, P. 1998, ApJ, 508, 844
75. Gould, A & Uza, O. 1998, ApJ, 494, 118
76. Gratton, R. G. et. al. 1997 ApJ, 491, 749
77. Grego, L. et al. 1999, presented at the AAS meeting (194.5807G)
78. Grogin, N. A. & Narayan, R. 1996, ApJ, 464, 92
79. Guinan, E. F. et al. 1998, ApJ, 509, L21
80. Hamilton, A. J. S. 1998, in *The Evolving Universe, ed. D. Hamilton*, (Kluwer, Dordrecht, 1998), p. 185
81. Hamuy, M. et al. 1996a, ApJ, 112, 2391
82. Hamuy, M. et al. 1996b, ApJ, 112, 2398
83. Hancock, S., Rocha, G., Lasenby, A. N. & Gutiérrez, C. M. 1998, MNRAS, 294, L1
84. Henry, J. P. & Arnaud, K. A. 1991, ApJ, 372, 410
85. Hanson, R. B. 1980, in *Star Clusters*, IAU Symposium 85, ed. J. E. Hesser (Reidel, Dordrecht), p. 71
86. Herrnstein, J. R. et al. 1999, Nature, 400, 539
87. Hu, W. & Sugiyama, N. 1995, Phys. Rev., D51, 2599
88. Hubble, E. P. 1925, Observatory, 48, 139
89. Iben, I. & Tuggle, R. S. 1975, ApJ, 197, 39
90. Impey, C. D. et al. 1998, ApJ, 509, 551
91. Jacoby, G. H., Ciardullo, R. & Ford, H. C. 1990a, ApJ, 356, 332
92. Jacoby, G. H., Walker, A. R. & Ciardullo, R. 1990b, ApJ, 365, 471
93. Jacoby, G. H., Ciardullo, R. & Harris, W. E. 1996, ApJ, 462,1
94. Jacoby, G. H. et al. 1992, PASP, 104, 599
95. Jensen, J. B., J. L.Tonry & Luppino, G. A. 1999, ApJ, 510, 71
96. Jha, S. et al. 1999, astro-ph/9906220, ApJ in press

97. Kaiser, N. 1987, MNRAS 227, 1
98. Kaiser, N. et al. 1998, astro-ph/9809268
99. Kauffmann, G., Guiderdoni, B. & White, S. D. M. 1994, MNRAS, 267, 981
100. Keeton, C. R. & Kochanek 1997, ApJ, 487, 42
101. Kelson, D. D. et al. 1999, astro-ph/9909222, ApJ in press
102. Kennicutt, R. C. et al. 1998, ApJ, 498, 181
103. Kochanek, C. S. 1996a, ApJ, 457, 228
104. Kochanek, C. S. 1996, ApJ, 466, 638
105. Kochanek, C. S. 1997, ApJ, 491, 13
106. Koopmans, L. V. E. & Fassnacht, C. D. 1999, astro-ph/9907258, ApJ, in press
107. Kovács, J & Jurcsik, J. 1996, ApJ, 466, L17
108. Kraan-Korteweg, R. C., Cameron, L. M. & Tammann, G. A. 1988, ApJ, 331, 620
109. Kundić, T., Cohen, J. G., Blandford, R. D. & Lubin, L. M. 1997a, AJ, 114, 507
110. Kundić, T. et al. 1997b, ApJ, 482, 75
111. Laney, C. D. & Stobie, R. S. 1994, MNRAS, 266, 441
112. Little, B. & Tremaine, S. 1987, ApJ, 320, 493
113. van Leeuwen, F. 1983, PhD thesis (Leiden University)
114. van Leeuwen, F. 1999, A& A, 341, L71
115. van Leeuwen, F. & Hansen-Ruiz, C. S. 1997, in *Hipparcos Venice '97*, ed. B. Battrick (ESA, Noordwijk, 1997), p. 689
116. van Leeuwen, F., Feast, M. W., Whitelock, P. A. & Yudin, B. 1997, MNRAS, 287, 955
117. Leibundgut, B. & Pinto, P. A. 1992, ApJ, 401, 49
118. Lineweaver, C. H. 1998, ApJ, 505, L69
119. Lovell, J. E. J. et al. 1998, ApJ, 508, L51
120. Lundqvist, P. & Sonneborn, G. 1997, in *SN1987A, Ten Years After*, ed. M. Phillips & N. Santzeff (ASP, San Francisco, 1997) to be published
121. Luri, X., Gómez, A. E., Torra, J., Figueras, F. & Mennessier, M. O. 1998, A& A, 335, L81
122. Lynden-Bell, D. et al. 1988, ApJ, 326, 19
123. Lyngå, G. 1987, *Catalogue of Open Cluster Data, 5th edition* (Lund University)
124. Madore, B. F. & Freedman, W. L. 1991, PASP, 103, 933
125. Madore, B. F. & Freedman, W. L. 1998, ApJ, 492, 110
126. Maoz, D. & Rix, H.-W. 1993, ApJ, 416, 425
127. Maoz, D. et al. 1993, ApJ, 402, 69
128. Maoz, E. et al. 1999, Nature, 401, 351
129. Mateo, M. & Hodge, P. 1986, ApJS, 60, 893
130. Mermilliod, J.-C., Turon, C., Robichon, N., Arenou, F. & Lebreton, Y. 1997, in *Hipparcos Venice '97*, ed. B. Battrick (ESA, Noordwijk, 1997), p. 643
131. Miyoshi, M. et al. 1995, Nature, 373, 127
132. Mochejska, B. J., Macri, L. M., Sasselov, D. D. & Stanek, K. Z. 1999, astro-ph/9908293
133. Myers, S. T., Baker, J. E., Readhead, A. C. S., Leitch, E. M. & Herbig, T. 1997, ApJ, 485, 1
134. Mould, J. R. et al. 1999, astro-ph/9909260, ApJ, in press
135. Narayanan, V. K. & Gould, A. 1999, ApJ, 523, 328
136. Ostriker, J. P. & Steinhardt, P. J. 1995, Nature, 377, 600
137. Oukbir, J. & Blanchard, A. 1992, A& A, 262, L21
138. Paczyński, B. 1997, in *The Extragalactic distance scale*, eds. M. Livio, M. Donahue, & N. Panagia (Cambridge University Press, Cambridge), p.273
139. Paczyński, B. & Stanek, K. Z. 1998, ApJ, 494, L219
140. Panagia, N., Gilmozzi, R., Macchetto, F., Adore, H.-M. & Kirshner, R. P. 1991, ApJ, 380, L23
141. Panagia, N. , Gilmozzi, R. & Kirshner, R. 1997, in *SN1987A, Ten Years After*, ed. M. Phillips & N. Santzeff (ASP, San Francisco, 1997) to be published

142. Paturel, G. et al. 1997, in Hipparcos Venice '97, ed. B. Battrick (ESA, Noordwijk, 1997), p. 629.
143. Peacock, J. A. 1997, MNRAS, 284, 885
144. Peacock, J. A. & Dodds, S. J. 1994, MNRAS, 267, 1020
145. Peebles, P. J. E. 1980, *The Large-Scale Structure of the Universe* (Princeton University Press, Princeton)
146. Peebles, P. J. E. 1984, ApJ, 284, 439
147. Peebles, P. J. E. 1995, ApJ, 449, 52
148. Peebles, P. J. E. 1999, in *Formation of Structure in the Universe*, eds Dekel, A. & Ostriker, J. P. (Cambridge University Press, Cambridge), p. 435
149. Pen, U.-L. 1998, ApJ, 498, 60
150. Peng et al. 1998, cited in Bridges, T. astro-ph/9811136
151. Perlmutter, S. et al. 1997, ApJ, 483, 565
152. Perlmutter, S. et al. 1999, ApJ, 517, 565
153. Perryman, M. A. C. et. al. 1998, A& A, 331, 81
154. Phillips, M. M. 1993, ApJ, 413, L105
155. Phillips, M. M. et al. 1999, astro-ph/9907052
156. Pierce, M. J. & Tully, R. B. 1988, ApJ, 330, 579
157. Pierce, M. J. et al. 1994, Nature, 371, 385
158. Pinsonneault, M. H., Stauffer, J., Soderblom, D. R., King, J. R. & Hanson, R. B. 1998, ApJ, 504, 170
159. Pskovskiĭ, Yu. P. 1984, Astron. Zh., 61, 1125 (Sov. Astron. 28, 658)
160. Refsdal, S. 1964, MNRAS, 128, 307
161. Reichart, D. E. et al. 1999, ApJ, 518, 521
162. Reid, I. N. 1997, AJ, 114, 161
163. Renzini, A. 1991, in *Observational Tests of Cosmological Inflation*, ed. T. Shanks et al. (Kluwer, Dordrecht), p. 131
164. Riess, A. G., Press, W. H. & Kirshner, R. P. 1995, ApJ, 438, L17
165. Riess, A. G., Filippenko, A. V., Li, W. & Schmidt, B. P. 1999, astro-ph/9907038, ApJ, in press
166. Riess, A. G. et al. 1998, AJ, 116, 1009
167. Saha, A. et al. 1999, 522, 802
168. Sakai, S. et al. 1999, astro-ph/9909269
169. Sandage, A. 1961, ApJ, 133, 355
170. Sandage, A. 1988, ARA& A, 26, 561
171. Sandage, A. 1993a, ApJ, 402, 3
172. Sandage, A. 1993b, AJ, 106, 703
173. Sandage, A., Bell, R. A. & Tripicco, M. J. 1999, ApJ, 522, 250
174. Sandage, A. & Tammann, G. A. 1968, ApJ, 151, 531
175. Sandage, A. & Tammann, G. A. 1982, ApJ, 256, 339
176. Sandage, A. & Tammann, G. A. 1997, in *Critical Dialogues in Cosmology*, ed. N. Turok (World Scientific, Singapore, 1997), p. 130
177. Sandage, A., Saha, A. Tammann, G. A., Panagia, N. & Macchetto, D. 1992, ApJ, 401, L7
178. Schechter, P. L. et al. 1997, ApJ, 475, L85
179. Schmidt, B. P., Kirshner, R. P. & Eastman, R. G. 1992, ApJ, 395, 366
180. Schmidt, B. P. et al. 1994, ApJ, 432, 42
181. Schmidt, B. P. et al. 1998, ApJ, 507, 46
182. Schommer, R. A., Olszewski, E. W. & Aaronson, M. 1984, ApJ, 285, L53
183. Schutz, B. F. 1986, Nature, 323, 310
184. Scott, P. F. et al. 1996, ApJ, 461, L1
185. Sekiguchi, M. & Fukugita, M. 1998, Observatory, 118, 73
186. Small, T. A., Ma, C.-P., Sargent, W. L. W. & Hamilton, D 1998, ApJ, 492, 45
187. Sonneborn, G. et al. 1997, ApJ, 477, 848
188. Squires, G., Kaiser, N., Fahlman, G., Babul, A. & Woods, D. 1996, ApJ 469, 73

189. Stanek, K. Z. & Garnavich, P. M. 1998, ApJ, 503, L131
190. Stanek, K. Z., Zaritsky, D. & Harris, J. 1998, ApJ, 500, L141
191. Stothers, R. B. 1988, ApJ, 329, 712
192. Strauss, M. A. & Willick, J. A. 1995, Phys. Rep., 261, 271
193. Suntzeff, N. B. et al. 1999, ApJ, 117, 1175
194. Szabados, L. in Hipparcos Venice '97, ed. B. Battrick (ESA, Noordwijk, 1997), p. 657
195. Tammann, G. A. 1999, in *Cosmological Parameters and the Evolution of the Universe* (IAU Symposium 183), ed. K. Sato (Kluwer, Dordrecht), p. 31
196. Tammann, G. A. & Leibundgut, B. 1990, A& A, 236, 9
197. Tegmark, M. 1999, ApJ, 514, L69
198. Thomsen, B., Baum, W. A., Hammergren, M. & Worthey, G. 1997, ApJ, 483, L37
199. Tonry, J. L., Blakeslee, J. P., Ajhar, E. A. & Dressler, A. 1997, ApJ, 475, 399
200. Tonry, J. L., Blakeslee, J. P., Ajhar, E. A. & Dressler, A. 1999, astro-ph/9907062
201. Tonry, J. L. & Franx, M. 1998, ApJ, 515, 512
202. Tonry, J. L. & Schneider, D. P. 1988, AJ, 96, 807
203. Torres, G., Stefanik, R. P. & Latham, D. W. 1997, ApJ, 474, 256
204. Tully, R. B. 1999, in *Cosmological Parameters and the Evolution of the Universe* (IAU Symposium 183), ed. K. Sato (Kluwer, Dordrecht), p. 54
205. Tully, R. B. & Fisher, J. R. 1977, A& A, 54, 661
206. Tully, R. B. et al. 1998, ApJ, 115, 2264
207. Tyson, J. A. & Fischer, P. 1995, ApJ, 446, L55
208. Udalski, A. 1998, Acta Astron., 48, 113
209. Udalski, A. et al. 1998a, Acta Astron., 48, 1
210. Udalski, A. et al. 1998b, ApJ, 509, L25
211. VandenBerg, D. A. & Bridges, T. J. 1984, ApJ, 278, 679
212. VandenBerg, D. A., Bolte, M. & Stetson, P. B. 1996, ARA& A, 34, 461
213. de Vaucouleurs, G. 1981, in *10th Texas Symp. on Relativistic Astrophysics*, ed. Ramaty, R. & Jones, F. C., Ann. N.Y. Acad. Sci. 375, 90
214. Viana, P. T. P. & Liddle, A. R. 1999, astro-ph/9902245 in *Cosmological Constraints from X-Ray Clusters* (to be published)
215. Weinberg, S. 1972, *Gravitation and Cosmology* (J. Wiley, New York)
216. White, D. A. & Fabian, A. C. 1995, MNRAS, 273, 72
217. White, S. D. M., Efstathiou, G. & Frenk, C. S. 1993a, MNRAS, 262, 1023
218. White, S. D. M., Navarro, J. F., Evrard, A. E. & Frenk, C. S. 1993b, Nature, 366, 429
219. Whitelock, P. A., van Leeuwen, F. & Feast. M. W. 1997, in *Hipparcos Venice '97*, ed. B. Battrick (ESA, Noordwijk, 1997), p. 213
220. Willick, J. A. et al. 1997, ApJS, 109, 333
221. Yasuda, N., Fukugita, M. & Okamura, S. 1997, ApJS, 108, 417
222. Zaldarriaga, M., Spergel, D. N. & Seljak, U. 1997, ApJ, 488, 1
223. de Zeeuw, P. T. et al. 1997, in *Hipparcos Venice '97*, ed. B. Battrick (ESA, Noordwijk, 1997), p. 495
224. Zaritsky, D., Smith, R., Frenk, C. & White, S. D. M. 1997, ApJ, 478, 39

OBSERVATIONAL TESTS FOR THE COSMOLOGICAL PRINCIPLE AND WORLD MODELS

OFER LAHAV
Institute of Astronomy
Madingley Road, Cambridge CB3 0HA, UK

Abstract. We review observational tests for the homogeneity of the Universe on large scales. Redshift and peculiar velocity surveys, radio sources, the X-Ray Background, the Lyman-α forest and the Cosmic Microwave Background are used to set constraints on inhomogeneous models and in particular on fractal-like models. Assuming the Cosmological Principle and the FRW metric, we estimate cosmological parameters by joint analysis of peculiar velocities, the CMB, cluster abundance, IRAS and Supernovae. Under certain assumptions the best fit density parameter is $\Omega_m = 1 - \lambda \approx 0.4$. We present a new method for joint estimation by combining different data sets in a Bayesian way, and utilising 'Hyper-Parameters'.

1. Introduction

The Cosmological Principle was first adopted when observational cosmology was in its infancy; it was then little more than a conjecture, embodying 'Occam's razor' for the simplest possible model. Observations could not then probe to significant redshifts, the 'dark matter' problem was not well-established and the Cosmic Microwave Background (CMB) and the X-Ray Background (XRB) were still unknown. If the Cosmological Principle turned out to be invalid then the consequences to our understanding of cosmology would be dramatic, for example the conventional way of interpreting the age of the Universe, its geometry and matter content would have to be revised. Therefore it is important to revisit this underlying assumption in the light of new galaxy surveys and measurements of the background radiations.

Like with any other idea about the physical world, we cannot prove a model, but only falsify it. Proving the homogeneity of the Universe is in

R.G. Crittenden and N.G. Turok (eds.), Structure Formation in the Universe, 131–142.

particular difficult as we observe the Universe from one point in space, and we can only deduce directly isotropy. The practical methodology we adopt is to assume homogeneity and to assess the level of fluctuations relative to the mean, and hence to test for consistency with the underlying hypothesis. If the assumption of homogeneity turns out to be wrong, then there are numerous possibilities for inhomogeneous models, and each of them must be tested against the observations.

Despite the rapid progress in estimating the density fluctuations as a function of scale, two gaps remain:

(i) It is still unclear how to relate the distributions of galaxies and mass (i.e. 'biasing'); (ii) Relatively little is known about fluctuations on intermediate scales between these of local galaxy surveys ($\sim 100h^{-1}$ Mpc) and the scales probed by COBE ($\sim 1000h^{-1}$ Mpc).

Here we examine the degree of smoothness with scale by considering redshift and peculiar velocities surveys, radio-sources, the XRB, the Ly-α forest, and the CMB. We discuss some inhomogeneous models and show that a fractal model on large scales is highly improbable. Assuming an FRW metric we evaluate the 'best fit Universe' by performing a joint analysis of cosmic probes.

2. Cosmological Principle(s)

Cosmological Principles were stated over different periods in human history based on philosophical and aesthetic considerations rather than on fundamental physical laws. Rudnicki (1995) summarized some of these principles in modern-day language:

- The Ancient Indian: *The Universe is infinite in space and time and is infinitely heterogeneous.*
- The Ancient Greek: *Our Earth is the natural centre of the Universe.*
- The Copernican CP: *The Universe as observed from any planet looks much the same.*
- The Generalized CP: *The Universe is (roughly) homogeneous and isotropic.*
- The Perfect CP: *The Universe is (roughly) homogeneous in space and time, and is isotropic in space.*
- The Anthropic Principle: *A human being, as he/she is, can exist only in the Universe as it is.*

We note that the Ancient Indian principle can be viewed as a 'fractal model'. The Perfect CP led to the steady state model, which although more symmetric than the CP, was rejected on observational grounds. The Anthropic Principle is becoming popular again, e.g. in explaining a non-zero cosmological constant. Our goal here is to quantify 'roughly' in the definition of the generalized CP, and to assess if one may assume safely the Friedmann-Robertson-Walker (FRW) metric of space-time.

3. Probes of Smoothness

3.1. THE CMB

The CMB is the strongest evidence for homogeneity. Ehlers, Garen and Sachs (1968) showed that by combining the CMB isotropy with the Copernican principle one can deduce homogeneity. More formally the EGS theorem (based on Liouville theorem) states that "If the fundamental observers in a dust spacetime see an isotropic radiation field, then the spacetime is locally FRW". The COBE measurements of temperature fluctuations $\Delta T/T = 10^{-5}$ on scales of 10° give via the Sachs Wolfe effect ($\Delta T/T = \frac{1}{3}\Delta\phi/c^2$) and Poisson equation rms density fluctuations of $\frac{\delta\rho}{\rho} \sim 10^{-4}$ on $1000\,h^{-1}$ Mpc (e.g. Wu, Lahav & Rees 1999; see Fig 3 here), i.e. the deviations from a smooth Universe are tiny.

3.2. GALAXY REDSHIFT SURVEYS

Figure 1 shows the distribution of galaxies in the ORS and IRAS redshift surveys. It is apparent that the distribution is highly clumpy, with the Supergalactic Plane seen in full glory. However, deeper surveys such as LCRS show that the fluctuations decline as the length-scales increase. Peebles (1993) has shown that the angular correlation functions for the Lick and APM surveys scale with magnitude as expected in a universe which approaches homogeneity on large scales.

Existing optical and IRAS (PSCz) redshift surveys contain $\sim 10^4$ galaxies. Multifibre technology now allows us to measure redshifts of millions of galaxies. Two major surveys are underway. The US Sloan Digital Sky Survey (SDSS) will measure redshifts to about 1 million galaxies over a quarter of the sky. The Anglo-Australian 2 degree Field (2dF) survey will measure redshifts for 250,000 galaxies selected from the APM catalogue. About 80,000 2dF redshifts have been measured so far (as of December 1999). The median redshift of both the SDSS and 2dF galaxy redshift surveys is $\bar{z} \sim 0.1$. While they can provide interesting estimates of the fluctuations on scales of hundreds of Mpc's, the problems of biasing, evolution and K-correction, would limit the ability of SDSS and 2dF to 'prove' the Cosmological Principle. (cf. the analysis of the ESO slice by Scaramella et al 1998 and Joyce et al. 1999).

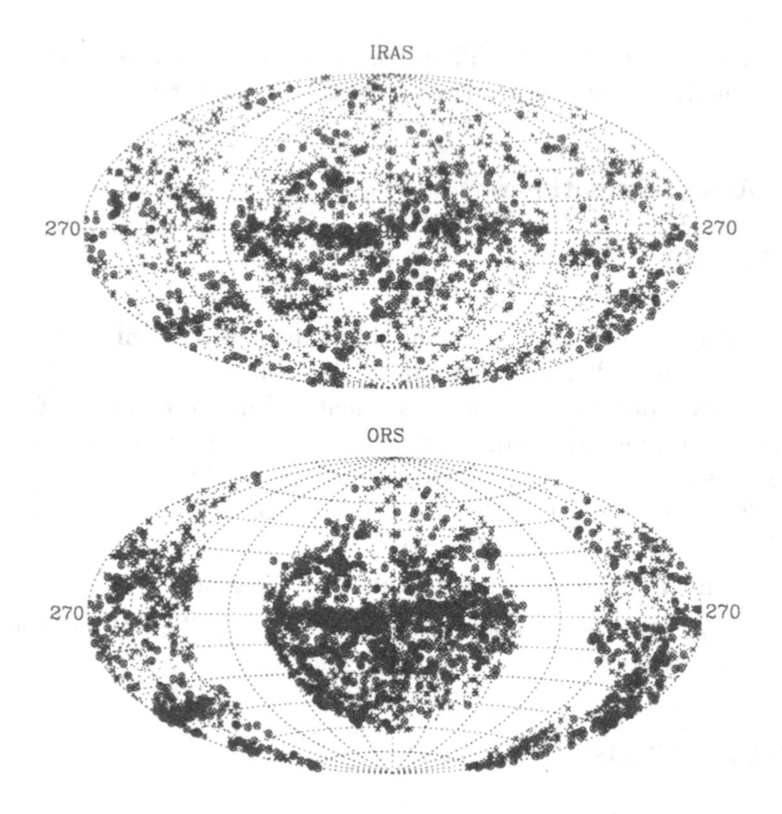

Figure 1. The distribution of galaxies projected on the sky in the IRAS and ORS samples. This is an Aitoff projection in Supergalactic coordinates, with $L = 90°, B = 0°$ (close to the Virgo cluster) in the centre of the map. Galaxies within 2000 km/sec are shown as circled crosses; galaxies between 2000 and 4000 km/sec are indicated as crosses, and dots mark the positions of more distant objects. Here we include only catalogued galaxies, which is why the Zone of Avoidance is so prominent in these two figures. (Plot by M. Strauss, from Lahav et al. 1999).

3.3. PECULIAR VELOCITIES

Peculiar velocities are powerful as they probe directly the mass distribution (e.g. Dekel et al. 1999). Unfortunately, as distance measurements increase with distance, the scales probed are smaller than the interesting scale of transition to homogeneity. Conflicting results on both the amplitude and coherence of the flow suggest that peculiar velocities cannot yet set strong constraints on the amplitude of fluctuations on scales of hundreds of Mpc's. Perhaps the most promising method for the future is the kinematic Sunyaev-Zeldovich effect which allows one to measure the peculiar velocities of clusters out to high redshift.

The agreement between the CMB dipole and the dipole anisotropy of relatively nearby galaxies argues in favour of large scale homogeneity. The IRAS dipole (Strauss et al 1992, Webster et al 1998, Schmoldt et al 1999) shows an apparent convergence of the dipole, with misalignment angle of only 15°. Schmoldt et al. (1999) claim that 2/3 of the dipole arises from within a $40\,h^{-1}$ Mpc, but again it is difficult to 'prove' convergence from catalogues of finite depth.

3.4. RADIO SOURCES

Radio sources in surveys have typical median redshift $\bar{z} \sim 1$, and hence are useful probes of clustering at high redshift. Unfortunately, it is difficult to obtain distance information from these surveys: the radio luminosity function is very broad, and it is difficult to measure optical redshifts of distant radio sources. Earlier studies claimed that the distribution of radio sources supports the 'Cosmological Principle'. However, the wide range in intrinsic luminosities of radio sources would dilute any clustering when projected on the sky (see Figure 2). Recent analyses of new deep radio surveys (e.g. FIRST) suggest that radio sources are actually clustered at least as strongly as local optical galaxies (e.g. Cress et al. 1996; Magliocchetti et al. 1998). Nevertheless, on the very large scales the distribution of radio sources seems nearly isotropic. Comparison of the measured quadrupole in a radio sample in the Green Bank and Parkes-MIT-NRAO 4.85 GHz surveys to the theoretically predicted ones (Baleisis et al. 1998) offers a crude estimate of the fluctuations on scales $\lambda \sim 600h^{-1}$ Mpc. The derived amplitudes are shown in Figure 3 for the two assumed Cold Dark Matter (CDM) models. Given the problems of catalogue matching and shot-noise, these points should be interpreted at best as 'upper limits', not as detections.

3.5. THE XRB

Although discovered in 1962, the origin of the X-ray Background (XRB) is still unknown, but is likely to be due to sources at high redshift (for review see Boldt 1987; Fabian & Barcons 1992). The XRB sources are probably located at redshift $z < 5$, making them convenient tracers of the mass distribution on scales intermediate between those in the CMB as probed by COBE, and those probed by optical and IRAS redshift surveys (see Figure 3).

The interpretation of the results depends somewhat on the nature of the X-ray sources and their evolution. By comparing the predicted multipoles to those observed by HEAO1 (Lahav et al. 1997; Treyer et al. 1998; Scharf et al. 1999) we estimate the amplitude of fluctuations for an assumed shape of the density fluctuations (e.g. CDM models). Figure 3 shows the amplitude of fluctuations derived at the effective scale $\lambda \sim 600h^{-1}$ Mpc probed by

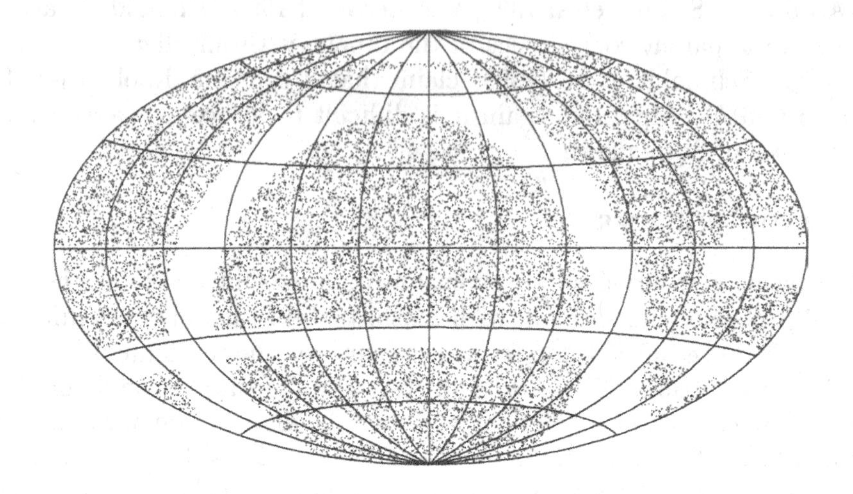

Figure 2. The distribution of radio source from the 87GB and PMN surveys projected on the sky. This is an Aitoff projection in Equatorial coordinates (from Baleisis et al. 1998).

the XRB. The observed fluctuations in the XRB are roughly as expected from interpolating between the local galaxy surveys and the COBE CMB experiment. The rms fluctuations $\frac{\delta\rho}{\rho}$ on a scale of $\sim 600h^{-1}$Mpc are less than 0.2 %.

3.6. THE LYMAN-α FOREST

The Lyman-α forest reflects the neutral hydrogen distribution and therefore is likely to be a more direct trace of the mass distribution than galaxies are. Unlike galaxy surveys which are limited to the low redshift Universe, the forest spans a large redshift interval, typically $1.8 < z < 4$, corresponding to comoving interval of $\sim 600\,h^{-1}$ Mpc. Also, observations of the forest are not contaminated by complex selection effects such as those inherent in galaxy surveys. It has been suggested qualitatively by Davis (1997) that the absence of big voids in the distribution of Lyman-α absorbers is inconsistent with the fractal model. Furthermore, all lines-of-sight towards quasars look statistically similar. Nusser & Lahav (1999) predicted the distribution of the flux in Lyman-α observations in a specific truncated fractal-like model. They found that indeed in this model there are too many voids compared with the observations and conventional (CDM-like) models for structure formation. This too supports the common view that on large scales the

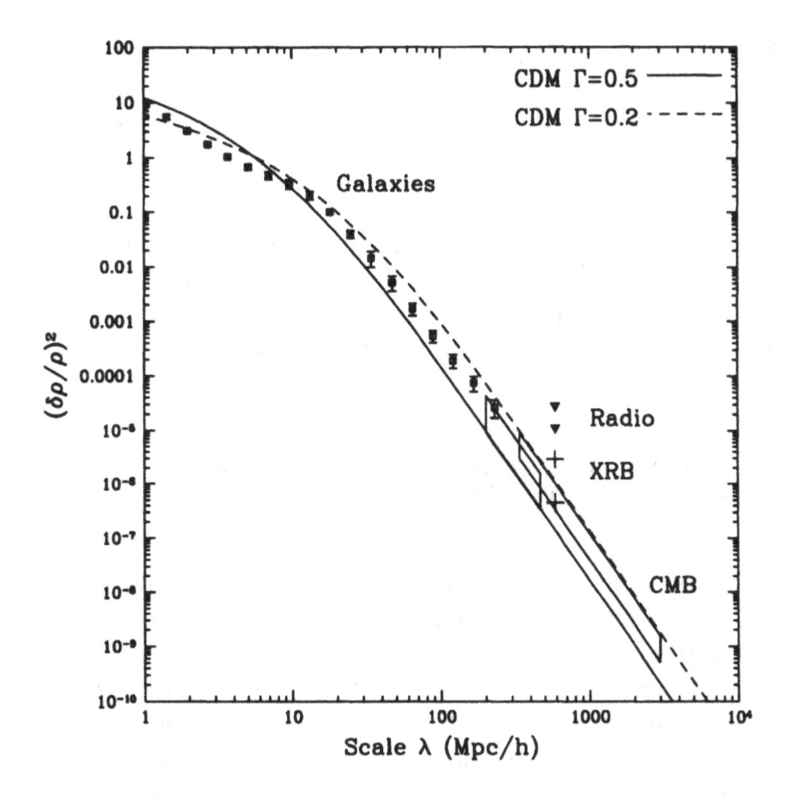

Figure 3. A compilation of density fluctuations on different scales from various observations: a galaxy survey, deep radio surveys, the X-ray Background and Cosmic Microwave Background experiments. The measurements are compared with two popular Cold Dark Matter models (with normalization $\sigma_8 = 1$ and shape parameters $\Gamma = 0.2$ and 0.5). The Figure shows mean-square density fluctuations $(\frac{\delta\rho}{\rho})^2 \propto k^3 P(k)$, where $k = 1/\lambda$ is the wavenumber and $P(k)$ is the power-spectrum of fluctuations. The open squares at small scales are estimates from the APM galaxy catalogue (Baugh & Efstathiou 1994). The elongated 'boxes' at large scales represent the COBE 4-yr (on the right) and Tenerife (on the left) CMB measurements (Gawiser & Silk 1998). The solid triangles and crosses represent amplitudes derived from the quadrupole of radio sources (Baleisis et al. 1998) and the quadrupole of the XRB (Lahav et al. 1997; Treyer et al. 1998). Each pair of estimates corresponds to assumed shape of the two CDM models. (A compilation from Wu, Lahav & Rees 1999).

Universe is homogeneous.

4. Is the Universe Fractal ?

The question of whether the Universe is isotropic and homogeneous on large scales can also be phrased in terms of the fractal structure of the Universe. A fractal is a geometric shape that is not homogeneous, yet preserves the property that each part is a reduced-scale version of the whole. If the matter in the Universe were actually distributed like a pure fractal on all scales then the Cosmological Principle would be invalid, and the standard model in trouble. As shown in Figure 3 current data already strongly constrain any non-uniformities in the galaxy distribution (as well as the overall mass distribution) on scales $> 300\,h^{-1}$ Mpc.

If we count, for each galaxy, the number of galaxies within a distance R from it, and call the average number obtained $N(< R)$, then the distribution is said to be a fractal of correlation dimension D_2 if $N(< R) \propto R^{D_2}$. Of course D_2 may be 3, in which case the distribution is homogeneous rather than fractal. In the pure fractal model this power law holds for all scales of R.

The fractal proponents (Pietronero et al. 1997) have estimated $D_2 \approx 2$ for all scales up to $\sim 500\,h^{-1}$ Mpc, whereas other groups have obtained scale-dependent values (for review see Wu et al. 1999 and references therein).

Estimates of D_2 from the CMB and the XRB are consistent with $D_2 = 3$ to within 10^{-4} on the very large scales (Peebles 1993; Wu et al. 1999). While we reject the pure fractal model in this review, the performance of CDM-like models of fluctuations on large scales have yet to be tested without assuming homogeneity *a priori*. On scales below, say, $30\,h^{-1}$ Mpc, the fractal nature of clustering implies that one has to exercise caution when using statistical methods which assume homogeneity (e.g. in deriving cosmological parameters). We emphasize that we only considered one 'alternative' here, which is the pure fractal model where D_2 is a constant on all scales.

5. More Realistic Inhomogeneous Models

As the Universe appears clumpy on small scales it is clear that assuming the Cosmological Principle and the FRW metric is only an approximation, and one has to average carefully the density in Newtonian Cosmology (Buchert & Ehlers 1997). Several models in which the matter in clumpy (e.g. 'Swiss cheese' and voids) have been proposed (e.g. Zeldovich 1964; Krasinski 1997; Kantowski 1998; Dyer & Roeder 1973; Holz & Wald 1998; Célérier 1999; Tomita 1999). For example, if the line-of-sight to a distant object is 'empty' it results in a gravitational lensing de-magnification of the object. This modifies the FRW luminosity-distance relation, with a clumping factor as another free parameter. When applied to a sample of SNIa the density parameter of the Universe Ω_m could be underestimated if FRW is used (Kantowski 1998; Perlmutter et al. 1999). Metcalf & Silk (1999) pointed out that this effect can be used as a test for the nature of the dark matter, i.e. to test if it is smooth or clumpy.

6. A 'Best Fit Universe': a Cosmic Harmony ?

Several groups (e.g. Eisenstein, Hu & Tegmark 1998; Webster et al. 1998; Gawiser & Silk 1998; Bridle et al. 1999) have recently estimated cosmological parameters by joint analysis of data sets (e.g. CMB, SN, redshift surveys, cluster abundance and peculiar velocities) in the framework of FRW cosmology. The idea is that cosmological parameters can be better estimated by the complementary of the different probes.

While this approach is promising and we will see more of it in the next generation of galaxy and CMB surveys (2dF/SDSS/MAP/Planck) it is worth emphasizing a 'health warning' on this approach. First, the choice of parameters space is arbitrary and in the Bayesian framework there is freedom in choosing a prior for the model. Second, the 'topology' of the parameter space is only helpful when 'ridges' of 2 likelihood 'mountains' cross each other (e.g. as in the case of the CMB and the SN). It is more problematic if the joint maximum ends up in a 'valley'. Finally, there is the uncertainty that a sample does not represent a typical patch of the FRW Universe to yield reliable global cosmological parameters.

6.1. COSMOLOGICAL PARAMETERS

Webster et al. (1998) combined results from a range of CMB experiments, with a likelihood analysis of the IRAS 1.2Jy survey, performed in spherical harmonics. This method expresses the effects of the underlying mass distribution on both the CMB potential fluctuations and the IRAS redshift distortion. This breaks the degeneracy e.g. between Ω_m and the bias parameter. The family of CDM models analysed corresponds to a spatially-flat Universe with an initially scale-invariant spectrum and a cosmological constant λ. Free parameters in the joint model are the mass density due to all matter (Ω_m), Hubble's parameter ($h = H_0/100$ km/sec), IRAS light-to-mass bias (b_{iras}) and the variance in the mass density field measured in an $8h^{-1}$ Mpc radius sphere (σ_8). For fixed baryon density $\Omega_b = 0.02/h^2$ the joint optimum lies at $\Omega_m = 1 - \lambda = 0.41 \pm 0.13$, $h = 0.52 \pm 0.10$, $\sigma_8 = 0.63 \pm 0.15$, $b_{iras} = 1.28 \pm 0.40$ (marginalised 1-sigma error bars). For these values of Ω_m, λ and H_0 the age of the Universe is ~ 16.6 Gyr.

The above parameters correspond to the combination of parameters $\Omega_m^{0.6}\sigma_8 = 0.4 \pm 0.2$. This is quite in agreement from results form cluster abundance (Eke et al. 1998), $\Omega_m^{0.5}\sigma_8 = 0.5 \pm 0.1$. By combining the abundance of clusters with the CMB and IRAS Bridle et al. (1999) found $\Omega_m = 1 - \lambda = 0.36$, $h = 0.54$, $\sigma_8 = 0.74$, and $b_{iras} = 1.08$ (with error bars similar to those above).

On the other hand, results from peculiar velocities yield higher values (Zehavi & Dekel 1999), $\Omega_m^{0.6}\sigma_8 = 0.8 \pm 0.1$. By combining the peculiar velocities (from the SFI sample) with CMB and SN Ia one obtains overlapping likelihoods at the level of $2 - sigma$ (Bridle et al. 2000). The best fit parameters are $\Omega_m = 1 - \lambda = 0.42$, $h = 0.63$, and $\sigma_8 = 1.24$.

6.2. HYPER-PARAMETERS

A complication that arises in combining data sets is that there is freedom in assigning the relative weights of different measurements. A Bayesian approach to the problem utilises 'Hyper Parameters' (Lahav et al. 2000).

Assume that we have 2 independent data sets, D_A and D_B (with N_A and N_B data points respectively) and that we wish to determine a vector of free parameters $\mathbf{w}$ (such as the density parameter $\Omega_{\rm m}$, the Hubble constant H_0 etc.). This is commonly done by minimising

$$\chi^2_{\rm joint} = \chi^2_A + \chi^2_B , \tag{1}$$

(or maximizing the sum of log likelihood functions). Such procedures assume that the quoted observational random errors can be trusted, and that the two (or more) χ^2s have equal weights. However, when combining 'apples and oranges' one may wish to allow freedom in the relative weights. One possible approach is to generalise Eq. 1 to be

$$\chi^2_{\rm joint} = \alpha\chi^2_A + \beta\, \chi^2_B , \tag{2}$$

where α and β are 'Lagrange multipliers', or 'Hyper-Parameters' (hereafter HPs), which are to be evaluated in a Bayesian way. There are a number of ways to interpret the meaning of the HPs. A simple example of the HPs is the case that

$$\chi^2_A = \sum \frac{1}{\sigma_i^2} [x_{{\rm obs},i} - x_{{\rm pred},i}(\mathbf{w})]^2 , \tag{3}$$

where the sum is over N_A measurements and corresponding predictions and errors σ_i. Hence by multiplying χ^2 by α each error effectively becomes $\alpha^{-1/2}\sigma_i$. But even if the measurement errors are accurate, the HPs are useful in assessing the relative weight of different experiments. It is not uncommon that astronomers discard measurements (i.e. by assigning $\alpha = 0$) in an ad-hoc way. The procedure we propose gives an objective diagnostic as to which measurements are problematic and deserve further understanding of systematic or random errors.

If the prior probabilities for $\ln(\alpha)$ and $\ln(\beta)$ are uniform then one should consider the quantity

$$-2\,\ln P(\mathbf{w}|D_A, D_B) = N_A \ln(\chi^2_A) + N_B \ln(\chi^2_B) \tag{4}$$

instead of Eq. 1.

More generally for M data sets one should minimise

$$-2 \ln P(\mathbf{w}|\text{data}) = \sum_{j=1}^{M} N_j \ln(\chi_j^2), \qquad (5)$$

where N_j is the number of measurements in data set $j = 1, ..., M$. It is as easy to calculate this statistic as the standard χ^2. The corresponding HPs can be identified as $\alpha_{\text{eff},j} = N_j/\chi_j^2$ (where the χ_j^2's are evaluated at the values of the parameters $\mathbf{w}$ that minimise eq. 4) and they provide useful diagnostics on the reliability of different data sets. We emphasize that a low HP assigned to an experiment does not necessarily mean that the experiment is 'bad', but rather it calls attention to look for systematic effects or better modelleing. The method is illustrated (Lahav et al. 2000) by estimating the Hubble constant H_0 from different sets of recent CMB experiments (including Saskatoon, Python V, MSAM1, TOCO and Boomerang).

7. Discussion

Analysis of the CMB, the XRB, radio sources and the Lyman-α which probe scales of $\sim 100 - 1000\,h^{-1}$ Mpc strongly support the Cosmological Principle of homogeneity and isotropy. They rule out a pure fractal model. However, there is a need for more realistic inhomogeneous models for the small scales. This is in particular important for understanding the validity of cosmological parameters obtained within the standard FRW cosmology.

Joint analyses of the CMB, IRAS, SN, cluster abundance and peculiar velocities suggests $\Omega_m = 1 - \lambda \approx 0.4$. With the dramatic increase of data, we should soon be able to map the fluctuations with scale and epoch, and to analyze jointly LSS (2dF, SDSS) and CMB (MAP, Planck) data, taking into account generalized forms of biasing.

Acknowledgments I thank my collaborators for their contribution to the work presented here.

References

1. Baleisis, A., Lahav, O., Loan, A.J. & Wall, J.V. 1998, MNRAS, 297, 545
2. Baugh C.M. & Efstathiou G. 1994, MNRAS , 267, 323
3. Boldt, E. A. 1987, Phys. Reports, 146, 215
4. Bridle, S.L., Eke, V.R., Lahav, O., Lasenby, A.N., Hobson, M.P., Cole, S., Frenk, C.S., & Henry, J.P. 1999, MNRAS, in press (astro-ph/9903472)
5. Bridle, S.L., Zehavi, I., Dekel, A., Lahav, O., Hobson, M.P. & Lasenby, A.N., 2000, in preparation
6. Buchert T & Ehlers, J. 1997, A&A, 320, 1
7. Célérier, M.N. 1999, submitted to A&A (astro-ph/9907206)
8. Cress C.M., Helfand D.J., Becker R.H., Gregg. M.D. & White, R.L. 1996, ApJ, 473, 7

9. Davis, M. 1997, *Critical Dialogues in Cosmology*, World Scientific, ed. N. Turok, pg. 13.
10. Dekel, A. et al., 1999, ApJ, in press (astro-ph/9812197)
11. Dyer, C.C. & Roeder, R.C. 1973, ApJ, 180, L31
12. Ehlers, J., Geren, P & Sachs, R.K. 1968, J Math Phys, 9(9), 1344, 1968
13. Eisenstein, D.J., Hu, W. & Tegmark, M. 1998 (astro-ph/9807130)
14. Eke, V.R., Cole, S., Frenk, C.S. & Henry, J.P. 1998, MNRAS, 298, 1145
15. Fabian, A. C. & Barcons, X. 1992, ARAA, 30, 429
16. Gawiser, E. & Silk, J., 1998, Science, 280, 1405
17. Holz, D.E. & Wald, R.M. 1998, Phys Rev D, 58, 063501
18. Joyce, M., Montuori, M., Sylos-Labini F. & Pietronero, L., 1999, A&A, 344, 387
19. Kantowski, R. 1998, ApJ, 507, 483
20. Krasinski, A. 1997, *Inhomogeneous Cosmological Models*, Cambridge University Press, Cambridge
21. Lahav O., Piran T. & Treyer M.A. 1997, MNRAS, 284, 499
22. Lahav, O., Santiago, B.X., Webster, A.M., Strauss, M.A., Davis, M., Dressler, A. & Huchra, J.P. 1999, MNRAS, in press (astro-ph/9809343)
23. Lahav, O., Bridle, S.L., Hobson, M.P., Lasenby, A.L., Sodré, L. 2000, submitted to MNRAS (astro-ph/9912105)
24. Magliocchetti, M., Maddox, S.J., Lahav, O.& Wall, J.V. 1998, MNRAS, 300, 257
25. Metcalf, R. B. , Silk, J. 1999, ApJ L, 519, L1
26. Nusser, A. & Lahav, O. 1999, submitted to MNRAS (astro-ph/991017)
27. Peebles, P. J. E. 1993, *Principles of Physical Cosmology*, Princeton University Press, Princeton.
28. Perlmutter et al. 1999, ApJ, 517, 565
29. Pietronero, L., Montuori M., & Sylos-Labini, F. 1997, in *Critical Dialogues in Cosmology*, World Scientific, ed. N. Turok, pg. 24
30. Rudnicki, K. 1995, *The cosmological principles*, Jagiellonian University, Krakow
31. Scaramella, R. et al. 1998, A&A, 334, 404
32. Scharf, C.A., Jahoda, K., Treyer, M., Lahav, O., Boldt, E. & Piran, T., et al., 1999, submitted to ApJ (astro-ph/9908187)
33. Schmoldt, I. et al. 1999, MNRAS, 304, 893
34. Strauss M.A. et al., 1992, ApJ, 397, 395
35. Tomita, K. 1999 (astro-ph/9906027)
36. Treyer, M., Scharf, C., Lahav, O., Jahoda, K., Boldt, E. & Piran, T. 1998, ApJ, 509, 531
37. Webster, M.A., Lahav, O., & Fisher, K.B. 1998, MNRAS, 287, 425
38. Webster, M., Hobson, M.P., Lasenby, A.N., Lahav, O., Rocha, G. & Bridle, S.L. 1998, ApJ, 509, L65
39. Wu, K.K.S., Lahav, O. & Rees, M.J. 1998, Nature, 397, 225
40. Zehavi,I & Dekel, A. 1999, Nature, 401, 252
41. Zeldovich, Ya, B. 1964, Soviet Astron, 8, 13

QUINTESSENCE AND COSMIC ACCELERATION

PAUL J. STEINHARDT
Department of Physics
Princeton University

1. Introduction[1]

Quintessence [1, 3] and the cosmological constant [4] are unanticipated and unwanted energy components from the point-of-view of both cosmologists and high energy physicists. Yet, confirming either would undoubtedly be one of the most important discoveries in both fields and would produce new links between the two.

For cosmology, the discovery of a new energy component would finally balance the energy budget, making the total energy content of the universe equal to the critical density predicted by inflation [5, 6, 7, 8]. The fact that the energy component has negative pressure and causes the universe to accelerate has a subtle but numerically significant impact on the past evolution of the universe and large-scale structure formation, resolving numerous difficulties with the standard cold dark matter model. The dramatic impact is on the future history of the universe. For decades, the common view was that the universe is decelerating due to the self- gravitation of matter and radiation. The only issue seemed to be whether the deceleration is sufficient to halt the expansion and cause the universe to contract to a big crunch, or whether the deceleration is too meager and the universe continues to expand at a slower and slower rate. Now, the evidence indicates that neither scenario is correct. The expansion is speeding up, driven by a mysterious form of dark energy that will ultimately overwhelm the universe.

For fundamental physics, quintessence or a cosmological constant represents new, ultra-low energy phenomena beyond the standard model. If firmly established by future observations, the discovery will be recognized as a fantastic, valuable hint about the ultimate, unified theory. The fact

[1]Reprinted from Proceedings of the Pritzker Symposium, U. of Chicago Press

R.G. Crittenden and N.G. Turok (eds.), Structure Formation in the Universe, 143–176.

that the component can be probed observationally is an added bonus, especially since many predictions of unified theories entail high energies beyond the realm of experiment.

Moreover, just as the Copernican revolution changed forever the view of our place in the universe, the discovery of cosmic acceleration will change the view of our place in time. In the static universe picture, what we see in the universe today is representative of the universe as it always has been and always will be. In the big bang picture, the universe has been undergoing steady evolution from a simple, uniform, cosmic soup of elementary particles to ever more complex structure, in close analogy to biological evolution. The view of the universe emerging today is that the universe as we know it is only a brief interlude between two periods of cosmic acceleration powered by negative pressure, inflation at early times and now acceleration once again. Life, the stars, the galaxies, and large-scale structure are completely ephemeral phenomena in the course of cosmic history, a momentary spark in an accelerating universe.

The evidence for cosmological acceleration is presented in Sec. II. We show how three distinct types of observations currently indicate cosmic acceleration. In Sec. III, we turn to the two competing theoretical explanations for explaining what powers the cosmic acceleration: either an inert vacuum density (or cosmological constant) or a dynamical, quintessence component. We focus particularly on the progress that has been made in developing quintessence from a rather artificial and ill-defined concept into a promising and well-motivated possibility. Perhaps the most important motivation for considering quintessence is the cosmic coincidence problem: why has cosmic acceleration begun at this particular moment in cosmic history? If acceleration had begun a little earlier, structure would never have formed in the universe, and, if acceleration had begun a little later, we would not detect it today. For this timing of the acceleration to occur, it must be that the matter density and dark energy density just happen to coincide (nearly) today even though they decrease at substantially different rates as the universe expands. The cosmological constant proposal offers little insight into the coincidence. However, some very promising ideas have emerged from the study of quintessence – tracker fields and creeper fields – which partially address the coincidence problem. In Sec. IV, the current observational status of quintessence will be summarized, and the future prospects for distinguishing quintessence from a cosmological constant will be discussed. In Sec. V, we outline some of the remaining theoretical challenges.

2. The evidence for cosmic acceleration

The most impressive aspect of the case for cosmic acceleration is that three separate lines of evidence have arisen which simultaneously lead us to the same startling conclusion [9]. Although the supernovae results [10, 11, 12, 13] are what first captured the attention of the broad scientific community, strong evidence already existed beforehand [19] and other kinds of measurements may ultimately provide the most reliable test in the future.
Direct evidence of accelerated expansion: Accelerated expansion produces a systematic deviation from the linear Hubble law at large red shift. Attempts to measure the deceleration parameter and higher order non-linearities have been a goal of cosmology for decades, and promising techniques have been explored, only to be foiled by unforeseen evolutionary effects. The latest approach, using Type IA supernovae as standard candles, appears promising from both a theoretical and empirical view at present. The results of the Supernovae Cosmology Project [10, 14, 15] are summarized in Figure 1; a competing group, the High-z Supernovae group [11, 12, 16, 17], uses somewhat different methods and achieves a similar result. Both groups find that distant supernovae are significantly fainter (by nearly half a magnitude) compared to a sample of nearby supernovae than would be expected in a cosmological model with $\Omega_m = 1$, such as the standard cold dark matter (SCDM) model. The SCP group reports $\Omega_m = 0.28 \pm 0.8 \pm 0.5$ assuming a flat universe ($\Omega_m + \Omega_\Lambda = 1$) [10]. (Allowing non-zero curvature, the constraint is $0.8\Omega_m - 0.6\Omega_\Lambda = -0.2 \pm 0.1$.) ($\Omega_i$ is the ratio of the energy density in component i to the critical density, $\rho_c = 8\pi G/3H_0^2$, where $H_0 = 100h$ km s^{-1} Mpc^{-1}. We use $i = m$ for the total (baryonic and nonbaryonic) matter density, $i = b$ for the baryon density, $i = r$ for the radiation density, $i = \Lambda$ for the cosmological constant or vacuum density, and $i = Q$ for quintessence.) Although the results are impressive, one should recall the sorry history of past attempts at long-distance, standard candles: in a nutshell, the initial, small statistical errors are ultimately replaced by large systematic uncertainties. In the case of Type IA supernovae, the most worrisome aspects are that the luminosity of supernovae may evolve with red shift in such a way as to mimic the predictions of accelerated expansion [18] or that dust at large red shift may make supernovae appear fainter than expected for a decelerating universe. Considerable efforts are already underway to test these possibilities.

If confirmed, cosmic acceleration can be interpreted as evidence for a substantial cosmic energy component with negative pressure. According to Einstein's theory of general relativity, the scale factor $a(t)$, which represents the expansion of the universe as a function of time, satisfies the differential

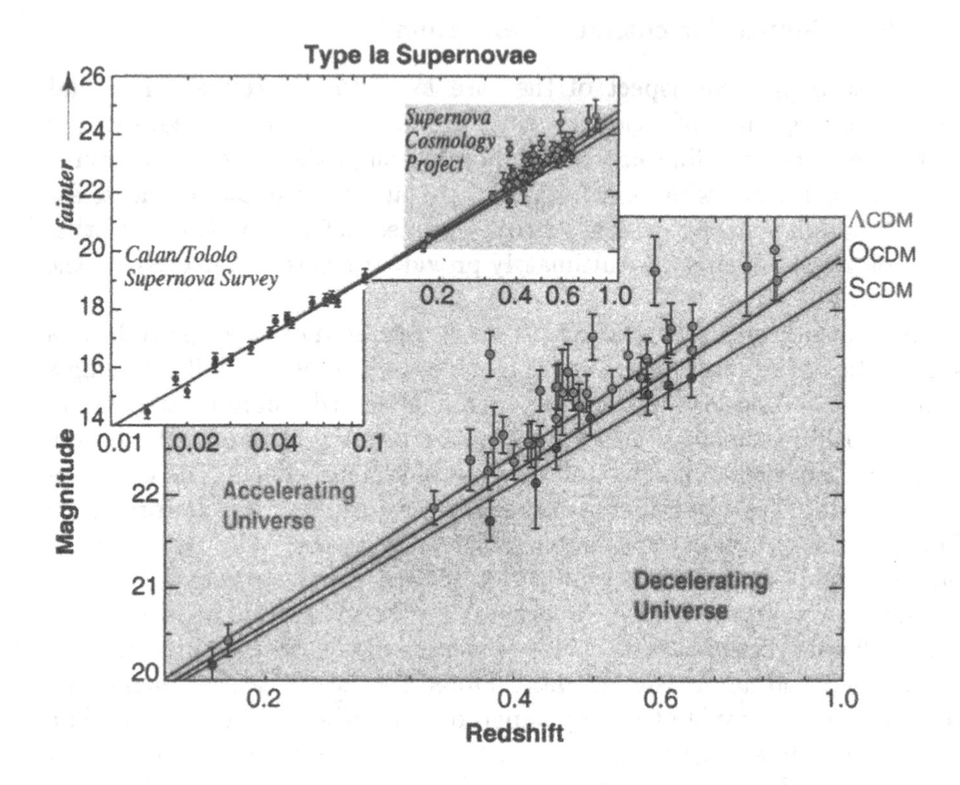

Figure 1. Observed brightness (magnitude) vs. red shift for Type Ia supernovae observed at low red shift by the Calan-Tololo Supernova Survey and at high red shift by the Supernova Cosmology Project (SCP) (with 1σ error bars) compared with model expectations. The standard cold dark matter model (SCDM) has $\Omega_m = 1$; OCDM is the best-fit open model and has $\Omega_m = 0.3$; and ΛCDM is a flat model with $\Omega_m = 0.3$ and $\Omega_m = 0.7$. The effect of a cosmological constant accelerating the expansion rate (as in ΛCDM) is seen as a relative 'dimming' of the distant SNIa compared to decelerating models. Similar results have been found by the High-Z Supernovae team (Riess *et al.*).

equation:

$$\ddot{a} = -\frac{4\pi G}{3}(\rho + 3p)a \tag{1}$$

where G is Newton's constant, ρ is the energy density, and p is the pressure. Baryonic and nonbaryonic cold dark matter are essentially pressureless, and radiation has positive pressure, $p = \rho/3$. If the universe contains only these energy components, then, according to Eq. (1), the universe is decelerating, $\ddot{a} < 0$. Note that this equation does not include explicitly the spatial curvature, so deceleration occurs whether the universe is open, flat, or closed if the pressure is non-negative. If the universe is found to be accelerating, there must be an energy component ρ_Q with negative pressure p_Q such that $\rho_{tot} + 3p_{tot} < 0$, or $p_{tot} < -\rho_{tot}/3 < 0$, where ρ_{tot} is the total energy density. Since $\rho_Q + p_Q \geq 0$ for any physically plausible negative pressure component ρ_Q (the positive energy condition), then ρ_Q must be at least one-third the total energy density, $\rho_Q > \rho_{tot}/3$ (assuming all other components have non-negative pressure) in order for $\ddot{a}$ to be greater than zero.

Evidence for a flat, low-density universe — "Cogito ergo sum": A strong case for a negative pressure component already existed and was presented forcefully several years ago [19], well before the supernovae data indicated an accelerating universe [15]. The argument relies on combining three observed features of the universe and an argument I entitle "cogito, ergo sum." First, the argument rests on the observation that the total mass density of the universe is less than the critical density predicted for a flat universe. While the observations were controversial for many decades, today at least eight different methods of constraining the mass density exist: cluster abundance, cluster abundance evolution, the mass-to-light (M/L) test, $\Gamma \equiv \Omega_m h$ as determined by large-scale structure, baryon fraction based on x-ray observations of clusters, gravitational lensing of massive clusters, and the age of the universe as inferred from globular clusters compared to the Hubble age determined from measurements of the Hubble parameter [21, 22, 9, 20]. Remarkably, all eight methods agree that the mass density is less than half the critical value [19, 20]. It is difficult to imagine so many different measurements with different systematic uncertainties reversing themselves to recover a good fit to an Einstein-de Sitter ($\Omega_m = 1$) universe.

The second assumed feature is that the universe is flat. Some would argue that flatness is a necessary condition based on confidence in inflationary cosmology (for which other evidence exists) or based on the classical flatness-problem argument [23]. Fortunately, the issue can be decided by observation rather than relying on theoretical arguments alone. The key observational test is the angular scale or, equivalently, the multipole moment (ℓ) of the first acoustic peak in the cosmic microwave background (CMB) temperature anisotropy power spectrum [26]. The power spectrum is the Legendre transform of the CMB temperature angular autocorrelation function. The shape of the power spectrum as a function of ℓ is an extraordinarily sensitive test of cosmological models and their parameters. A prominent feature is a series of peaks resulting from acoustic oscillations of the baryon-photon cosmic fluid. See Figure 2. The oscillations are caused by density perturbations, such as those created during inflation. In the case of inflation, if the density can be decomposed into a sum of Fourier modes with different comoving wavelengths, then comoving wavelengths longer than the Hubble horizon are frozen at some amplitude. When the sound horizon grows to be comparable to the wavelength, the mode be-

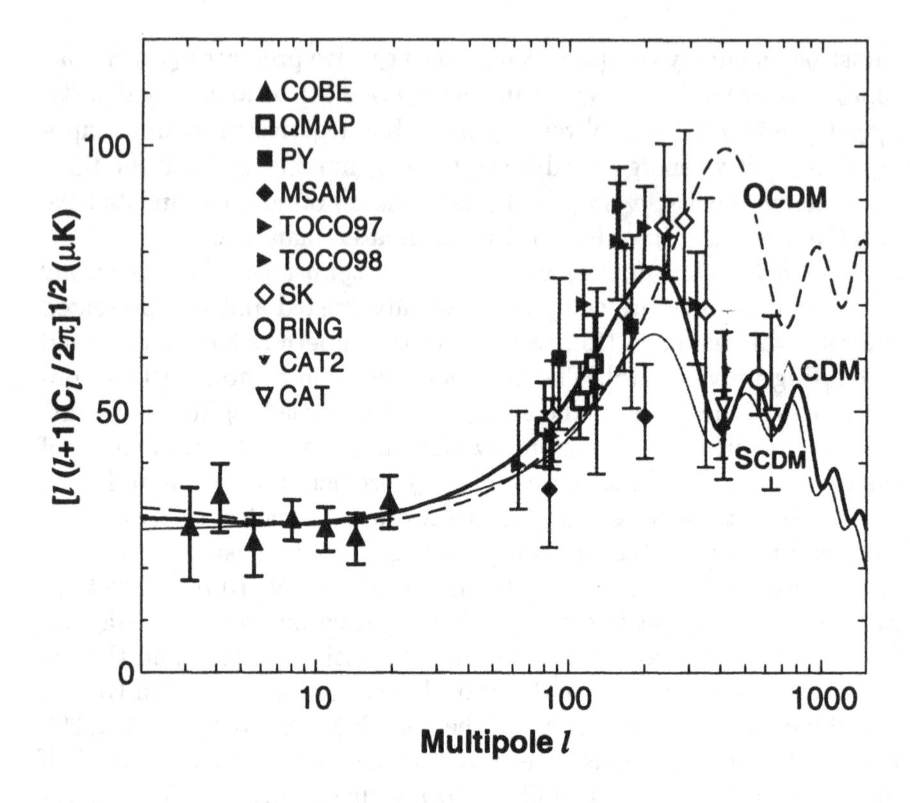

Figure 2. The cosmic microwave background temperature anisotropy power spectrum is shown as a function of angular scale. The multipole ℓ corresponds roughly to an angular scale of π/ℓ radians. Flat models ($\Omega_m + \Omega_\Lambda = 1$), such as the standard cold dark matter (SCDM) model with $\Omega_m = 1$ and the best-fit ΛCDM model with $\Omega_m = 0.3$, produce an acoustic peak at $\ell =\sim 200$ (about one degree on the sky). Shown also is the predicted anisotropy power spectrum for the best-fit open (OCDM) model with adiabatic fluctuations.

gins to oscillate. The oscillation is due to a combination of gravity, which causes the amplitude to grow as baryons are drawn together in regions of high density, and the pressure of the baryon-photon fluid which pushes the baryons apart when the amplitude is too high. In measuring the temperature anisotropy on different angular scales, one is probing different modes at different stages of compression and rarefaction. The first acoustic peak corresponds to the mode undergoing its first compression; that is, the mode whose wavelength is equal to the sound horizon at recombination. The magnitude of the sound horizon is relatively insensitive to most cosmological parameters, and, so, can serve as a kind of "standard ruler." If space is flat, then it is straightforward to show that the angle subtended by this standard ruler on the last scattering surface as seen on the sky today is about 1 degree (or $\ell_{flat} \approx 220$). If space is curved (open or closed), the path of

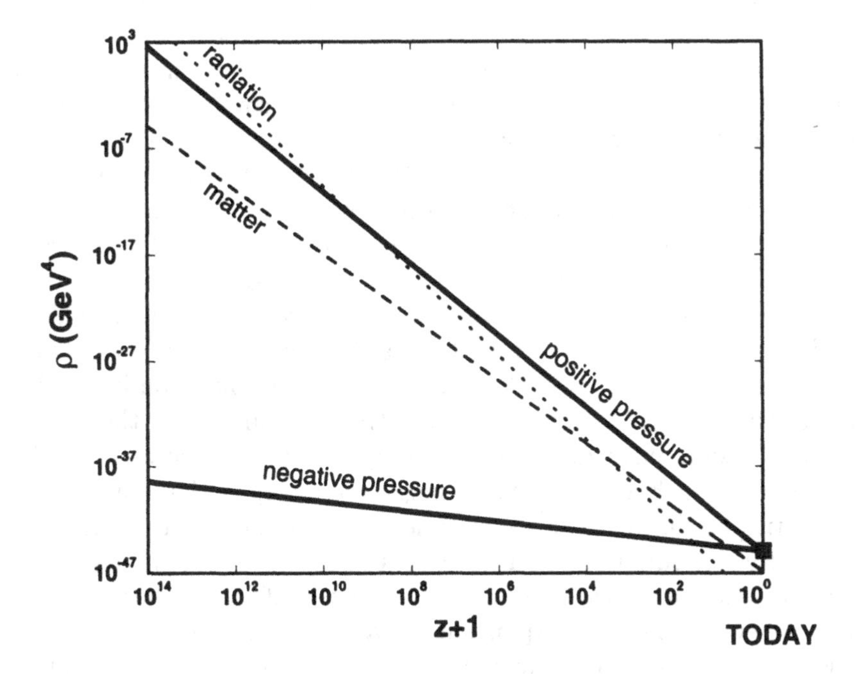

Figure 3. Pressure versus red shift plot indicating the evolution of the matter and radiation density compared to a dark energy component with either positive or negative pressure. According to observations, the dark energy dominates the matter and radiation energy density today (solid square). If the dark energy has positive pressure in the past, as well. Since the matter density never dominates, it is impossible to form large scale structure. On the other hand, dark energy density with negative pressure becomes negligible compared to the matter density in the past, allowing a finite range of time between $z = 10^4$ and $z \approx 1$ where matter dominates and structure can form.

light from the last scattering surface to our detectors is distorted so as to change the apparent subtended angle. For example, in an open universe, the sound horizon subtends a smaller angle so that the maximum of the first acoustic peak lies at a substantially larger value of $\ell \approx \ell_{flat}(\Omega_m)^{-1/2}$. Hence, measuring the angle is the most promising method for determining whether the universe is flat. Already five years ago, with less CMB data in hand than we have today, the best-fit (adiabatic) open model did not fit the combination of CMB and cluster abundance constraints at an acceptable level [19]; the misfit is statistically much more significant today [9].

If subcritical density and flatness are accepted as properties of our universe, then the only other feature one must assume to prove cosmic acceleration is that we exist: *cogito, ergo sum.* I think, therefore, I am. And, if

I am, there must be an earth for me to stand on, a sun to shine over me, a galaxy to make the sun, and clustering matter on cosmic scales to make the galaxy.

How do the three features — subcritical mass density, flatness, and clustered matter – combine to imply a negative pressure component? Consider Figure 3, which shows a plot of energy density versus red shift (z). The figure shows the matter and radiation density falling at different rates owing to the fact that they have different equations-of-state, $w \equiv p/\rho$. Radiation has $w = 1/3$ and matter has $w = 0$. For constant w, the scale factor grows as $a(t) = t^{2/[3(1+w)]}$ and the energy density decreases as $\rho \propto a^{-3(1+w)}$. Now suppose that the matter and radiation density are less than half the critical density, and the universe is flat, as current data suggests. Then, there must be an additional dark energy component that dominates the universe today. If the component has positive pressure, then the energy density in this component decreases more rapidly than the matter density so that the slope in Figure 3 is more negatively steeped than the matter density. Extrapolating backwards in time, the component dominates the matter density not only today, but forever in the past. With gravity alone, structure growth is a delicate balance between the effect of inhomogeneity drawing matter together and expansion spreading the matter. Only during the period when $\Omega_m \approx 1$ does the first effect win and structure grow appreciably. In the positive pressure case, since matter never dominates, it is not possible to form the non-linear structures (galaxies, clusters, etc.) observed throughout the universe beginning from the tiny fluctuations observed by the COBE satellite experiment [28]. (In this case, structure could not grow by gravitational instability alone, but would require an additional long-range force or some other new physics.) On the other hand, if the pressure is negative, the energy density decreases more slowly than the matter density. Extrapolating backwards in time, a negative pressure component that dominates the universe today becomes subdominant to the matter density at some time in the past. If the component has a *sufficiently* negative w, the matter will dominate for a long enough period to form the observed structure via gravitational stability beginning from the tiny fluctuations measured in the cosmic microwave background anisotropy. To account for the observed structure, sufficiently negative means $w < -0.33$ [19, 20], which corresponds to the regime in which the expansion of the universe is accelerating. Hence, we see that evidence of subcritical matter density and flatness, combined with existence of structure, is sufficient to prove the case for a negative pressure component and, an accelerating universe. (The argument as presented here assumes w is constant or changing slowly; a more detailed discussion is required to dispose of cases where w is changing rapidly for a component, as in the case of decaying dark matter. This is left as an exercise for the reader.)

Evidence for a high acoustic peak in the cosmic microwave background power spectrum: The most recent and, statistically, the weakest evidence for a negative pressure component is based on measuring the height of the first acoustic (Doppler) peak in the CMB temperature power spectrum [24, 25], We pointed out above that the position of the acoustic peak as a function of multipole number (neglecting its height) is a measure of the flatness of the universe, and so supports the previous argument for negative pressure [26]. Now we point out that recent observations suggest that the amplitude of the peak is substantially higher than the value predicted for the standard cold dark matter (CDM) model [9, 27]. Various factors can account for the discrepancy: a negative pressure component, higher than anticipated baryon density, lower than anticipate Hubble parameter, and positive spectral tilt are all examples [24, 25]. Given what is already known about constraints on the baryon density and Hubble constant from other observations, a negative pressure component is the most likely explanation for the peak height. Improvements in measurements over the next few years based on long duration balloon experiments (such as *BOOMERANG*) and the MAP satellite experiment will dramatically reduce the current uncertainties. Perhaps the CMB test for a negative pressure component will be the most compelling ultimately.

The cosmic triangle: The observational evidence for negative pressure can be summarized visually in a "cosmic triangle" diagram [9], as described in Figure 4. The triangle is based on the Friedmann equation,

$$H^2 \equiv \left(\frac{\dot{a}}{a}\right)^2 = \frac{8\pi G}{3}\rho_m + \frac{8\pi G}{3}\rho_\Lambda - \frac{k}{a^2}, \tag{2}$$

where H is the Hubble parameter, ρ_m is the matter density, ρ_Λ is the vacuum density, and $k = 0, \pm 1$ is the spatial curvature. Dividing through by H^2, one finds

$$1 = \Omega_m + \Omega_\Lambda + \Omega_k \tag{3}$$

where $\Omega_m \equiv (8\pi G/3H^2)\rho_m$, $\Omega_\Lambda \equiv (8\pi G/3H^2)\rho_\Lambda$, and $\Omega_k = -k/a^2H^2$. Note that Ω_k is defined to include the negative sign so that $\Omega_k < 0$ for a closed universe and > 0 for an open universe. The Friedmann equation has been converted to a sum rule in which Ω_i represents the fractional contribution of component i to the expansion of the universe. Because the energy densities in the two components decrease at different rates as the universe expands, the fractional contributions may change; however, the sum rule must be obeyed at all times.

Every point in the triangle has the property that the perpendicular distances to the three edges of the triangle sum to unity; so, if the three distances correspond to the three Ω_i, every point in the triangle obeys the

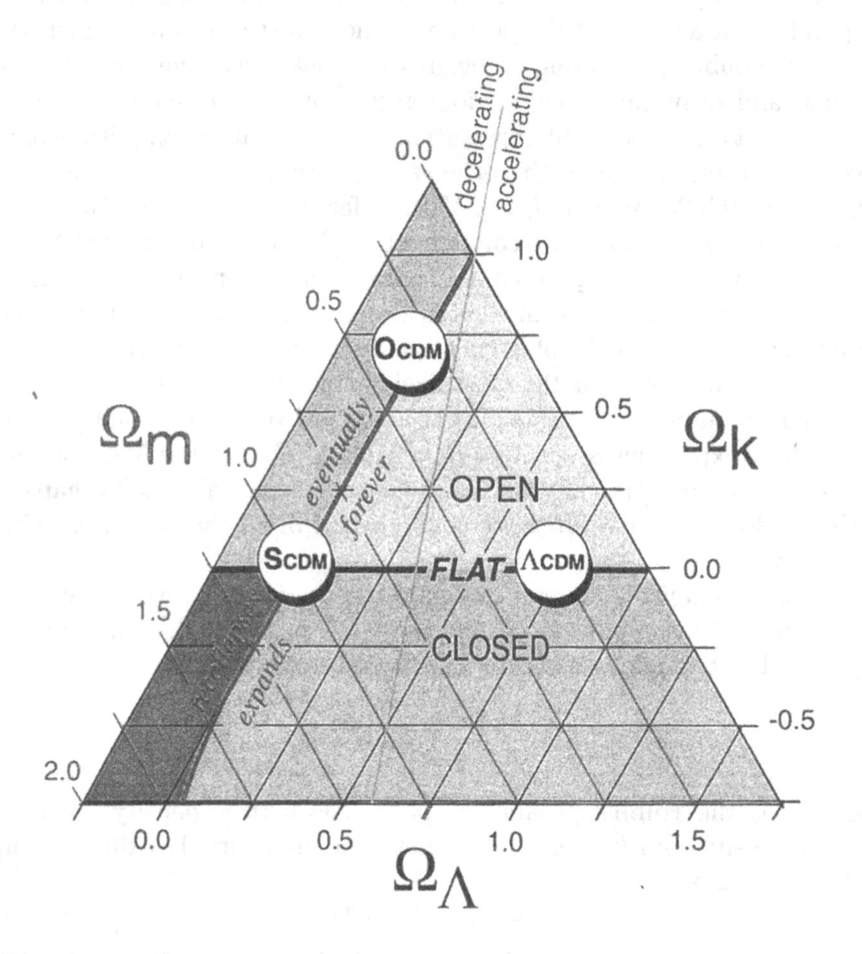

Figure 4. The triangle plot represents the three key cosmological parameters – Ω_m, Ω_Λ, and Ω_k – where each point in the triangle satisfies the sum rule $\Omega_m + \Omega_\Lambda + \Omega_k = 1$. The central horizontal line (marked Flat) corresponds to a flat universe ($\Omega_m + \Omega_\Lambda = 1$), separating an open universe from a closed one. The diagonal line on the left (nearly along the $\Lambda = 0$ line) separates a universe that will expand forever (approximately $\Omega_\Lambda > 0$) from one that will eventually recollapse (approximately $\Omega_\Lambda < 0$). And the light-gray, nearly vertical line separates a universe with an expansion rate that is currently decelerating from one that is accelerating. The location of three key models are highlighted: standard cold-dark-matter (SCDM) is dominated by matter ($\Omega_m = 1$) and no curvature or cosmological constant; flat (ΛCDM), with $\Omega_m = 1/3$, $\Omega_\Lambda = 2/3$, and $\Omega_k = 0$; and Open CDM (OCDM), with $\Omega_m = 1/3$, $\Omega_\Lambda = 0$ and curvature $\Omega_k = 2/3$.

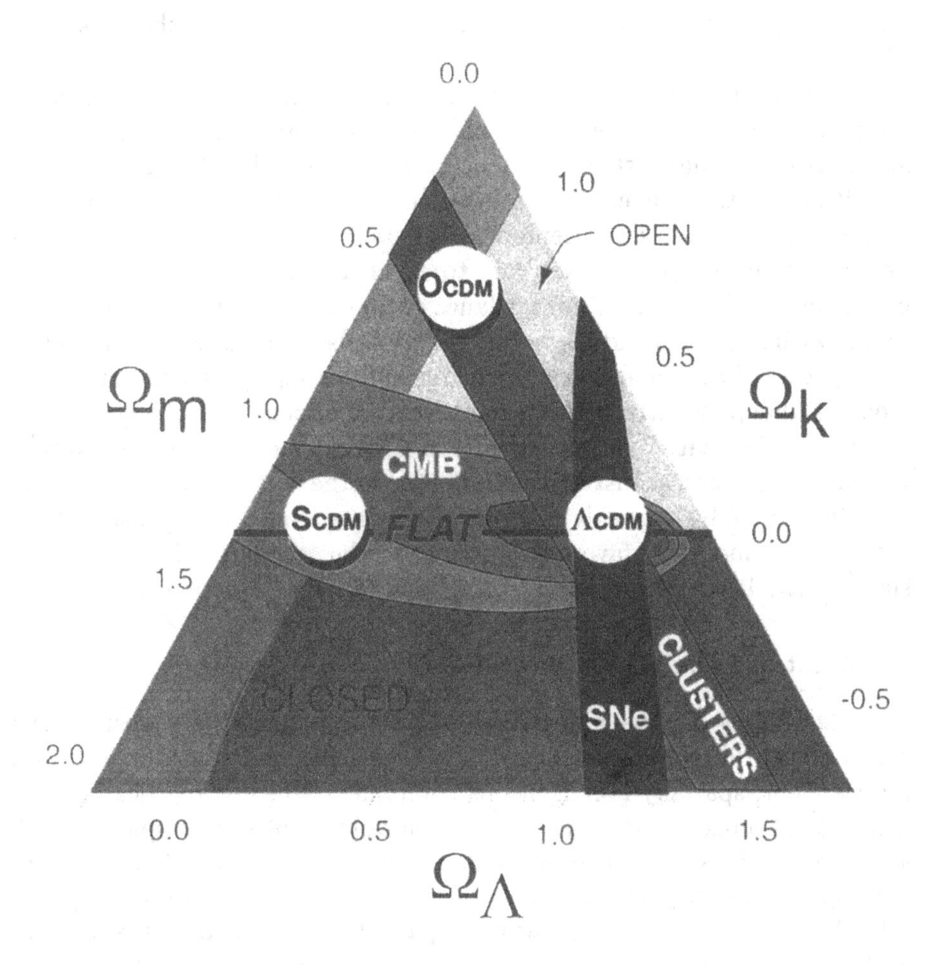

Figure 5. Current observational constraints are represented on the cosmic triangle plot. The tightest constraints from measurements at low red shift (clusters, including the mass-to-light method, baryon fraction, and cluster abundance evolution), intermediate red shift (supernovae), and high red shift (CMB) are shown by the three bands. The bands for cluster and supernovae measurements represent 1σ uncertainties; 1σ, 2σ and 3σ uncertainties are shown for the CMB.

sum rule. The evolution of the universe corresponds to a trajectory in the cosmic triangle plot in Figure 4. A central goal of observational cosmology is to determine the point corresponding to the present universe.

The current measurements of the CMB temperature anisotropy power spectrum are shown in Figure 2. In the plot, the CMB data is compared with the predicted power spectra for the best fit standard, open and Λ cold dark matter models. The data strongly favors flat models over open models and moderately favors models with Λ (or quintessence) over the Einstein-de Sitter model.

Figure 5 illustrates the constraints from measurements of cluster abundance, supernovae, and the cosmic microwave background on the parameter-space. Because the cluster abundance, supernova and CMB are measuring conditions in the universe in different red shift regimes, the contours of maximum likelihood are oriented at different angles. As a result, the independent measurements combine to form an overconstrained region of concordance. The current concordance region is centered near the ΛCDM model with $\Omega_m \approx 0.3$ and $\Omega_\Lambda \approx 0.7$. Figure 5 also illustrates how the cosmic acceleration of the universe can be deduced from supernovae measurements alone or, independently, from the combination of cluster abundance (which constrains the matter density) and the CMB (which constrains flatness). Because of the high acoustic peak indicated in the most recent CMB results, a likelihood analysis reveals that the CMB data alone points to the ΛCDM model being modestly favored compared to the standard CDM model and significantly favored compared to the open (adiabatic) model.

3. What are the the explanations?

The candidates for the negative pressure component are a cosmological constant [4, 19] or quintessence [1, 3]. A cosmological constant is a time-independent, spatially homogeneous component which is physically equivalent to a non-zero vacuum energy: each volume of empty space has the same energy density at each moment in time. The pressure of vacuum density equals precisely the negative of the energy density, or $w \equiv p/\rho = -1$. Quintessence is a time-varying, spatially inhomogeneous component with negative pressure, $-1 < w < 0$. Formally, vacuum energy density is quintessence in the limit $w \to -1$, although the two forms of energy are quite distinct physically. Quintessence is a dynamical component whereas vacuum density is inert.

The term "quintessence" was introduced historically in an attempt to resolve a different problem of acceleration. Namely, in ancient times, the centripetal acceleration of the moon was inconsistent with the ancient physical world-view. According to this view, the universe consists of four constituents: earth, air, fire and water. An important property of earth is that it is the densest and so everything made of earth falls to the center of universe (from which one concludes that the Earth is the center of the universe). The moon is problematic in that it has mountain and valley feature like the Earth, and yet the moon does not fall to the center. One proposed explanation was that there is a fifth element, or "quintessence," which permeates space and keeps the moon suspended, but which otherwise does not interact with the other four components. Millenia later, the term is being re-introduced to resolve another problem of cosmic acceleration. Cosmological models composed of four basic elements, baryons, leptons, photons and cold dark matter cannot explain the apparent acceleration in the universe on a grander scale, and perhaps quintessence is the cause.

The prime example of quintessence discussed in the literature [1, 3] is a scalar field Q slowly rolling down a potential $V(Q)$. The pressure of a scalar field is the difference between the kinetic and potential energy, $p = \frac{1}{2}\dot{\phi}^2 - V$. "Slow-roll," a condition in which the kinetic energy is less than the potential energy, produces negative pressure. The notion that a scalar field can produce negative pressure and cosmic acceleration was already established with "new inflation," [6, 7] in which the scalar inflaton field undergoes slow-roll and drives inflation. The difference is that the energy scale for quintessence is much smaller and the associated time-scale is much longer compared to inflation. Although we will consider only the scalar field example in the rest of the paper, other forms of quintessence are possible. For example, a tangled web of non-Abelian cosmic strings [37, 38] produces a negative pressure with $w = -1/3$, and a network of domain walls has $w = -2/3$.

3.1. QUINTESSENCE: NOT A TIME-VARYING COSMOLOGICAL CONSTANT!

Quintessence is sometimes referred to as a "time-varying cosmological constant" or "smooth component," based on its average effect on the expansion of the universe. Discussions of a time-varying cosmological constant in the smooth approximation date back at least as far as the papers by the Russian physicist, Bronstain in 1933 [39], and the idea has been revisited frequently over the intervening decades [1, 2, 3]. However, treating quintessence in this manner oversimplifies the concept to a point where some of the most difficult theoretical challenges and intriguing possibilities are lost.

A good analogy is the description of inflation as a "de Sitter phase," a finite period in which the universe is dominated by a cosmological constant. The description captures some of the gross features of inflation, such as the superluminal expansion. Yet, a de Sitter phase of finite duration is a physical contradiction. Once one appreciates that inflation must be of finite duration, it then becomes clear that a dynamical component is required instead of a cosmological constant, and immediately issues arise: what is the nature of the dynamical component? how did inflation begin? how did it end? what happened to the energy that drove the inflationary expansion? By pursuing these issues, one encounters one of the great surprises of inflation: the existence of tiny density fluctuations following inflation as a result of the stretching of quantum fluctuations from microscopic scales to cosmic scales [40, 41, 42, 43]. One is also led to consider the associated problem of tuning required to insure that the density fluctuations have an acceptable amplitude. Describing inflation as a de Sitter phase misses these important features.

Similarly, describing quintessence as a time-varying cosmological constant is a physical contradiction. Nor can quintessence be properly considered a smooth component. An energy component which is varying in time but spatially homogeneous contradicts the equivalence principle. See Section IIIA. The moment one realizes that full dynamics must be specified, a host of theoretical issues arise, just as in the case of inflation). This more serious treatment is only recent. The pioneering work was done by N. Weiss, C. Wetterich, and B. Ratra and J. Peebles in the late 1980's [3], and, more recently, a systematic march through these issues has appeared in a series of papers by R. Caldwell and students R. Dave, L. Wang, I. Zlatev, and G. Huey [1, 20, 44, 45, 46, 50, 51, 52, 53]. As a result of their work, many of the most worrisome aspects of quintessence are well-understood and under control.

3.2. SOME BASIC THEORETICAL ISSUES AND THEIR RESOLUTION

This subsection raises basic questions that arise when one considers the dynamical aspects of a quintessence component and the answers that have been found in recent studies.

Can quintessence be perfectly smoothly distributed? No. A time-varying but smoothly distributed component is inconsistent with the equivalence principle [1, 44, 54]. If the scalar field $Q(\tau, x)$ is separated into spatially homogeneous and inhomogeneous pieces, $Q(\tau, \vec{x}) = Q_o(\tau) + \delta Q(\tau, \vec{x})$, where τ is the conformal time, then, the Fourier transform of the fluctuating component obeys the wave equation [44]

$$\delta Q_k'' + 2aH\delta Q_k' + (k^2 + a^2 V_{,QQ})\delta Q_k = -\frac{1}{2}h_k' Q_o' \tag{4}$$

where the prime denotes $\partial/\partial\tau$, the index k indicates the Fourier transform amplitude, $V_{,QQ}$ is the second derivative of the quintessence potential V with respect to Q, and h is the trace of the synchronous gauge metric perturbation. Even if one sets $\delta Q(\tau, \vec{x}) = 0$ initially, $\delta Q(\tau, \vec{x})$ cannot remain zero if the source term on the right-hand-side is non-zero. For quintessence, the right-hand-side is non-zero because Q_o' is non-zero (the field is rolling) and, so long as one considers practical models where matter clusters, h_k' is non-zero. Hence, Q cannot remain perfectly homogeneous.

If $w = p/\rho < 0$, is the sound speed imaginary? Does this mean that inhomogeneities suffer catastrophic gravitational collapse? No. The sound speed squared is $dp/d\rho$, which may be positive even though $w = p/\rho$ is negative. In many cases (including slow-rolling scalar fields), the sound speed is a function of the wavenumber, k. At small wavenumbers, corresponding to superhorizon scales, the sound speed may be imaginary formally, but this has no physical significance since superhorizon modes do not propagate. At wavenumbers corresponding to subhorizon scales, the sound speed is real and positive [1, 44, 54]. For example, for the scalar field, the dispersion term in (4) indicates a k-dependent sound speed $v_k^2 = (1 - a^2 V_{,QQ}/k^2)^{-1}$. Recall that quintessence has negative pressure because Q is rolling slowly. The slow-roll means that the curvature of the potential is less than the square of the Hubble damping scale, $H^2 > V_{,QQ}$. For modes inside the horizon, $k^2/a^2 > H^2$, the sound speed is relativistic on scales much smaller than the horizon ($v_k \to 1$) and becomes non-relativistic on large scales comparable to the horizon ($v_k \to 0$). Formally, the sound-speed becomes imaginary on superhorizon scales, but that has no physical significance because superhorizon modes do not propagate.

To what extent do scalar fields span the possible ways w can change as a function of red shift? Scalar fields completely span the space of possibilities for $w(a)$. Given the evolution of $w(a)$, we may reconstruct the equivalent potential and field evolution, $V(Q[a])$, using the parametrized system of integral equations [35, 36]

$$V(a) = \frac{3H_0^2\Omega_Q}{16\pi G}[1 - w(a)] \exp\left(3[\log\frac{a_0}{a} + \int_a^{a_0}\frac{d\tilde{a}}{\tilde{a}}\, w(\tilde{a})]\right) \tag{5}$$

$$Q(a) = \sqrt{\frac{3H_0^2\Omega_Q}{8\pi G}} \int_{a_0}^{a} d\tilde{a}\frac{\sqrt{1 + w(\tilde{a})}}{\tilde{a}H(\tilde{a})} \exp\left(\frac{3}{2}[\log\frac{a_0}{\tilde{a}} + \int_{\tilde{a}}^{a_0}\frac{d\hat{a}}{\hat{a}}\, w(\hat{a})]\right) \tag{6}$$

Note that we implicitly require the Einstein-FRW equation for $H(a)$ to evaluate $Q(a)$, so that the form of the potential which yields a particular equation-of-state depends on all components of the cosmological fluid, not just the quintessence. For simple $w(a)$, these equations can be combined to give an analytic expression for $V(Q)$. Otherwise, $V(Q)$ can computed numerically and approximated by a fit.

Hence, even if quintessence does not consist of a physical scalar field, studying scalar fields suffices to study all possible equations-of-state. The equation-of-state is sufficient to specify the background evolution. To include the perturbations due to quintessence, one also needs to determine the anisotropic stress. If quintessence is composed of strings [37, 38] or tensor fields [48], say, the anisotropic stress cannot be mimicked by a scalar field. However, the differences in the perturbative effects are typically small and so a scalar field can be used as a first approximation.

Doesn't quintessence introduce nearly an infinite number of free parameters in the choice of $V(Q)$? (If so, the concept has no useful, predictive power.) Yes and no. The situation is very similar to inflation. Once one determines that inflation must be driven by a dynamical energy component, such as a rolling scalar field, the concern arises that there are an infinite number of choices for the inflaton potential energy. In practice, though, there are only a few degrees of freedom relevant to observations because the observables depend only on the behavior of the inflaton over the last 60 e-folds before the end of inflation. During the last 60 e-folds, the inflaton traverses only a very small range of the potential which can be parameterized by a few degrees of freedom. Hence, in practice, inflationary predictions only depend on a small number of parameters, which is why the theory has powerful predictive power.

Similarly, observable consequences of quintessence occur between red shift $z = 5$ and today when the Q field traverses only a small range of its potential. Thus, the possible potentials $V(Q)$ can be effectively parameterized by only one or two constants [50]. The constants might be chosen as V(Q) and $V'(Q)$ today. A more convenient choice is the effective (Ω_Q-weighted) equation of state

$$\bar{w} \equiv \int \Omega_Q(a) w(a) da / \int \Omega_Q(a) da, \tag{7}$$

and the effective first time-derivative $\dot{w}$,

$$\dot{\bar{w}}^2 \equiv \int dz\, \Omega_Q(z) [\dot{w}]^2 / \int dz\, \Omega_Q(z) \tag{8}$$

where $\dot{w} \equiv dw/d\ln z$. Most observations are only sensitive to $\bar{w}$ and, in some cases, $\dot{\bar{w}}$.

To what extent must the analysis of observations be modified because quintessence is spatially inhomogeneous? Conventional treatments of the mass power spectrum and the CMB temperature rely on the mass density being the only spatially inhomogeneous component. The Press-Shechter formalism [46], the computation of the linear mass power spectrum [1, 44], the nonlinear corrections [47], and the CMB temperature power spectrum [1, 44] all must be modified to properly incorporate the spatial inhomogeneities in quintessence. In the typical case, the modification is a small, quantitative correction. In extreme cases, the spatial inhomogeneities can produce anomalous bumps in $P(k)$ (not unlike the bumps reported in some large-scale structure surveys) and amplify peaks in the CMB power spectrum (not unlike some recent observations) [49].

Isn't quintessential cosmology sensitive to initial conditions for Q and $\dot{Q}$ as well as to the potential parameters? It depends. For many potentials discussed in the literature [1, 3], the initial value of Q and $\dot{Q}$ must be finely tuned to obtain the correct value of Ω_Q today. The tuning of the initial field expectation value is required in addition to tuning the potential parameters. Since the initial conditions for the field are hard to control, the scenario seems even more contrived than the the tuning of the cosmological constant. However, as discussed in the next subsection, a large class of potentials has been found for which the cosmology is insensitive to the initial Q and $\dot{Q}$ because there exist classical attractor solutions to the equations-of-motion which result in the same value of Ω_Q independent of the initial conditions [51, 52, 55].

Isn't quintessential cosmology sensitive to the initial conditions for δQ, the spatial fluctuations in Q? No. Without the source term in Eq. (4), perturbations in Q would remain small due to the highly relativistic nature of Q. With the source term, the fluctuations in Q do grow significantly at a rate determined by fluctuations in the metric which, in turn, are determined by the clustering matter component [1, 44, 54]. The amplitude of the perturbation as it enters the horizon depends principally on the source term and is insensitive to the initial conditions in Q itself. In particular, the numerical difference in predictions obtained assuming adiabatic initial conditions for δQ (as might be expected after inflation) versus smooth initial conditions is negligible [36].

Taken together, the answers to these questions go a long way to transforming quintessence from a seemingly arbitrary proposal with many free parameters and choices of initial conditions to a predictive scenario described by few parameters. What remains to be resolved is the cosmic coincidence problem.

3.3. THE COSMIC COINCIDENCE PROBLEM

The key problem posed by a negative pressure component and, in my view, the principle motivation for considering quintessence is the cosmic coincidence problem. The problem has two aspects whose resolution may require two different concepts. One puzzle is to explain why the energy density of the negative pressure component, $\rho_Q \sim (1 \text{ meV})^4$, is so tiny compared to typical particle physics scales. At present, some discount the current evidence for negative pressure simply because it requires a seemingly extraordinary fine-tuning. However, if the evidence described in Part II progresses and becomes overwhelmingly decisive in the next few years, which is technologically feasible, then the view will change. Instead of the small energy density being a problem for cosmology, it will become a new, fundamental parameter whose measured value must be explained by particle physics, just as particle physics is expected to explain ultimately the mass of the electron. This paper anticipates that day.

However, explaining the small value of the energy density is not enough. A second puzzle is to explain why the matter density and the energy density of the negative pressure component are comparable today. Throughout the history of the universe, the two densities decrease at different rates, so it appears that the initial conditions in the early universe have to be set with exponentially sensitive precision to arrange comparable energy densities today. For example, after inflation, the ratio of vacuum density to matter-radiation density would have to be tuned to be of order 10^{-100}. Since the ratio is inferred on the basis of extrapolating a cosmological model backwards in time, the solution to the initial conditions problem may lie in the domain of cosmology, rather than particle physics. That is, perhaps the tuning may be avoided perhaps by changing the cosmological model.

What would be ideal is an energy component that is initially comparable to the matter and radiation density, remains comparable during most of the history of the universe, and then jumps ahead late in the universe to initiate a period of cosmic acceleration. This is quite unlike a cosmological constant. However, recently, a large class of quintessence models with "runaway scalar fields" have been identified which have many of the desired properties.

3.4. RUNAWAY SCALAR FIELDS AND THE STICKING POINT THEOREM

Runaway scalar fields are promising candidates for quintessence. We use the term "runaway scalar fields" [56] to refer to cases in which the potential $V(Q)$, the curvature $V''(Q)$, and their ratio, V''/V all converge to zero as $Q \to \infty$. The potentials occur in string and M-theory models associated with the many moduli fields or with fermion condensates [57, 58, 59, 60, 61, 62]. The potentials are typically flat to perturbative order but have runaway potentials when non-perturbative effects are included. Inverse power-law potentials, general functionals with inverse powers or fractional powers of the field (or condensate) are examples of runaway potentials.

A runaway field has a simple but profound effect on cosmology. The ultimate fate of the universe is sealed: the universe is destined to undergo cosmic acceleration [56]. The prediction is as sure as if one had introduced a positive cosmological constant into the theory. The argument is based on a simple theorem we refer to as the "sticking point theorem": Given a runaway potential $V(Q)$, there is a largest value of the field Q_{sp} for which $V''(Q_{sp})/V(Q_{sp}) = 8\pi G/3$; that is, since V''/V converges to zero at large Q, there must be a point where the ratio reaches the particular value $8\pi G/3$. (One could imagine potentials where all Q satisfy $V''(Q)/V(Q) < 8\pi G/3$. Then, the sticking point theorem is trivially satisfied; see below.)

The sticking point theorem says that, for all $Q > Q_{sp}$, the rolling field is critically damped by the Hubble expansion and the field is frozen. The word "frozen" is used judiciously – one should imagine a frozen glacier which slowly flows downhill. In this case, the field flows so slowly downhill that the energy density decreases much more slowly than matter and radiation density. The field energy eventually overtakes the matter and radiation, driving the universe into cosmic acceleration.

The proof is trivial: For any $Q > Q_{sp}$, it must be that $V''(Q)/V(Q) < 8\pi G/3$, by the definition of Q_{sp}. Then, we have

$$V''(Q) < \frac{8\pi G}{3} V(Q) < \frac{8\pi G}{3}[\rho_{m,r} + \frac{1}{2}\dot{Q}^2 + V(Q)] = H^2, \tag{9}$$

where $\rho_{m,r}$ is the background matter-radiation density, which is positive. The right hand equation is the definition of the Hubble parameter according to the Friedmann equation. The chain of relations reduces to $V'' < H^2$, which is precisely the condition for the Q-field kinetic energy density to be overdamped by the Hubble expansion so as to force slow-roll and negative pressure [7]. Hence, the sticking point theorem assures us that, once the field rolls past the sticking point, Q slows to a crawl and acts as a negative pressure component. Since the potential is decreasing, there is nothing to stop Q from ultimately reaching and surpassing the sticking point. Beyond the sticking beyond, the potential energy of the Q-field is sufficient to create a Hubble damping that freezes the field, independent of the value of the matter and radiation density. It is just a matter of time before this energy density comes to dominate the universe. The corollary is that it is just a matter of time before the energy density of the runaway field overtakes the matter and radiation, and cosmic acceleration commences. For most runaway potentials, the sticking point corresponds to a large expectation value of the field and a small energy. This feature of runaway fields satisfies in a very rough, qualitative way the condition desired to explain why cosmic acceleration begins late in the history of the universe when the mean density is small. There remains the issue of why acceleration commences after 10 billion years rather than 10 million years or 10 trillion years, but the qualitative character of runaway fields seems, at least to this author, to be an attractive feature which may well be incorporated in the final answer.

3.5. TRACKER FIELDS AND TRACKING POTENTIALS

There are different ways the runaway field may come to surpass the sticking point, which leads to different cosmological scenarios. We discuss two examples: tracker models and creeper models.

Tracker fields [51, 52] are runaway fields in which the equation-of-motion has an attractor-like solution so that the evolution of Q is insensitive to the initial conditions for Q and $\dot{Q}$. in order to have an attractor solution, one requirement is that $\Gamma \equiv V''V/(V')^2$ exceed 5/6 and be nearly constant [52]. The only constraint on the initial energy density in the tracker field is that it be less than or equal to the initial matter-radiation energy and greater than the present-day matter density. (As we shall see, this condition is necessary in order for Q to converge to the attractor solution before the present epoch.) This constraint allows an extraordinary range of initial conditions for ρ_Q spanning over 100 orders of magnitude. The range includes the physically well-motivated possibility of equipartition between quintessence and matter-radiation. The term "tracker" refers to the fact that the cosmology follows the same evolutionary track independent of initial conditions. See Figure 6.

The attractor solution has the property that, beginning from the initial Q and $\dot{Q}$, Q rapidly converges to the point on the potential where $V'' \approx H^2$. The Hubble parameter H is determined by the matter and radiation density. As the universe expands and H decreases, Q moves down the potential so as to maintain the condition $V'' \approx H^2$. In this sense, the evolution of the Q-field controlled by the matter-radiation density rather than evolving independently according to its own potential. This is the distinctive feature of tracker fields. The controlled evolution continues until Q finally surpasses the sticking point. Then, its own potential energy density is sufficient to freeze the field and cause it to overtake the background energy.

What if the initial energy density is far below the value where $V'' = H^2$ but $Q < Q_{sp}$? We refer to this as the "undershoot" initial condition. Since V''/V is decreasing, lower energy density means that the initial V'' must be much less than H^2. The evolution of Q is overdamped by the Hubble expansion, and so its value remains nearly constant. The field has not surpassed the sticking point ($Q < Q_{sp}$), and the field is only frozen because of the matter-radiation density contribution to H. As the universe expands, the matter-radiation density eventually decreases to a point where $V'' \approx H^2$ and the field becomes unfrozen. The field begins to roll down the potential with V'' tracking H^2 just as it would if the field had begun on the attractor solution initially. Ultimately, Q passes the sticking point and cosmic acceleration begins.

What if the initial energy density begins above the value where $V'' = H^2$ (the "overshoot" initial condition)? See Figure 6. In this case, $V'' \gg H^2$ and the Hubble damping is irrelevant at first. The potential is so steep that the field rapidly accelerates to a condition where the its energy density is dominated by its kinetic energy. A field dominated by kinetic energy has $w = 1$ and decreases as $1/a^6$. The field energy rapidly falls compared to the matter and radiation at such a rate that it overshoots and falls below the attractor solution, $V'' \approx H^2$. Eventually, though, the Hubble damping red shifts away the kinetic energy and the field becomes frozen after a

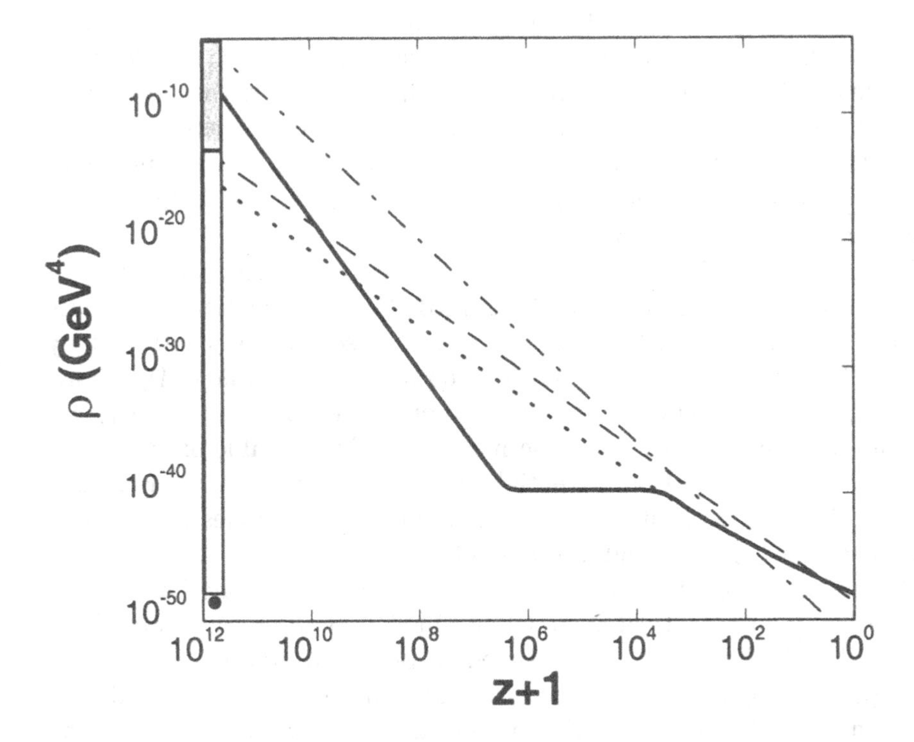

Figure 6. Energy density versus red shift for the evolution of a tracker field. For computational convenience, $z = 10^{12}$ has been arbitrarily chosen as the initial time when the field begins to roll. The white and grey bars represent the span of of allowed initial conditions for ρ_Q (which corresponds to 100 orders of magnitude if extrapolated back to inflation). The white bar on left represents the undershoot type of initial condition and the grey bar represents the overshoot. The solid black circle represents the unique initial condition required if the missing energy consists of vacuum energy density. The dotted curve is the attractor (tracker) solution. The solid thick curve illustrates the evolution beginning from an overshoot initial condition in which ρ_Q has a value greater than the tracker solution. Q rushes rapidly down the potential, overshooting the tracker solution and the matter density, and freezes. As the universe expands, H decreases to a point were $V'' \approx H^2$ and the field joins the tracker solution.

displacement of [51, 52]:

$$\Delta Q \approx \left(\frac{3}{4\pi}\Omega_{Qi}\right)^{1/2} M_p, \tag{10}$$

where Ω_{Qi} is the initial ratio of ρ_Q to the critical density and M_p is the Planck mass. (Typically, the initial value of Q is small compared to ΔQ, so $Q \approx \Delta Q$ when the field is frozen.) Now the field is frozen at a point where $V'' \ll H^2$, the initial condition of the undershoot case. The frozen values of $Q \approx \Delta Q$ and V'' are independent of the potential (since the kinetic energy dominates during the as Q rapidly falls), but they are dependent on the initial value of ρ_Q, as shown in Eq. (10) above. Nevertheless, this is irrelevant for cosmology since the energy density in Q is much smaller than matter and radiation density during the frozen period. As in the undershoot case, the field remains frozen until H^2 decreases to a point where it matches V'' and the field joins the tracker solution. By the time ρ_Q grows to influence cosmology, Q is on the same evolutionary track as if Q and $\dot{Q}$ had begun on the tracker solution in the first place.

The tracker solutions lead to a new prediction: a relationship between Ω_m and the equation-of-state for Q, w_Q [51, 52]. The relationship occurs because, for any given potential, the attractor solution is controlled by only one free parameter, which can be chosen to be the value of Ω_m today (assuming a flat universe). Consequently, once the potential and Ω_m are fixed, no freedom remains to choose independently the value of w_Q today. There is some variation from potential form to potential form; see Figure 7. But, the variation is limited and, most importantly, includes a forbidden region between $w_Q = -1$ and $w_Q \approx -3/4$.

To construct models with w_Q between -1 and $-3/4$, one has to consider rather poorly motivated and fine-tuned potentials. An example is $V(Q) \sim 1/Q^\alpha$ where $\alpha \ll 0.1$ is tuned to be a tiny fractional power so that $V(Q)$ is designed to be nearly like a cosmological constant. Adding a true cosmological constant to $V(Q$ would also allow $w_Q \to -1$, but then there is no point to having the tracker field. The Ω_Q-w_Q relation we present is intended as a prediction that distinguishes cases where there is no true cosmological constant from cases where there is.

3.6. CREEPING QUINTESSENCE

An alternative possibility, "creeping quintessence," [56] occurs for the very same tracker potentials if the initial energy density in the Q-field greatly exceeds the matter-radiation density. Then, as in the overshoot case discussed in the previous subsection, the field begins with $V'' \gg H^2$ and rapidly accelerates down the potential until the kinetic energy dominates the potential energy. As before, the field kinetic energy red shifts away as Q rolls a distance [52, 56]:

$$\Delta Q = \sqrt{\frac{3}{4\pi}} \left(1 + \frac{1}{2} \ln \left(\frac{\rho_{Qi}}{\rho_{Bi}}\right) M_p\right). \tag{11}$$

Note that this expression is different from Eq. (10) because here $\rho_{Qi} > \rho_{Bi}$, where ρ_{Bi} is the initial background matter-radiation density, and the field rolls farther than overshoot case for the tracker field. The Q-field rolls so far that it overshoots not only the initial matter density, but also the sticking point before it finally freezes. See Figure 8. Now, the field is frozen and

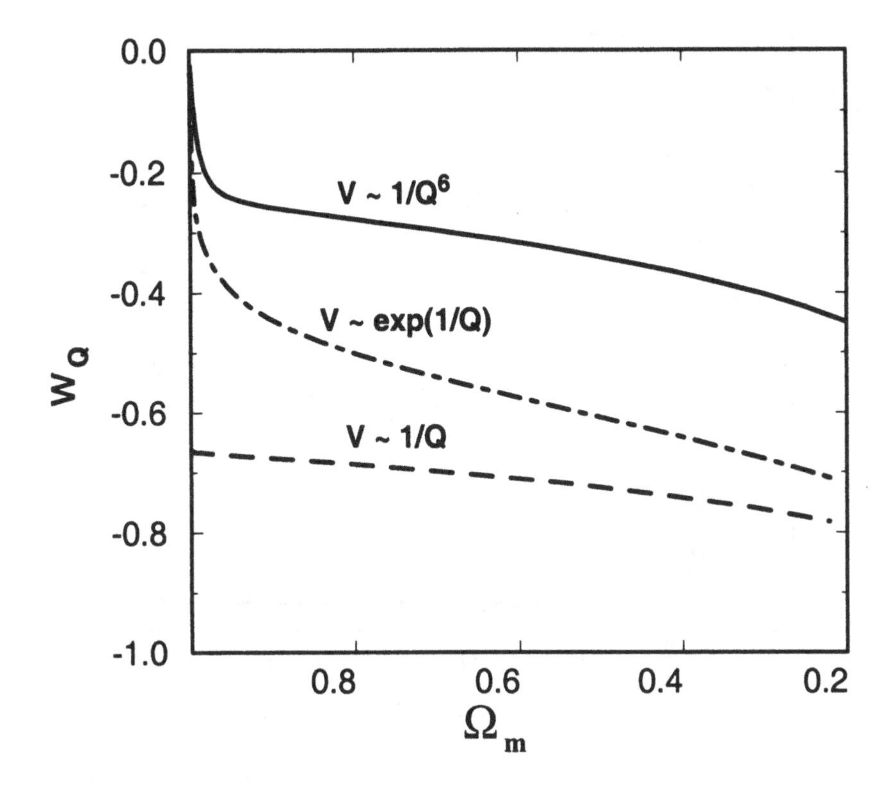

Figure 7. The Ω_Q-w_Q relation can be illustrated on a plot of w_Q versus $\Omega_m = 1 - \Omega_Q$. The predictions are shown for various potentials. Note that the forbidden region at values of w_Q less than -0.75.

remains frozen forever, never joining the tracker solution. Because the field overshoots the sticking point by a significant margin, the energy density is tiny compared to the matter-radiation density, and equation of state is $w \approx -1$. The field creeps down the potential for the remainder of the history of the universe, with ρ_Q eventually overtaking the matter-radiation energy density and inducing cosmic acceleration.

The creeping case differs in several ways from the tracker potential. First, the creeping scenario is possible for a somewhat wider range of potentials. Since it never utilizes the tracker features, the condition on Γ can be dropped. All that is required is a sticking point at sufficiently low energy. Second, the scenario is more sensitive to initial conditions because the value of ρ_Q at which Q freezes is dependent on the degree to which the field overdominates the background density. However, one observes from Eq. (11) that the value at which Q freezes depends only logarithmically on ρ_{Qi}/ρ_{Bi}, which seems to be a relatively mild sensitivity to initial conditions [56].

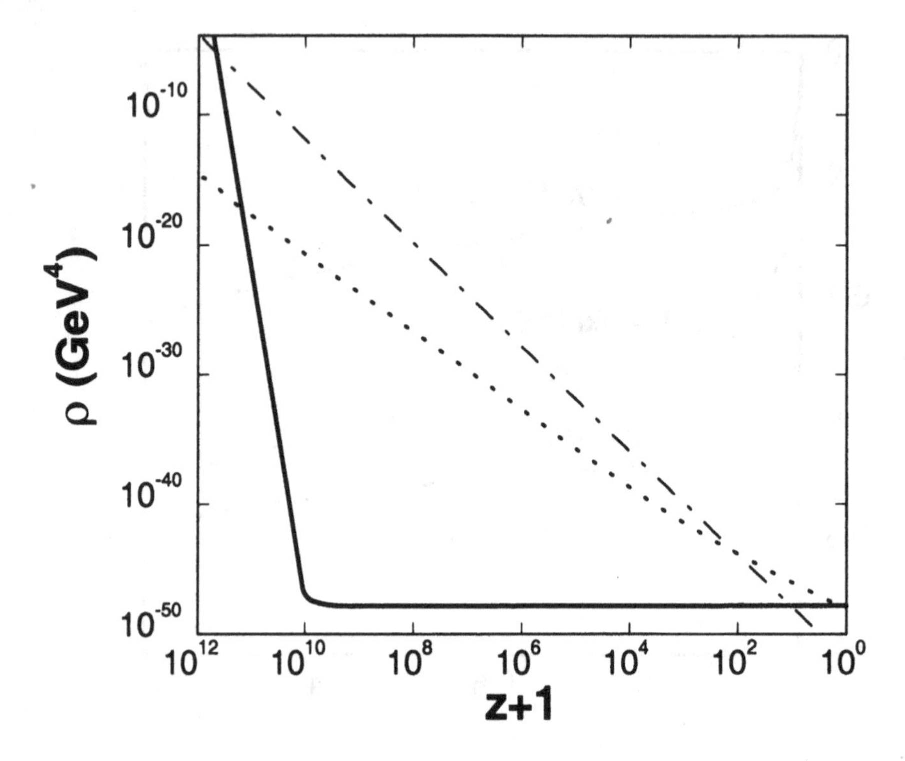

Figure 8. A plot showing the case for creeping quintessence in which the field overshoots the sticking point and freezes early in the history of the universe. Compare to Figure 6, in which the energy density eventually joins a tracker solution. Here, once frozen, the field rolls very slowly for the rest of the history of the universe. Consequently, w is very close to -1.

The disadvantage of creeping quintessence is that w is very close to -1 today so that distinguishing it from a vacuum density is difficult. There is no significant difference in terms of cosmic evolution, astrophysics, or the cosmic microwave background. In most quantum field theories, Q couples through quantum corrections to other interactions, and time-variation in Q results in time-varying constants. Here, the field is moving so slowly that any variation in coupling constants is exponentially small! In addition to cosmic acceleration, the ultra-slow evolution of couplings constants is an intriguing consequence of this scenario that would be very difficult to detect directly.

3.7. WHY IS QUINTESSENCE BEGINNING TO DOMINATE TODAY?

We have argued that tracker models and, with somewhat less precision, creeper models produce nearly the same cosmic evolutionary track independent of the initial conditions for Q. This is one of the properties desired

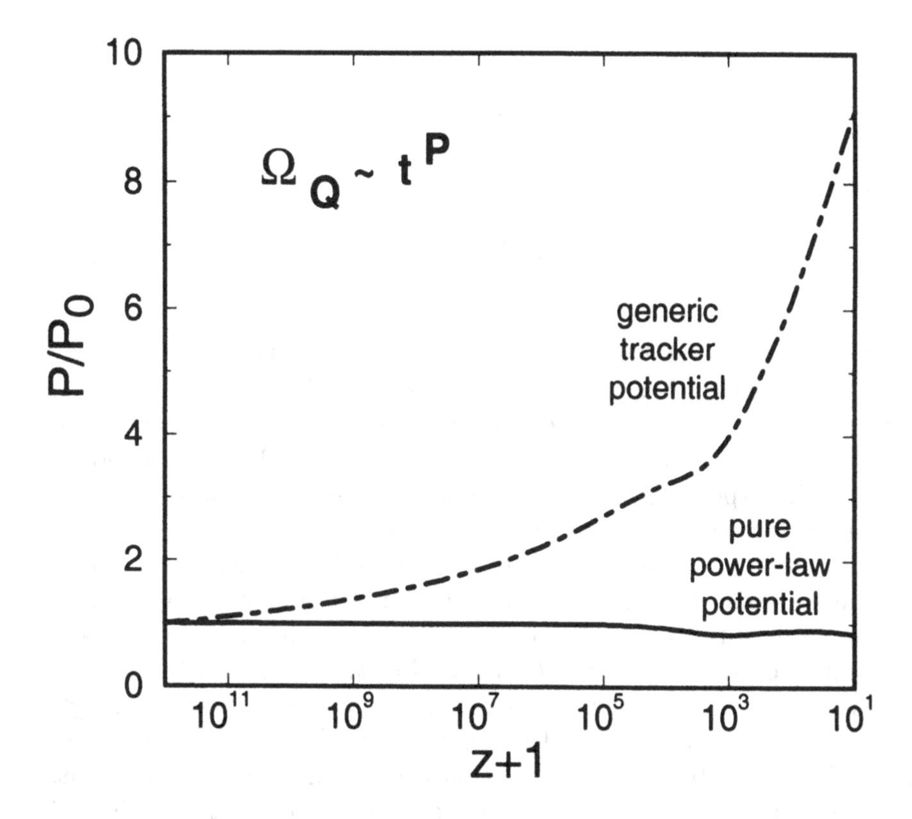

Figure 9. A plot of P/P_0 versus t, where $\Omega_Q \propto t^P$ and P_0 is the initial value of P. The plot compares pure inverse power-law ($V \sim 1/Q^\alpha$) potentials for which P is constant with a generic potential (*e.g.*, $V \sim \exp(1/Q)$) for which P increases with time.

to address the cosmic coincidence problem. What remains is to determine why the track turns out to be one where Q has begun to dominate recently. The time when Q overtakes the matter density is determined by M for a quintessence potential $V(Q) = M^4 f(Q/M)$. The tuning of M might be viewed as similar to the tuning of Λ in the case of a cosmological constant.

However, there is more to the issue because different forms of $f(Q/M)$ produce different families of tracker solutions which overtake the matter density at different times. To consider this issue, we want to change our point-of-view. Up to this point, we have imagined fixing M so that $\Omega_Q = 1 - \Omega_m$ has the measured value today. This amounts to considering one tracker solution for each $V(Q)$. Now we want to consider the entire family of tracker solutions (as a function of M) for each potential form $f(Q/M)$ and consider whether Ω_Q is more likely to dominate late in the universe for one one family of solutions or another.

In general, Ω_Q is proportional to $a^{3(w_B - w_Q)} \propto t^{2(w_B - w_Q)/(1+w_B)}$, where

[51]

$$w_B - w_Q = \frac{2(\Gamma - 1)(w_B + 1)}{1 + 2(\Gamma - 1)}. \tag{12}$$

Hence, we find $\Omega_Q \propto t^P$ where

$$P = \frac{4(\Gamma - 1)}{1 + 2(\Gamma - 1)}. \tag{13}$$

For two special cases ($V \sim 1/Q^\alpha$ and $V \sim \exp(\beta Q)$), $\Gamma - 1$ is nearly constant, and, hence, P is nearly constant as well. The interpretation is that Ω_Q grows as the same function of time throughout the radiation- and matter-dominated epochs. So, there is no tendency for Ω_Q to grow slowly at first and then speed up later. See Figure 9. The same situation occurs for Ω_Λ for models with a cosmological constant.

However, for more general quintessence potentials, P increases as the universe ages. Consider first a potential which is the sum of two inverse power-law terms with exponents $\alpha_1 < \alpha_2$. The term with the larger power is dominant at early times when Q is small, but the term with the smaller power dominates at late times as Q rolls downhill and obtains a larger value. Hence, the effective value of α decreases and $\Gamma - 1 \propto 1/\alpha$ increases; the result is that P increases at late times. For more general potentials, such as $V \sim \exp(1/Q)$, the effective value of α decreases continuously and P increases with time. Figure 9 illustrates the comparison in the growth of P.

How does this relate to why Ω_Q dominates late in the universe? Because an increasing P means that Ω_Q grows more rapidly as the universe ages. Figure 10 compares a tracker solution for a pure inverse power-law potential ($V \sim 1/Q^6$) model with a tracker solution for a generic potential with a combination of inverse powers (in this case, $V \sim \exp(1/Q)$), where the two solutions have been chosen to begin at the same value of Ω_Q. (The start time has been chosen arbitrarily at $z = 10^{17}$ for the purposes of this illustration.) Following each curve to the right, there is a dramatic (10 orders of magnitude) difference between the time when the first solution (solid line) meets the background density versus the second solution (dot-dashed line). Beginning from the same Ω_Q, the first tracker solution dominates well before matter-radiation equality and the second (generic) example dominates well after matter-domination.

Hence, an intriguing conclusion is that the generic quintessence potential has a family of solutions in which Ω_Q tends to dominate late in the history of the universe and induces a recent period of accelerated expansion. Although the trend towards late domination is an improvement over models with cosmological constant or pure power-law or exponential $V(Q)$, we have not

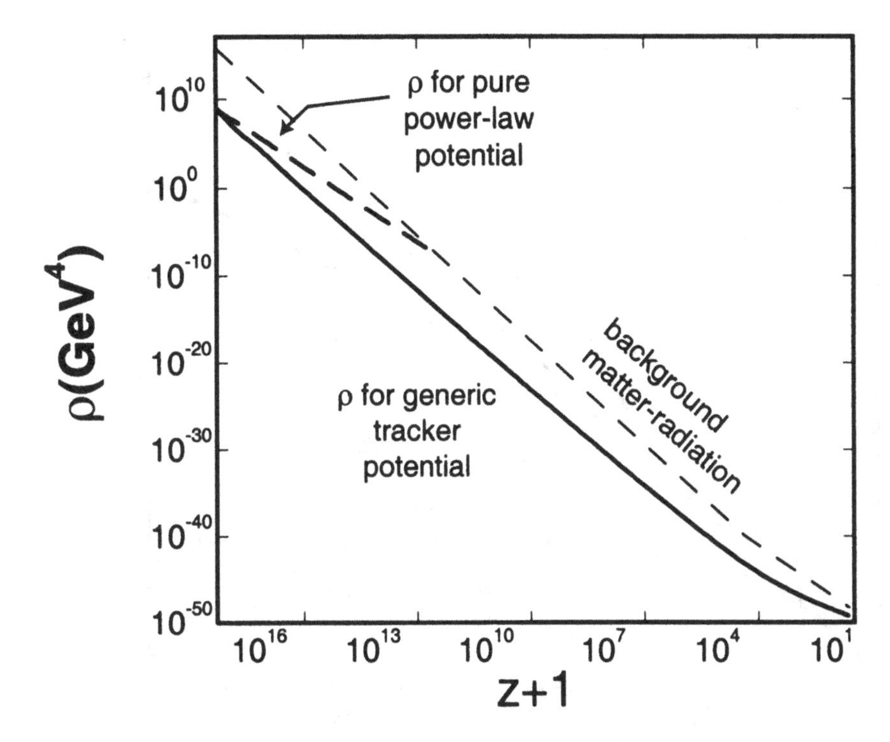

Figure 10. A plot comparing two tracker solutions for the special case of a pure power-law ($V \sim 1/Q^6$) potential (solid line) and a general potential composed of a combination of inverse power-law contributions ($V \sim \exp(1/Q)$) (dot-dash). The dashed line is the background density. The two tracker solutions were chosen to have the same energy density initially. The tracker solution for the generic example ($V \sim \exp(1/Q)$) reaches the background density much later than for the pure inverse-power law potential. By this measure, Ω_Q is more likely to dominate late in the history of the universe in the generic case than for a pure power-law potential.

answered the quantitative question: why is Q dominating after 15 billion year and not, say, 1.5 billion years or 150 billion years. Yet further ideas are required.

4. Current constraints and future tests

The observable consequences of quintessence are due to its effect on the expansion of the universe and its inhomogeneous spatial distribution. Figure 11 compares the expansion rate of universe for an Einstein-de Sitter model ($\Omega_m = 1$) with three models with $\Omega_m = 0.3$: an open model, a flat model with quintessence ($w = -0.7$), and a flat model with a cosmological constant. The expansion rate is decelerating for the Einstein-de Sitter and open models and accelerating for the models with cosmological constant and quintessence. For a given Ω_Q, the cosmological constant has more

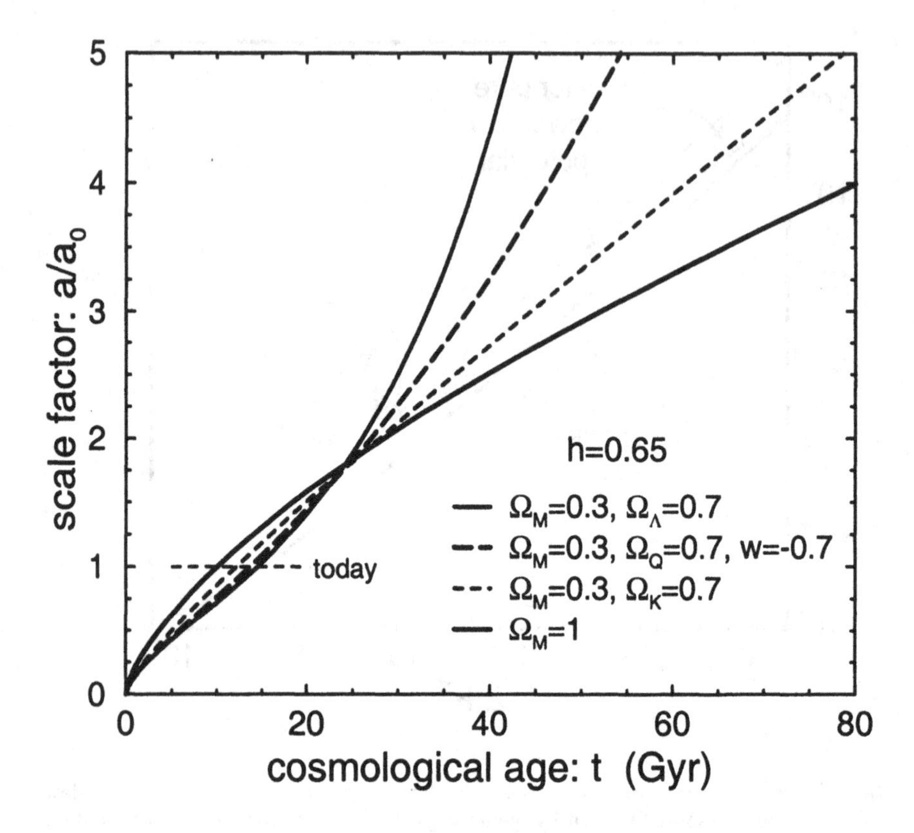

Figure 11. The scale factor vs. time for an Einstein-de Sitter model ($\Omega_m = 1$) and three models with $\Omega_m = 0.3$: an open model, a flat model with quintessence ($w = -0.7$), and a flat model with a cosmological constant.

negative pressure and, hence, induces a greater acceleration. The present epoch corresponds to scale factor $a/a_0 = 1$. The figure shows that the age of the universe increases with acceleration. The open model and the models with quintessence or cosmological constant predict ages consistent with estimates based on the ages of globular clusters [21, 22].

Because the spatial inhomogeneities in the quintessence typically have small amplitudes compared to the density perturbations, they are difficult to detect. The largest effect is on the CMB anisotropy and the mass power spectrum on large scales, which is incorporated in computations of the power spectra for quintessence models [1]. Figure 12 shows the predictions for the CMB temperature anisotropy power spectrum for a series of quintessence models with different choices of w, Ω_m and the spectral index, n, of the primordial density fluctuation distribution. The results are compared to an Einstein-de Sitter model and a model with cosmological constant. As with a cosmological constant, quintessence increases the height of the first acoustic peak compared to the Einstein-de Sitter model. The differences in height are due to a combination of differences in equation-of-state, tilt, and normalization. Note the tiny difference in the shape of the CMB power spectra at small ℓ for the quintessence models compared to the ΛCDM model. The effect is primarily due to the spatial inhomogeneities in the quintessence field [1, 36], as well as differences in the integrated Sachs-Wolfe contribution. Although the spatial fluctuations in the quintessence field make a large contribution to the CMB anisotropy at small ℓ, the difference in the CMB power spectrum shape is nearly impossible to detect. The quintessence fluctuations become important, though, when the amplitude of the CMB anisotropy is correlated with the amplitude of the mass density fluctuations on large scales. For example, consider a quintessence model and a ΛCDM with nearly indistinguishable CMB anisotropy power spectra. The two models have nearly the same CMB anisotropy on large scales, but, for one model, spatial fluctuations in the quintessence makes a significant contribution. Consequently, the contribution of mass fluctuations is different for the two models. By measuring the mass power spectrum directly and comparing to the two model predictions, the presence or absence of spatial fluctuations in the quintessence field can be determined.

Based on these effects, Wang *et al.* [20] have completed a comprehensive survey of cosmological tests of quintessence models which shows that a substantial range of Ω_m and w is consistent with current observations. The allowed region in the Ω_m-w plane is shown in Figure 13. For a given combination of Ω_m and w to be included, there must exist a choice of the Hubble parameter, spectral tilt, and baryon density such that the model satisfies all current tests at the 2-σ level or better. (See the original paper for a full discussion and an alternative plot using conventional likelihood analysis which shows similar results.)

In the future, it should be possible to distinguish the Λ or creeper regime from the tracker regime with improved CMB anisotropy measurements combined with other cosmological observations [50]. The wider the difference between the actual w and -1, the easier it is to distinguish Q from Λ. Measurements of the CMB power spectrum constrain models to a degeneracy curve in the Ω_m-w plane; along the degeneracy curve, changes in Ω_m, w and h combine to make the power spectrum indistinguishable [50]. The effects of gravitational lensing on the CMB, which may be detectable by measuring the spectrum at very small angles, reduce but do not totally eliminate the degeneracy [50]. Measurements of the Hubble parameter, deceleration parameter, or matter density can break the degeneracy, as can, perhaps, measurements of gravitational lensing arcs or improved luminosity-red shift measurements for supernovae [50, 63, 64, 65, 66]. Davis has suggested a more direct approach in which measuring velocity disper-

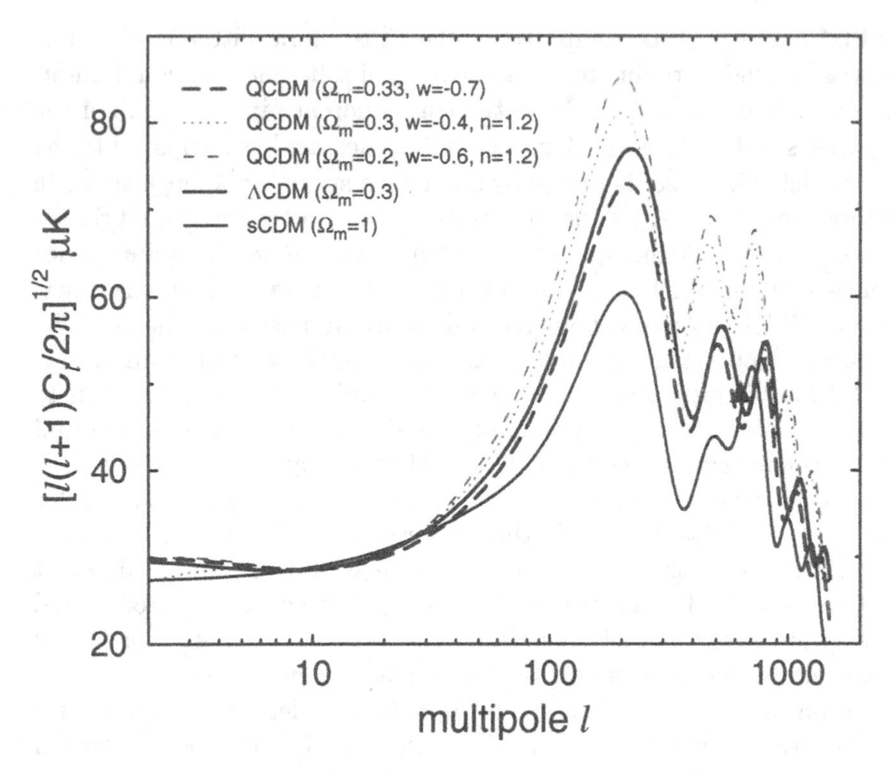

Figure 12. The CMB power spectrum for an Einstein-de Sitter model ($\Omega_m = 1$), a flat model with cosmological constant (ΛCDM), and a sequence of flat models with quintessence (QCDM). The ΛCDM and QCDM models are consistent all current cosmological observations.

sion (line-widths) of clusters as function of red shift is used as a cosmic barometer.

5. Theoretical Challenges

The quintessence scenario began as a somewhat arbitrary, poorly motivated and ill-defined concept with many free parameters. However, the range of possibilities has begun to be explored and regimes of more plausible and predictive models have been identified. Runaway fields are an especially promising possibility. Runaway fields appear to be endemic to many unified theories, including string theory and M-theory, and, as shown here, they lead inevitably to cosmic acceleration at low energies.

Significant theoretical challenges remain. Obviously, one goal is to identify the quintessence scalar field or to replace the field by some dynamical mechanism. Perhaps a more satisfying solution to the coincidence problem can be found which explains why quintessence could not have dominated at

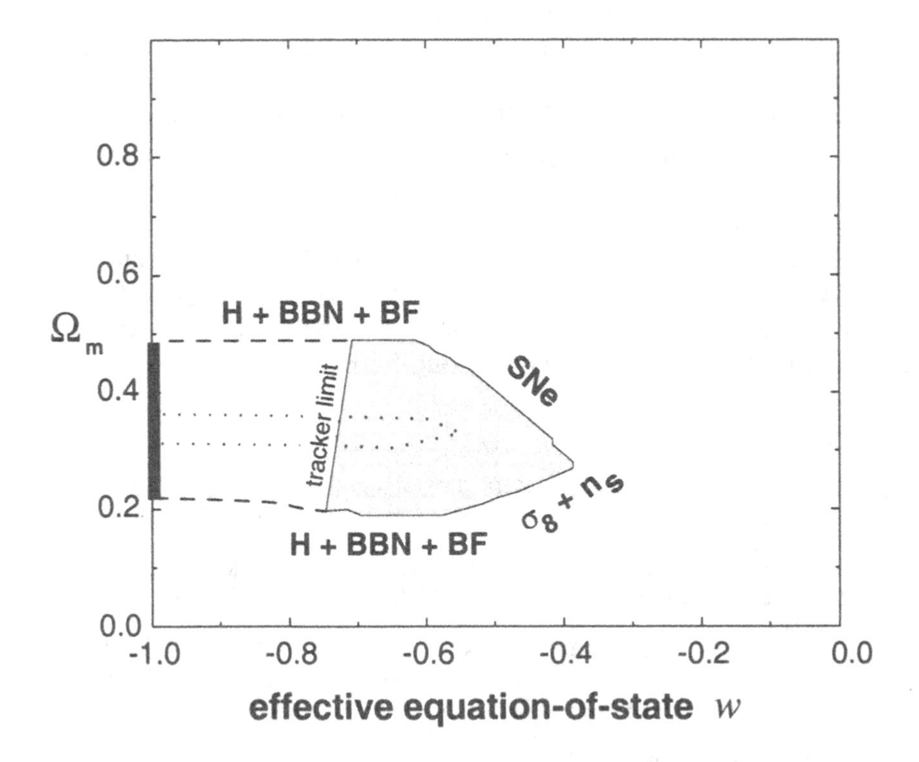

Figure 13. The region of the Ω_m-w plane consistent with current observations and theoretical constraints. The black strip on left corresponds to the allowed range for models with a cosmological constant or creeping quintessence. The grey region on right corresponds to tracker models. The gap spanning $-1.0 < w < -0.75$ is allowed observationally, but does not fit within the tracker or creeper scenarios. (This range appears to require fine-tuning of initial conditions or more exotic potentials.) The dotted region is the 1-sigma best-fit according to maximum likelihood analysis. The markings around the allowed region indicate the measurements which delimit each boundary: Hubble constant (H), big bang nucleosynthesis (BBN), baryon fraction (BF), supernovae searches (SNe), cluster abundance (σ_8), power spectrum tilt (n_s).

a much earlier epoch. A second issue is whether quintessence induces fifth force effects and time-varying constants which are inconsistent with known experimental constraints [67]. For generic, light scalar fields, a dimensional estimate suggests that the effects would exceed current experimental limits [68]. To avoid the problem, a dimensionless parameter expected to be of order unity would instead have to be set to less than 10^{-5}. Whether this tuning is problematic is a matter of judgment; one must recall that the problem which quintessence addresses is at the 10^{100} level, so perhaps the tuning is acceptable. Or, perhaps the tuning is not needed because the scalar field has suppressed couplings for reasons of symmetry. A third issue is the time-honored question concerning the cosmological constant: if quintessence accounts for the current cosmic acceleration, why is the vacuum density zero ($\Lambda = 0$) or, perhaps, non-zero but much smaller than the quintessence energy density. Finally, the quintessence scenario must be integrated into fundamental theory. In many ways, the list of challenges are similar to the challenges for inflationary cosmology, the cosmic acceleration of the past. Perhaps nature is giving us a message.

I would like to thank the the staff of the Isaac Newton Institute for Mathematical Sciences for the their kind hospitality and support during the NATO conference. This research was supported by the US Department of Energy grant DE-FG02-91ER40671 (Princeton).

References

1. R. R. Caldwell, R. Dave and P. J. Steinhardt, *Phys. Rev. Lett.* **80**,, 1582(1998). R. R. Caldwell, R. Dave and P. J. Steinhardt, *Phys. Rev. Lett.* **80**,, 1582(1998).
2. A.D. Dolgov, *Pis'ma ZhETF* **41**, 345 (1985). M. Ozer and M.O. Taha, *Phys. Lett. B***171**, 363 (1986); *Nucl. Phys. B***287**, 776 (1987); *Phys.Rev.D***45**, 997 (1992).
3. N. Weiss, *Phys. Lett. B* **197**, 42 (1987); B. Ratra and J.P.E. Peebles, *Astrophys. J.* ,, 325,L17 (1988); C.Wetterich, *Nucl. Phys. B***302**, 668 (1988), and *Astron. Astrophys.* **301**, 32 (1995); J.A. Frieman, *et al. Phys. Rev. Lett.* **75**, 2077 (1995); K. Coble, S. Dodelson, and J. Frieman, *Phys. Rev. D* **55**, 1851 (1995); P.G. Ferreira and M. Joyce, *Phys. Rev. Lett.* **79**, 4740 (1997); *Phys. Rev. D* **58**, 023503 (1998). 1582
4. A. Einstein, *Sitz. Preuss. Akad. Wiss.* **142**, (1917).
5. A.H. Guth, *Phys. Rev. D***23**, 347 (1981).
6. A. D. Linde, *Phys. Lett.* **108B**, 389 (1982).
7. A. Albrecht and P. J. Steinhardt, *Phys. Rev. Lett.* **48**, 1220 (1982).
8. For an introduction to inflationary cosmology, see A. H. Guth and P. J. Steinhardt, "The Inflationary Universe" in **The New Physics**, ed. by P. Davies, (Cambridge U. Press, Cambridge, 1989) pp. 34-60.
9. For a recent review, see N. Bahcall, J.P. Ostriker, S. Perlmutter and P.J. Steinhardt, *Science*, 284, 1481 (1999).
10. S. Perlmutter, *et al.*, LBL-42230 (1998), astro-ph/9812473; S. Perlmutter, *et al.*, *Astrophys. J.* **(in**, press),astro-ph/9812133.
11. A.G. Riess, *et al.*, *Astrophys. J.* **116**,, 1009(1998).
12. P. M. Garnavich, *et al.*, *Astrophys. J.* **509**,, 74(1998).
13. S. Perlmutter, M. S. Turner and M. White, *Astrophys. J.* **(submitted)**,, astro-ph/9901052,(1999).
14. S. Perlmutter, *et al.*, *Astrophys. J.* **483**,, 565(1997).
15. S. Perlmutter, *et al.*, 1995. LBL-38400, page I.1 (1995); also published in Thermonuclear Supernovae, P. Ruiz-Lapuente, R. Canal and J.Isern, editors, Dordrecht: Kluwer, page 749 (1997)
16. P. Garnavich, *et al.*, *Astrophys. J.* **493**,, L53(1998)
17. B.P. Schmidt, *et al.*, *Astrophys. J.* **507**,, 46(1998)
18. A.G. Riess, A.V. Filippenko, W. Li, B.P. Schmidt, astro-ph/9907038.stro-ph/9907038
19. J. P. Ostriker and P. J. Steinhardt, *Nature* **377**, 600 (1995).

20. L. Wang, R. Caldwell, J.P. Ostriker and P.J. Steinhardt, *Astrophys. J.* **(submitted)**,, astro-ph/9901388,(1999).
21. B. Chaboyer, *et al. Astrophys. J.* **494**,, 96(1998).
22. M. Salaris & A. Weiss, *Astron. Astrophys.* **335**, 943 (1998).
23. R.H. Dicke and P.J.E. Peebles, in *General Relativity: An Einstein Centenary Survey*, ed. by S.W. Hawking and W. Israel, (Cambridge: Cambridge U. Press, 1979), pp. 504-517.
24. For an introduction to the CMB power spectrum and its interpretation see, P.J. Steinhardt, "Cosmology at the Crossroads", *Particle and Nuclear Astrophysics and Cosmology in the Next Millenium,* ed. by E.W. Kolb and P. Peccei (World Scientific, Singapore, 1995), pp. 51-72.
25. The physics of CMB anisotropy generation is discussed in W. Hu, N. Sugiyama and J. Silk, *Nature* **386**, 37 (1997).
26. M. Kamionkowski, D. N. Spergel, N. Sugiyama, *Astrophys. J.* **426**,, L57(1994).
27. A. Miller *et al.*, astro-ph/9906421
28. G.F. Smoot, *et al.*, *Astrophys. J.* **396**,, L1(1992); C. L. Bennett *et al.*, *Astrophys. J.* **464**,, L1(1996).
29. S.R. Platt *et al.*, *Astrophys. J.* **475**,, L1(1997).
30. E. S. Cheng, *et al.*, *Astrophys. J.* **488**,, L59(1997).
31. M. J. Devlin, *et al.*, *Astrophys. J.* **509**,, L69(1998); T. Herbig, *et al.*, *Astrophys. J.* **509**,, L73(1998); A. Oliveira-Costa, *et al.*, *Astrophys. J.* **509**,, L77(1998)
32. B. Netterfield, M.J. Devlin, N. Jarosik, L. Page and E.L. Wallack, *Astrophys. J.* **474**,, 47(1997).
33. P. F. Scott, *et al.*, *Astrophys. J.* **461**,, L1(1996).
34. E. M. Leitch, A.C.S. Readhead, T.J. Pearson, S.T. Myers and S. Gulkis, *Astrophys. J.* **518**,, (1999),astro-ph/9807312, (1998).
35. R. Caldwell and P.J. Steinhardt, in *Proceedings of the Non-sleeping Universe Conference*, ed. by T. Lago and A. Blanchard (Kluwer Academic, 1998).
36. R. Dave, U. of Penn. Ph.D. thesis (1999).
37. A. Vilenkin, *Phys. Rev. Lett.* **53**, 1016 (1984).
38. D. Spergel & U.-L. Pen, *Astrophys. J.* **491**,, L67(1996).
39. M.P. Bronstein, *Phys. Zeit. der Sowjetunion* **3**, 73 (1933).
40. J. Bardeen, P. J. Steinhardt and M. S. Turner, *Phys. Rev. D* **28**, 679 (1983).
41. A. H. Guth and S.-Y. Pi, *Phys. Rev. Lett.* **49**, 1110 (1982).
42. A. A. Starobinskii, *Phys. Lett. B* **117**, 175 (1982).
43. S. W. Hawking, *Phys. Lett. B***115**, 295 (1982).
44. P.J. Steinhardt and R. Caldwell, "Introduction to Quintessence," in **Cosmic Microwave Background and Large Scale Structure of the Universe**, ed. by Y.-I. Byun and K.-W. Ng, (Astronomical Society of the Pacific, 1998), pp. 13-21.
45. R. Caldwell and P.J. Steinhardt, *Phys. Rev. D* **57** 6057 (1998).
46. L. Wang and P.J. Steinhardt, *Astrophys. J.* **508**,, 483(1998).
47. R. Caldwell, R. Juskiewicz, P.J. Steinhardt and P. Bouchet, in preparation (1999).
48. W. Hu, D.J. Eisenstein, M. Tegmark, M. White, *Phys. Rev. D***59**, 023512 (1999).
49. I. Zlatev, L. Wang. R. Caldwell, and P.J. Steinhardt, to appear.
50. G. Huey, L. Wang, R. Dave, R. Caldwell, and P.J. Steinhardt, *Phys. Rev. D***59**, 063005 (1999).
51. I. Zlatev, L. Wang and P.J. Steinhardt, *Phys. Rev. Lett.* **82**, 896-899 (1999).
52. P.J. Steinhardt, L. Wang and I. Zlatev, *Phys. Rev. D***59**, 123504 (1999).
53. I. Zlatev and P.J. Steinhardt, to appear *Phys. Lett.* (1999).
54. R. Dave, Ph.D. thesis, U. of Penn. (1999).

55. B. Ratra and J.P.E. Peebles, *Astrophys. J.* ,, 325,L17 (1988); B. Ratra, and P.J.E. Peebles, *Phys. Rev. D***37**, 3406 (1988).
56. G. Huey and P.J. Steinhardt, in preparation.
57. I. Affleck, *et al.*, *Nucl Phys B* 256, 557 (1985).
58. C. Hill, and G.G. Ross, *Nuc Phys B***311**, 253 (1988).
59. C. Hill and G.G. Ross, *Phys Lett B***203**, 125 (1988).
60. P. Binetruy, M. K. Gaillard, and Y.-Y. Wu, *Nucl Phys B***481**, 109 (1996).
61. T. Barreiro, B. Carlos, and E. J. Copeland, *Phys. Rev. D* ,, 57,7354 (1998).
62. P. Binetruy, *Phys. Rev.D***60**, 063502 (1999).
63. S. Perlmutter, M.S. Turner, and M. White, *Phys. Rev. Lett.* **83**, 670 (1999).
64. G. Efstathiou, astro-ph/9904356.
65. S. Podariou and B. Ratra, KSUPT-99/6
66. T.D. Saini, S. Raychaudhury, V. Sahni, A.A. Starobinsky, astro-ph/9910231.
67. S. Carroll, *Phys. Rev. Lett.* **81** 3067 (1998).
68. For a review, see T. Damour, *Class. Quant. Grav.* **13**, 133 (1996); also, *Proceedings of the 5th Hellenic School of Elementary Particle Physics*, gr-qc/9606079.

Part III
THE COSMIC MICROWAVE SKY

CMB ANISOTROPIES AND THE DETERMINATION OF COSMOLOGICAL PARAMETERS

G EFSTATHIOU
Institute of Astronomy
Madingley Road
Cambridge CB3 OHA

1. Abstract

I review the basic theory of the cosmic microwave background (CMB) anisotropies in adiabatic cold dark matter (CDM) cosmologies. The latest observational results on the CMB power spectrum are consistent with the simplest inflationary models and indicate that the Universe is close to spatially flat with a nearly scale invariant fluctuation spectrum. We are also beginning to see interesting constraints on the density of CDM, with a best fit value of $\omega_c \equiv \Omega_c h^2 \sim 0.1$. The CMB constraints, when combined with observations of distant Type Ia supernovae, are converging on a Λ-dominated Universe with $\Omega_m \approx 0.3$ and $\Omega_\Lambda \approx 0.7$.

2. Introduction

The discovery of temperature anisotropies in the CMB by the COBE team (Smoot *et al.* 1992) heralded a new era in cosmology. For the first time COBE provided a clear detection of the primordial fluctuations responsible for the formation of structure in the Universe at a time when they were still in the linear growth regime. Since then, a large number of ground based and balloon borne experiments have been performed which have succeeded in defining the shape of the power spectrum of temperature anisotropies C_ℓ[1] up to multipoles of $\ell \sim 300$ clearly defining the first acoustic peak in the spectrum. Figure 1 shows a compilation of band power anisotropy

[1] The power spectrum is defined as $C_\ell = \langle |a_{\ell m}|^2 \rangle$, where the $a_{\ell m}$ are determined from a spherical harmonic expansion of the temperature anisotropies on the sky, $\Delta T/T = \sum a_{\ell m} Y_{\ell m}(\theta, \phi)$.

R.G. Crittenden and N.G. Turok (eds.), Structure Formation in the Universe, 179–189.

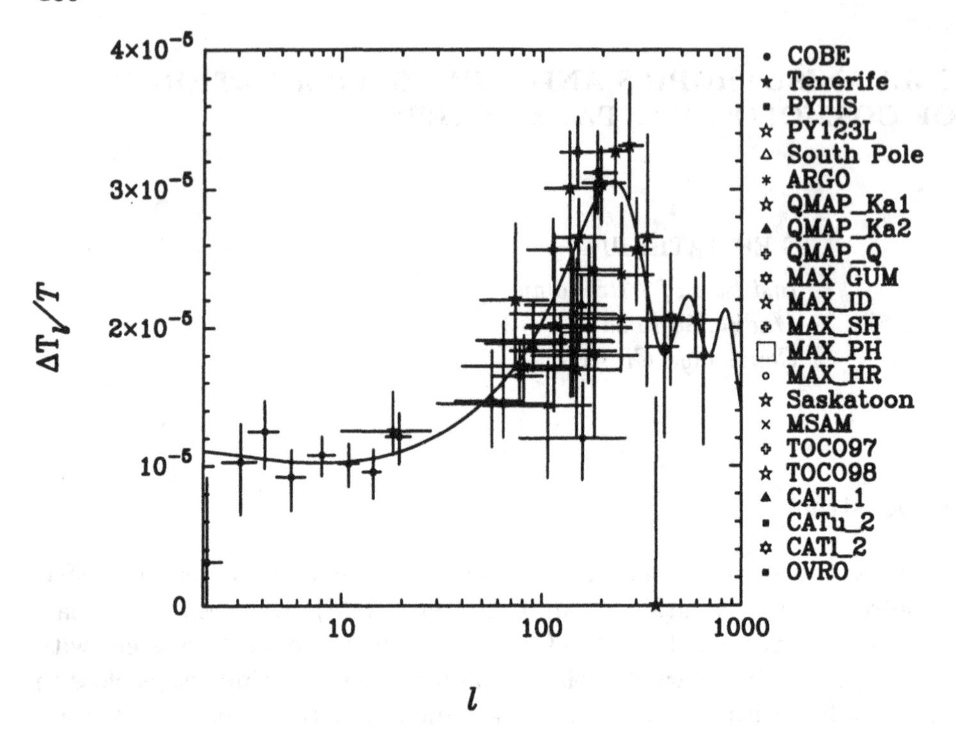

Figure 1. Current constraints on the power spectrum of CMB temperature anisotropies. The error bars in the vertical direction show 1σ errors in the band power estimates and the error bars in the horizontal direction indicate the width of the band. The solid line shows the best fit adiabatic CDM model with parameters $\omega_b = 0.019$, $\omega_c = 0.10$, $n_s = 1.08$, $Q_{10} = 0.98$, $\Omega_m = 0.225$, $\Omega_\Lambda = 0.775$.

measurements

$$\frac{\Delta T_\ell}{T} = \sqrt{\frac{1}{2\pi}\ell(\ell+1)C_\ell} \tag{1}$$

that is almost up to date at the time of writing. The horizontal error bars show the multipole range probed by each experiment. The recent results from the VIPER experiment (Peterson *et al.* 1999) and the Boomerang test flight (Mauskopf *et al.* 1999) are not plotted because the exact window functions are not yet publically available. Neither are the published results from the Python V experiment (Coble *et al.* 1999) which seem to be discrepant with the other experiments particularly in the multipole range $\ell \lesssim 100$. The points plotted in figure 1 are generally consistent with each other and provide strong evidence for a peak in the power spectrum at $\ell \sim 200$.

In this introductory article, I will review briefly the theory of CMB anisotropies in adiabatic models of structure formation and then discuss the implications of Figure 1 for values of cosmological parameters. The literature on the CMB anisotropies has grown enormously over the last few years and it is impossible to do the subject justice in a short article. General reviews of the CMB anisotropies are given by Bond (1996) and Kamionkowski and Kosowsky (1999). A recent review on constraining cosmological parameters from the CMB is given by Rocha (1999).

3. Basic Theory

Most of the key features of figure 1 can be understood using a simplified set of equations. The background universe is assumed to be spatially flat together with small perturbations h_{ij} so that the metric is

$$ds^2 = a^2(\tau)\,(\eta_{ij} + h_{ij})\,dx^i dx^j, \tag{2}$$
$$\eta_{ij} = (1,\ -1,\ -1,\ -1), \qquad \tau = \int dt/a.$$

We adopt the synchronous gauge, $h_{00} = h_{0i} = 0$, and ignore the anisotropy of Thomson scattering and perturbations in the relativistic neutrino component. With these assumptions, the equations governing the evolution of scalar plane wave perturbations of wavenumber k are

$$\dot{\Delta} + ik\mu\Delta + \Phi = \sigma_T n_e a\,[\Delta_0 + 4\mu v_b - \Delta] \tag{3a}$$
$$\Phi = -(3\mu^2 - 1)\dot{h}_{33} - (1 - \mu^2)\dot{h} \tag{3b}$$
$$\dot{v}_b + \frac{\dot{a}}{a}v_b = \sigma_T n_e a \frac{\bar{\rho}_\gamma}{\bar{\rho}_b}\left(\Delta_1 - \frac{4}{3}v_b\right), \tag{3c}$$
$$\dot{\delta}_b = \frac{1}{2}\dot{h} - ikv_b, \qquad \dot{\delta}_C = \frac{1}{2}\dot{h} \tag{3d}$$
$$\ddot{h} + \frac{\dot{a}}{a}\dot{h} = 8\pi G a^2(\bar{\rho}_b\delta_b + \bar{\rho}_c\delta_c + 2\bar{\rho}_\gamma\Delta_0) \tag{3e}$$
$$ik(\dot{h}_{33} - \dot{h}) = 16\pi G a^2(\bar{\rho}_b v_b + \bar{\rho}_\gamma\Delta_1). \tag{3f}$$

Here, Δ is the perturbation to the photon radiation brightness and Δ_0 and Δ_1 are its zeroth and first angular moments, δ_b and δ_c are the density perturbations in the baryonic and CDM components, v_b is the baryon velocity and $h = \mathrm{Tr}(h_{ij})$. Dots denote differentiation with respect to the conformal time variable τ. It is instructive to look at the solutions to these equations in the limits of large ($k\tau_R \ll 1$) and small ($k\tau_R \gg 1$) perturbations, where τ_R is the conformal time at recombination:

3.1. LARGE ANGLE ANISOTROPIES

In the limit $k\tau_R \ll 1$, Thomson scattering is unimportant and so the term in square brackets in the Boltzmann equation for the photons can be ignored.

In the matter dominated era $h_{33} = h \propto \tau^2$ and so equation (3a) becomes

$$\dot{\Delta} + ik\mu\Delta = 2\mu^2\dot{h} \tag{4}$$

with approximate solution

$$\Delta(k, \mu, \tau) \approx -\frac{2\ddot{h}(\tau_R)}{k^2} \exp\left(ik\mu(\tau_s - \tau)\right). \tag{5}$$

This solution is the Sachs-Wolfe (1967) effect. Any deviation from the evolution $\ddot{h}$ = constant, caused for example by a non-zero cosmological constant, will lead to additional terms in equation (5) increasing the large-angle anisotropies (sometimes referred to as the late-time Sachs-Wolfe effect, see *e.g.* Bond 1996). The CMB power spectrum is given by

$$C_\ell = \frac{1}{8\pi} \int_0^\infty |\Delta_\ell|^2 k^2 dk, \tag{6}$$

where the perturbation Δ has been expanded in Legendre polynomials,

$$\Delta = \sum_\ell (2\ell + 1)\Delta_\ell P_\ell(\mu). \tag{7}$$

Inserting the solution of equation (5) into equation (6) gives

$$C_\ell = \frac{1}{2\pi} \int_0^\infty \frac{|\ddot{h}|^2}{k^4} j_l^2(k\tau_0) k^2 \, dk, \tag{8}$$

and so for a power-law spectrum of scalar perturbations, $|h|^2 \propto k^{n_s}$, the CMB power spectrum is

$$C_\ell = C_2 \frac{\Gamma\left(\ell + \frac{(n_s - 1)}{2}\right) \Gamma\left(\frac{9 - n_s}{2}\right)}{\Gamma\left(\ell + \frac{(5 - n_s)}{2}\right) \Gamma\left(\frac{3 + n_s}{2}\right)} \tag{9}$$

giving the characteristic power-law like form, $C_\ell \propto \ell^{(n_s - 3)}$ at low multipoles ($\ell \lesssim 30$).

3.2. SMALL ANGLE ANISOTROPIES AND ACOUSTIC PEAKS

In the matter dominated era, equation (3a) becomes

$$\dot{\Delta} + ik\mu\Delta = \sigma_T n_e a \left[\Delta_0 + 4\mu v_b - \Delta\right] + 2\mu^2 \dot{h}, \tag{10}$$

and taking the zeroth and first angular moments gives

$$\dot{\Delta}_0 + ik\Delta_1 = \frac{2}{3}\dot{h} \tag{11a}$$

$$\dot{\Delta}_1 + ik\left(\frac{\Delta_0 + 2\Delta_2}{3}\right) = \sigma_T n_e a \left[\frac{4}{3} v_b - \Delta_1\right]. \tag{11b}$$

Prior to recombination, $\tau/\tau_c \gg 1$ where $\tau_c = 1/(\sigma_T n_e a)$ is the mean collision time, and so the matter is tightly coupled to the radiation. In this limit $\Delta_1 \approx 4/3 v_b$ from equation (3c) and Δ_2 in equation (11b) can be ignored. With these approximations, equation (11b) becomes

$$\dot{\Delta}_1 + \frac{ik\Delta_0}{3} = -\frac{\bar{\rho}_b}{\bar{\rho}_\gamma}\left[\frac{3}{4}\dot{\Delta}_1 + \frac{\dot{a}}{a}\Delta_1\right]. \tag{12}$$

Neglecting the expansion of the universe, equations (11a) and (12) can be combined to give a forced oscillator equation

$$\ddot{\Delta}_0 = -\frac{k^2}{3R}\Delta_0 + \frac{2}{3}\ddot{h}, \qquad R \equiv 1 + \frac{3\bar{\rho}_b}{4\bar{\rho}_\gamma}, \tag{13}$$

with solution

$$\Delta_0(\tau) = \left(\Delta_0(0) - \frac{2R\ddot{h}}{k^2}\right)\cos\frac{k\tau}{\sqrt{3R}} + \frac{\sqrt{3R}}{k}\dot{\Delta}_0(0)\sin\frac{k\tau}{\sqrt{3R}} + \frac{2R\ddot{h}}{k^2}, \tag{14}$$

where $\Delta_0(0)$ and $\dot{\Delta}_0(0)$ are evaluated when the wave first crosses the Hubble radius, $k\tau \sim 1$. For adiabatic perturbations the first term dominates over the second because the perturbation breaks at $k\tau \sim 1$ with $\dot{\Delta}_0 \approx 0$. It is useful to define (gauge-invariant) radiation perturbation variables

$$\tilde{\Delta}_0 = \Delta_0 - \frac{2\ddot{h}}{k^2}, \qquad \tilde{\Delta}_1 = \Delta_1 + i\frac{2\dot{h}}{3k},$$

then the solution of equation (10) is

$$\tilde{\Delta}(k,\mu,\tau) = \int_0^\tau \sigma_T n_e a\left[\tilde{\Delta}_0 + 4\mu\left(v_b + \frac{i\dot{h}}{2k}\right)\right] e^{ik\mu(\tau'-\tau) - \int_{\tau'}^{\tau}[\sigma_T n_e a]d\tau''}d\tau'. \tag{15}$$

If $k\tau \gg 1$, the second term in the square brackets is smaller than the first by a factor of $k\tau$, and the solution of equation (15) gives a power spectrum with a series of modulated acoustic peaks spaced at regular intervals of $k_m r_s(a_r) = m\pi$, where r_s is the sound horizon at recombination

$$r_s = \frac{c}{\sqrt{3}H_0\Omega_m^{1/2}}\int_0^{a_r}\frac{da}{(a + a_{equ})^{1/2}}\frac{1}{R^{1/2}}, \tag{16}$$

(Hu and Sugiyama 1995). Here a_{equ} is the scale factor when matter and radiation have equal densities and a_r is the scale factor at recombination.

The multipole locations of the acoustic peaks in the angular power spectrum are given by

$$\ell_m = \alpha m\pi\frac{d_A(z_r)}{r_s} \tag{17}$$

where α is a number of order unity and d_A is the angular diameter distance to last scattering

$$d_A = \frac{c}{H_0|\Omega_k|^{1/2}} \sin_k(|\Omega_k|^{1/2} x) \tag{18a}$$

$$x \approx \int_{a_r}^{1} \frac{da}{[\Omega_m a + \Omega_k a^2 + \Omega_\Lambda a^4]^{1/2}} \tag{18b}$$

where $\Omega_k = 1 - \Omega_\Lambda - \Omega_m$ and $\sin_k \equiv \sinh$ if $\Omega_k > 0$ and $\sin_k = \sin$ if $\Omega_k < 0$.

The general dependence of the CMB power spectrum on cosmological parameters is therefore clear. The positions of the acoustic peaks depend on the geometry of the Universe via the angular diameter distance of equation (18) and on the value of the sound horizon r_s. The relative amplitudes of the peaks depend on the physical densities of the various constituents $\omega_b \equiv \Omega_b h^2$, $\omega_c \equiv \Omega_c h^2$, $\omega_\nu \equiv \Omega_\nu h^2$, *etc.* and on the scalar fluctuation spectrum (parameterized here by a constant spectral index n_s). Clearly, models with the same initial fluctuation spectra and identical physical matter densities ω_i will have identical CMB power spectra at high multipoles if they have the same angular diameter distance to the last scattering surface. This leads to a strong *geometrical degeneracy* between Ω_m and Ω_Λ (*e.g.* Efstathiou and Bond 1999 and references therein). The power spectrum on large angular scales (equation 9) is sensitive to the spectral index and amplitude of the power spectrum, geometry of the Universe and, for extreme values of Ω_k can break the geometrical degeneracy via the late-time Sachs-Wolfe effect. We will discuss briefly some of the constraints on cosmological parameters from the current CMB data in the next section. Before moving on to this topic, I mention some important points that cannot be covered in detail because of space limitations:

• Inflationary models can give rise to tensor perturbations with a characteristic spectrum that declines sharply at $\ell \gtrsim 100$ (see *e.g.* Bond 1996 and references therein). In power-law like inflation, the tensor spectral index n_t is closely linked to the scalar spectral index, $n_t \approx n_s - 1$, and to the relative amplitude of the tensor and scalar perturbations.

• The anisotropy of Thomson scattering causes the CMB anisotropies to be linearly polarized at the level of a few percent (see Bond 1996, Hu and White 1997, and references therein). Measurements of the linear polarization can distinguish between tensor and scalar perturbations and can constrain the epoch of reionization of the intergalactic medium (Zaldarriaga, Spergel and Seljak 1997).

• The main effect of reionization is to depress the amplitude of the power spectrum at high multipoles by a factor of $\exp(-2\tau_{opt})$ where τ_{opt} is the optical depth to Thomson scattering. In the 'best fit' CDM universe described in the next section ($\omega_b = 0.019$, $h = 0.65$, $\Omega_m = \Omega_c + \Omega_b \approx 0.3$

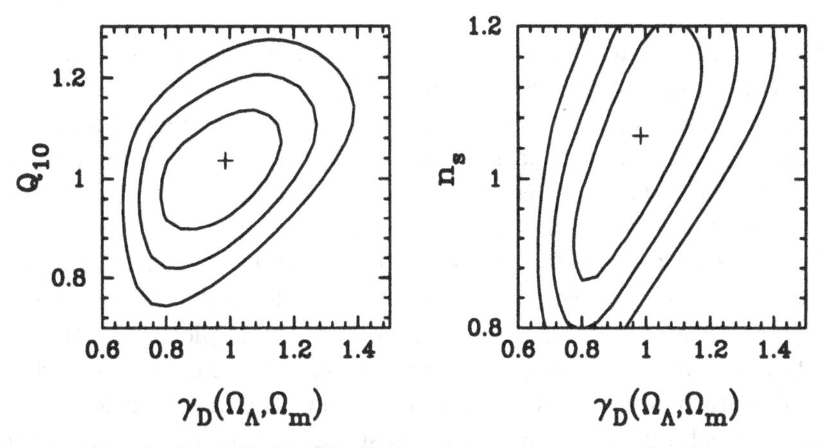

Figure 2. Marginalized likelihoods (1, 2 and 3σ contours) in the Q_{10}–γ_D and n_s–γ_D planes, where γ_D is the acoustic peak location parameter defined in equation 18.

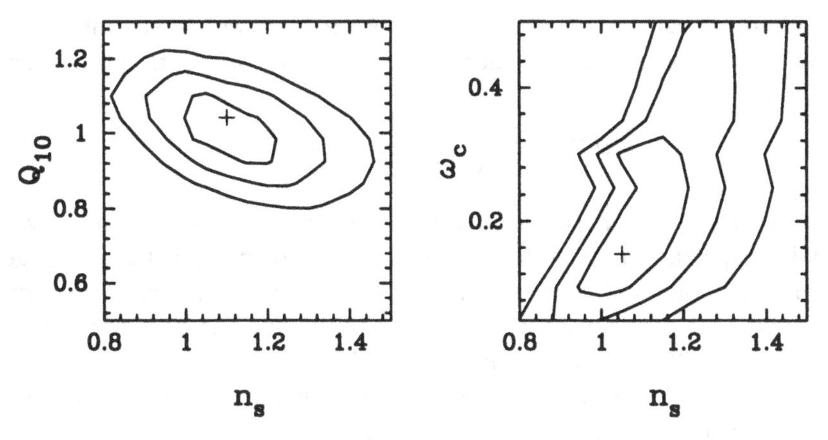

Figure 3. Marginalized likelihoods (1, 2 and 3σ contours) in the Q_{10}–n_s and ω_c–n_s planes. The crosses show where the likelihood function peaks.

and $\Omega_\Lambda \approx 0.7$) and a reionization redshift of $z_{reion} \approx 20$ (a plausible value) $\tau_{opt} \approx 0.2$ which is significant. There is a reasonable chance that we might learn something about the 'dark ages' of cosmic history from precision measurements of the CMB.

4. Cosmological Parameters from the CMB

In this section, we review some of the constraints on cosmological parameters from the CMB data plotted in figure 1. The analysis is similar to that presented in Efstathiou *et al.* (1999, hereafter E99), in which we map the full likelihood function in 5 parameters Ω_Λ, Ω_m, ω_c, n_s and Q_{10} (the amplitude of $\sqrt{C_\ell}$ at $\ell = 10$ relative to that inferred from COBE). The baryon density is constrained to $\omega_b = 0.019$, as determined from primordial nucleosynthesis and deuterium abundances measurements from quasar spectra (Burles and Tytler 1998). The results presented below are insensitive to modest variations ($\sim 25\%$) of ω_b and illustrate the main features of cosmological parameter estimation from the CMB. Recently, Tegmark and Zaldarriaga (2000) have performed a heroic 10 parameter fit to the CMB data, including a tensor contribution and finite optical depth from reionization. I will discuss the effects of widening the parameter space briefly below but refer the reader to Tegmark and Zaldarriaga for a detailed analysis.

The best fit model in this five parameter space is plotted as the solid line in figure 1. It is encouraging that the best fitting model has perfectly reasonable parameters, a spatially flat universe with a nearly scale invariant fluctuation spectrum and a low CDM density $\omega_c \sim 0.1$. Marginalised likelihood functions are plotted in various projections in the parameter space in figures 2, 3 and 5 (uniform priors are assumed in computing the marginalized likelihoods, as described in E99). Figure 2 shows constraints on the position of the first acoustic peak measured by the 'location' parameter

$$\gamma_D = \frac{\ell_D(\Omega_\Lambda, \Omega_m)}{\ell_D(\Omega_\Lambda = 0, \Omega_m = 1)}, \tag{19}$$

i.e. the parameter γ_D measures the location of the acoustic peak relative to that in a spatially flat model with zero cosmological constant. The geometrical degeneracy between Ω_m and Ω_Λ described in the previous section is expressed by $\gamma_D =$ constant. Figure 2 shows that the best fitting value is $\gamma_D = 1$ with a 2σ range of about ± 0.3. The position of the first acoustic peak in the CMB data thus provides powerful evidence that the Universe is close to spatially flat.

Figure 3 shows the marginalized likelihoods in the $Q_{10} - n_s$ and $\omega_c - n_s$ planes. The constraints on Q_{10} and n_s are not very different to those from the analysis of COBE alone (see *e.g.* Bond 1996). The experiments at higher multipoles are so degenerate with variations in other cosmological parameters that they do not help tighten the constraints on Q_{10} and n_s. The constraints on ω_c and n_s show an interesting result; if $n_s \approx 1$, then the best fit value of ω_c is about 0.1 with a 2σ upper limit of about 0.3. This constraint on ω_c comes from the height of the first acoustic peak, as shown in figure 4. In this diagram, the CMB data points have been averaged in 10 band-power estimates as described by Bond, Knox and Jaffe (1998). The solid curve shows the best-fit model as plotted in figure 1, which has $\omega_c = 0.1$. The dashed lines show models with $\omega_c = 0.25$ and $\omega_c = 0.05$ with the other parameters held fixed. Raising ω_c lowers the height of the peak and vice-versa. This result is not very sensitive to variations of ω_b in the neighbourhood of $\omega_b \sim 0.02$. Reionization and the addition of a tensor component can lower the height of the first peak relative to the anisotropies

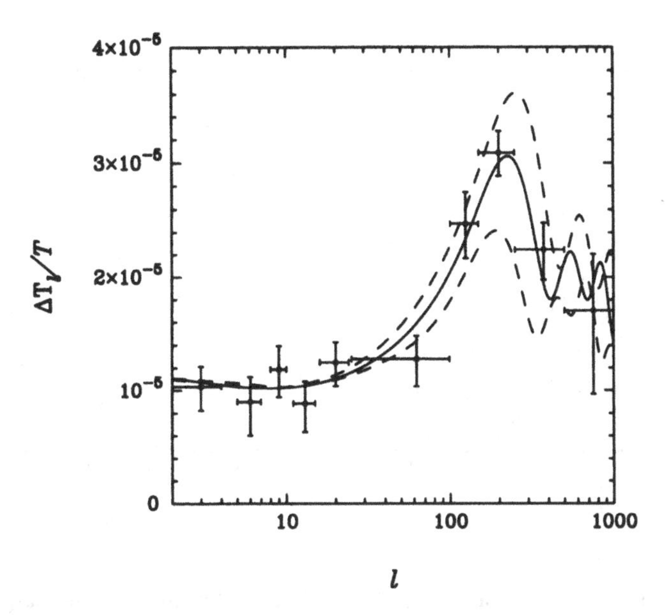

Figure 4. The crosses show maximum likelihood bandpower averages of the observations shown in figure 1 together with 1σ errors. The solid line shows the best fit adiabatic CDM model as plotted in figure 1 which has $\omega_c = 0.1$. The dashed lines show the effects of varying ω_c keeping the other parameters fixed. The upper dotted line shows $\omega_c = 0.05$ and the lower dashed line shows $\omega_c = 0.25$.

at lower multipoles and so the upper limits on ω_c are robust to the addition of these parameters. The CMB data have now reached the point where we have good constraints on the height of the first peak, as well as its location, and this is beginning to set interesting constraints on ω_c. The best fit value of $\Omega_m \approx 0.3$, derived from combining the CMB data with results from distant Type Ia supernovae (see figure 5) implies $\omega_c \approx 0.11$ for a Hubble constant of $h = 0.65$, consistent with the low values of ω_c favoured by the height of the first acoustic peak.

The left hand panel of figure 5 shows the marginalized likelihood for the CMB data in the Ω_Λ-Ω_m plane. The likelihood peaks along the line for spatially flat universes $\Omega_k = 0$ and it is interesting to compare with the equivalent figure in E99 to see how the new experimental results of the last year have caused the likelihood contours to narrow down around $\Omega_k = 0$. (See also Dodelson and Knox 1999 for a similar analysis using the latest CMB data). As is well known, the magnitude-redshift relation for distant Type Ia supernovae results in nearly orthogonal constraints in the Ω_Λ-Ω_m plane, so combining the supernovae and CMB data can break the geometrical degeneracy. The right hand panel in Figure 5 combines the CMB likelihood function derived here with the likelihood function of the supernovae sample of Perlmutter *et al.* (1999) as analysed in E99. The combined likelihood function is peaked at $\Omega_m \approx 0.3$ and $\Omega_\Lambda \approx 0.7$.

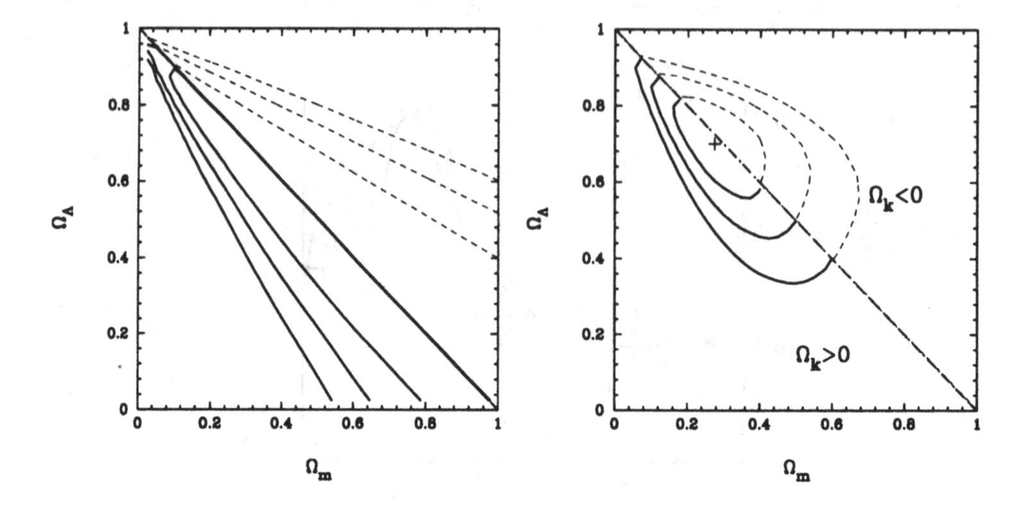

Figure 5. The figure to the left shows the 1, 2 and 3σ likelihood contours marginalized in the Ω_Λ and Ω_m plane from the observations plotted in figure 1. The figure to the right shows the CMB likelihood combined with the likelihood function for Type Ia supernovae of Permutter *et al.* (1999) as analyzed by E99. The dotted contours in both figures extend the CMBFAST (Seljak and Zaldarriaga 1996) computations into the closed universe domain using the approximate method described in E99.

It is remarkable how the CMB data and the supernovae data are homing in on a consistent set of cosmological parameters that are compatible with the simplest inflationary models and also with parameters inferred from a number of other observations (*e.g.* galaxy clustering, baryon content of clusters and dynamical estimates of the mean mass density, see Bahcall *et al.* 1999 for a review). It is also remarkable that the 'best fit' model requires a non-zero cosmological constant, a result that few cosmologists would have thought likely a few years ago.

The next few years will see a revolutionary increase in the volume and quality of CMB data. The results of the Boomerang Antarctic flight are awaited with great interest and should be of sufficient quality to render all previous analyses of cosmological parameters from the CMB obsolete. The polarization of the CMB has not yet been discovered, but a number of ground based and balloon borne experiments designed to detect polarization are under construction (Staggs, Gundersen and Church 1999). The MAP satellite, scheduled for launch in late 2000, will have polarization sensitivity and should determine the power spectrum C_ℓ accurately to about $\ell \sim 800$, defining the first three acoustic peaks. Further into the future, the Planck satellite, scheduled for launch in 2007, should determine the power spectrum to $\ell \gtrsim 2500$, provide sensitive polarization measurements and extremely accurate subtraction of foregrounds. Evidently, the era of precision cosmology is upon us and the next decade should see a dramatic improvement in our knowledge of fundamental cosmological parameters and in our understanding of the origin of fluctuations in the early Universe.

References

1. Bahcall N., Ostriker J.P., Perlmutter S., Steinhardt P.J., (1999) The cosmic triangle: revealing the state of the Universe, *Science*, **284**, 1481-1488.
2. Bond J.R., (1996) Theory and Observations of the Cosmic Microwave Background Radiation, in Cosmology and large scale structure, eds Schaeffer R., Silk J., Spiro M., Zinn-Justin J., Elsevier Science, Amsterdam, 469-666.
3. Bond J.R., Jaffe A.H., Knox L., (1998) Estimating the power spectrum of the cosmic microwave background, *Phys. Rev. D*,**57**, 2117-2137.
4. Burles S., Tytler D. (1998) The deuterium abundance towards QSO 1009+2956, *ApJ*, **507**, 732-744.
5. Coble K., *et al.* (1999) Anisotropy in the cosmic microwave background at degree angular scales, *ApJ*, **519**, L5-L8.
6. Dodelson S., Knox S., (1999) Dark energy and the CMB. astro-ph/9909454.
7. Efstathiou G., Bond J.R., (1999) Cosmic confusion: degeneracies among cosmological parameters derived from measurements of microwave background anisotropies, *MNRAS*, **304**, 75-97.
8. Efstathiou G., Bridle S.L., Lasenby A.N., Hobson M.P., Ellis R.S., (1999) Constraints on Ω_Λ and Ω_m from distant Type Ia supernovae and cosmic microwave background anisotropies, *MNRAS*, **303**, L47-52.
9. Hu W., Sugiyama N., (1995) Towards understanding CMB anisotropies and their implications, *Phys. Rev. D*, **51**, 2599-2630.
10. Kamionkowski M, Kosowsky A., (1999) The cosmic microwave background and particle physics, *Ann. Rev. Nucl. Part. Sci.*, in press. astro-ph/9904108.
11. Hu W., White M., (1997) A CMB polarization primer, *New Astronomy*, **2**, 323-344.
12. Mauskopf P.D., *et al.* (1999) Measurement of a peak in the cosmic microwave background power spectrum from the North American test flight of Boomerang, *ApJ*, submitted. astro-ph/9911444.
13. Perlmutter S. *et al.* (1999) Measurement of omega and lambda from 42 high-redshift supernovae, *ApJ*, **517**, 565-586.
14. Peterson J.B. *et al.* (1999) First results from VIPER: detection of small-scale anisotropy at 40 GHz. *ApJ*, in press. astro-ph/9910503.
15. Rocha G., (1999) Constraints on the cosmological parameters using CMB observations, to appear in the proceedings of the 'Early Universe and Dark Matter Conference', DARK98, Heidelberg. astro-ph/9907312.
16. Sachs R.K., Wolfe A.M., (1967) Perturbations of a cosmological model and angular variations of the microwave background, *ApJ*, **147**, 73-90.
17. Seljak U., Zaldarriaga M., (1996) A line-of-sight integration approach to cosmic microwave background anisotropies, *ApJ*, **469**, 437-444.
18. Smoot G.F., (1992) Structure in the COBE differential microwave radiometer first-year maps, *ApJ*, **396**, L1-L5.
19. Staggs S.T., Gundersen J.O., Church S.E., (1999) CMB polarization experiments. astro-ph/9904062.
20. Tegmark M., Zaldarriaga M., (2000) Current cosmological constraints from a 10 parameter CMB analysis *ApJ*, submitted. astro-ph/0002091.
21. Zaldarriaga M., Spergel D.N., Seljak U., (1997) Microwave background constraints on cosmological parameters, *ApJ*, **488**, 1-13.

MEASURING THE ANISOTROPY IN THE CMB: CURRENT STATUS & FUTURE PROSPECTS

L.A. PAGE
Dept of Physics, Princeton University
Princeton, New Jersey
http://physics.princeton.edu/~page

Abstract.
The CMB is perhaps the cleanest cosmological observable. Given a cosmology model, the angular spectrum of the CMB can be computed to percent accuracy. On the observational side, as far as we know, there is little that stands in the way between an accurate measurement and a rigorous confrontation with theory. In this article, we review the state of the data and indicate future directions.

The data clearly show a rise in the angular spectrum to a peak of roughly $\delta T_l = (l(l+1)C_l/2\pi)^{1/2} \approx 85\ \mu$K at $l \approx 200$ and a fall at higher l. In particular, δT_l at $l = 400$ is significantly less than at $l = 200$. This is shown by a combined analysis of data sets and by the TOCO data alone.

In the simplest open models with $\Omega_m = 0.35$, one expects a peak in the angular spectrum near $l = 400$. For spatially flat models, a peak near $l = 200$ is indicated and thus this model is preferred by the data. The combination of this, along with the growing body of evidence that $\Omega_m \approx 0.3$, suggests a cosmological constant is required. Further evidence for a cosmological constant is provided by the height of the peak. This conclusion is independent of the supernovae data.

1. Introduction

These notes are from two talks given at the Newton Institute in July 1999. The goal was to assess the status of CMB anisotropy measurements and give some indication of what the future holds. Given the extraordinarily rapid development of this field, this article is sure to be outdated soon after it appears. The program included a section on interferometers and

R.G. Crittenden and N.G. Turok (eds.), Structure Formation in the Universe, 191–213.

the data therefrom by Anthony Lasenby, on the physics of the CMB by George Efstathiou, and on data analysis by Dick Bond so I shall not discuss those matters here. Anthony Lasenby also covered work on ESA's Planck satellite.

My talks are biased toward the experiments I know best. Some of the experiments which I will discuss, in particular the MAT experiments (a.k.a. TOCO97[59] and TOCO98[42]), were done by a collaboration between Mark Devlin's group at the University of Pennsylvania and the Princeton group.

It would be stunning if the currently popular model survived to be our favorite model of the universe in a few years. Recent panoramic assessments of cosmological data [1] [60] suggest the universe is made of $\Omega_b \approx 0.05$, $\Omega_{cdm} \approx 0.3$, & $\Omega_\Lambda \approx 0.65$[1]. In other words, only 5% of the universe is made of something with which we are familiar.

There are three classes of observations that lead to the current picture. The supernovae data indicate that the universe is accelerating and thus need something like a cosmological constant to explain them. Secondly, the mass density, $\Omega_m = \Omega_b + \Omega_{cdm}$, as inferred from galactic velocities, cluster abundances, cluster x-ray luminosities, the S-Z effect in clusters, the cluster mass to light ratio, etc. is $\Omega_m \approx 0.35$. Thirdly, the CMB data suggest that the universe, within the context of adiabatic cold dark matter models, is spatially flat [1] [9]. The CMB data are improving rapidly. New data (TOCO97, TOCO98, CAT[2]), since [1], [9], strongly disfavor the nominal open spatial geometry models, and the case is getting tighter by the month [35]. We should point out that the position of the first peak does not *prove* the universe is spatially flat; there is enough wiggle room with the other parameters even within the limited context of adiabatic CDM models [34], but a spatially flat model is the simplest explanation when one assumes prior knowledge of other parameters such as H_0.

2. The temperature of the CMB

In 1990 John Mather and colleagues [39], using the Far infrared Spectrophotometer (FIRAS) aboard the COBE satellite, showed that the CMB is a blackbody emitter over the frequency range of 70 to 630 GHz. It is perhaps the best characterized blackbody. A recent analysis [40] gives the temperature as T$= 2.725 \pm 0.002$ K (95% confidence). The error, 2 mK, is entirely systematic and so it is difficult to assign a precise confidence limit. The statistical error is of order 7 μK. This measurement was quickly followed by the UBC rocket experiment [31] which found T$= 2.736 \pm 0.017$ K (1σ).

[1]The b subscript stands for baryons, cdm for cold dark matter, and Λ for a cosmological constant type term.

At frequencies greater than 90 GHz, the FIRAS measurement will not be bettered without another satellite. At lower frequencies, the measurements are less precise and there is plenty of room for improvement. The best long wavelength measurement is by Staggs et al. [54] at 11 GHz. They find $T = 2.730 \pm 0.014$ K. Deviations from a pure thermal spectrum are expected to show up near a few GHz. In addition, we know the universe was reionized at $z \approx 5$, so there should be remnant free-free emission, also at a few GHz. Unfortunately, near these frequencies, our Galaxy emits about 2 K making a 0.01% determination of the CMB temperature difficult, to say the least. Our picture of the spectrum of the CMB will not be complete until the long wavelength part of the spectrum is known though only a few groups are seriously considering these tough experiments.

3. The unbiased anisotropy spectrum

Figure 1 shows all the anisotropy data that has at least made it into preprint form (as of this writing, Oct. 99, all of it has been accepted for publication). Many of the data points have not been confirmed or are essentially unconfirmable; others have large calibration errors; some data sets comprise sets of correlated points; still others have foreground contamination. Despite this, the trend is clear. From the Sachs-Wolfe plateau discovered by COME/DMR [53] there is a rise to an amplitude of $\delta T_l \approx 85$ μK at $l \approx 200$ and a fall after that.

It is worth reviewing what sort of systematic checks we have between different experiments. At both large and small angular scales, the spectrum of the anisotropy is seen to be thermal. Also, at the largest angular scales, there is a clear correlation between DMR at 53 GHz and the FIRS data at 180 GHz [28]. In this analysis, the dust contribution to FIRS was subtracted though inclusion of it did not significantly alter the results: the dust is not correlated with the CMB. At smaller angular scales, SK at 35 GHz[43] saw the same signal as did the MSAM experiment at 200 GHz. In an analysis tour de force, Fixsen et al. [27] show that the COBE/FIRAS instrument–remember FIRAS is an absolute measurement–sees the same anisotropy as the COBE/DMR instrument which is a differential microwave radiometer. A plot of the cross correlation is consistent with a thermal spectrum from 90 to 300 GHz.

Outside of the above measurements, teams have not confirmed each others' findings. The reason is that it is difficult to match scan strategies from different instruments. This is one of the reasons that maps are desired. There are preliminary indications that the SK and QMAP maps agree [22] as do the maps from two seasons of PYTHON [13].

Figure 2 show all the data from Figure 1 binned into ten l-space bins.

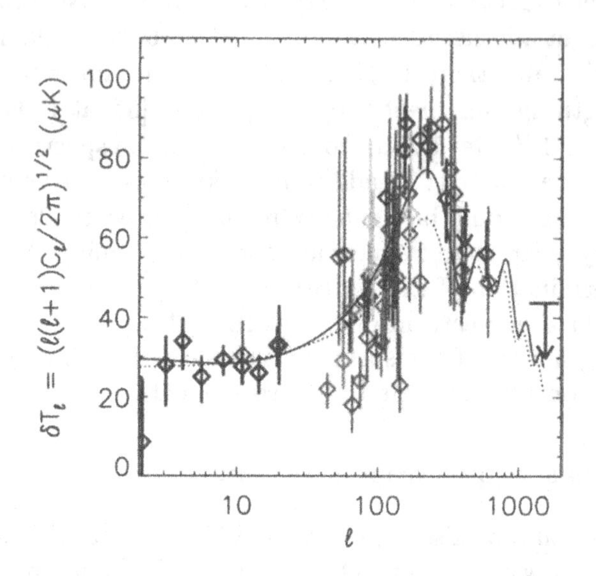

Figure 1. Unbiased sampling of data. The solid line on top is a model from Wang *et al.* (1999) with $\Omega_b = 0.05$, $\Omega_{cdm} = 0.3$, $\Omega_\Lambda = 0.65$, and $h = 0.65$. The dotted curve is the "standard cold dark matter" model, which is inconsistent with many non-CMB observations, with $\Omega_b = 0.05$, $\Omega_{cdm} = 0.95$, $\Omega_\Lambda = 0.0$, and $h = 0.5$. For the PYTHON 5 data we use the data from K. Coble's thesis (Coble 1999) rather than those from the paper.

The plot is remarkable and gives us faith in the hot big bang model. The rise of the angular spectrum and the location of the peak for a spatially flat universe were predicted well in advance of the measurements. It is also satisfying that these data have shown that a number of alternative models simply do not work. For instance, large classes of isocurvature models do not fit the data (but by no means is the isocurvature mechanism excluded), simple open models do not fit the data [35], [25], and a broad class of defect models do not fit the data [45].

4. The observational setting & foreground emission

The CMB is a 2-D random field in temperature with variance of order (115 μK)^2. If one could measure the anisotropy with sharp filters in l-space, one would find the *rms* variations for l between $2 < l < 40$ to be 54 μK, between $40 < l < 400$ to be 88 μK, and between $400 < l < 1500$ to be 53 μK. So far, the data are consistent with a Gaussian temperature distribution.

Characterizing the anisotropy is challenging because one wishes to mea-

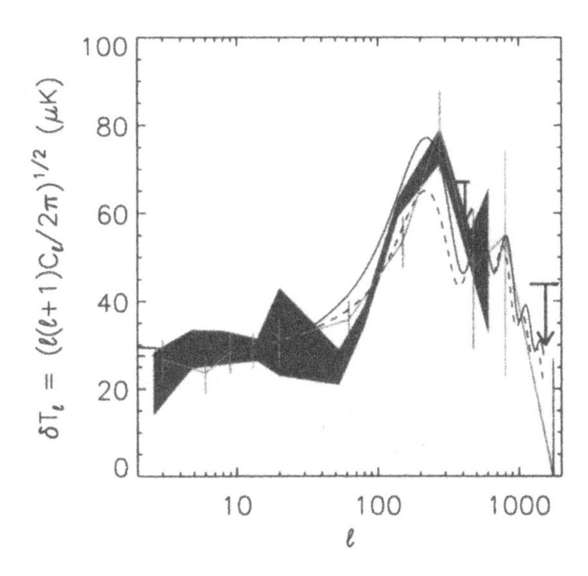

Figure 2. All the data from Figure 1 binned into ten logarithmically spaced bins. There is no accounting for calibration error, correlations, etc. The upper limits are at 95% confidence. The width of the blue swath is the statistical weight of the data that land in the corresponding bin. The orange line is a more sophisticated analysis by Bond *et al.* (1999) that uses a subset of the data in Figure 1.

sure accurately micro-Kelvin variations from an experiment sitting on, or just above, a 300 K Earth. Nature, though, has been kind. The CMB is the brightest thing in the sky between 0.6 and 600 GHz; and fluctuations from emission from our Galaxy (for galactic latitudes $|b| > 20°$) are smaller than the fluctuations intrinsic to the CMB [57], as shown in Figure 3. We do not yet know if we shall be so fortunate with the polarization, but low frequency measurements suggest this may be the case [17].

We may get a sense of the scale of the corrections for foreground emission from the SK data. SK observed near the North Celestial Pole at $b = 25°$. The contribution to the original data set from foreground emission is 4% at 40 GHz [20]. It turns out the contamination was not due to free-free emission, Haslam-like synchrotron emission, or extra galactic sources, but rather was due to a component correlated with interstellar dust emission. The favorite current explanation is that this component is due to radiation by spinning dust grains [26]. This component was not expected when the experiment was conceived. In the QMAP experiment, the contamination at Ka ($\approx$ 25 GHz) is $\approx$ 8% [23]; no significant contribution to the Q band data was measured. Coble *et al.* [12], looking in the Southern Hemisphere at high Galactic latitudes, found effectively no contamination at 40 GHz.

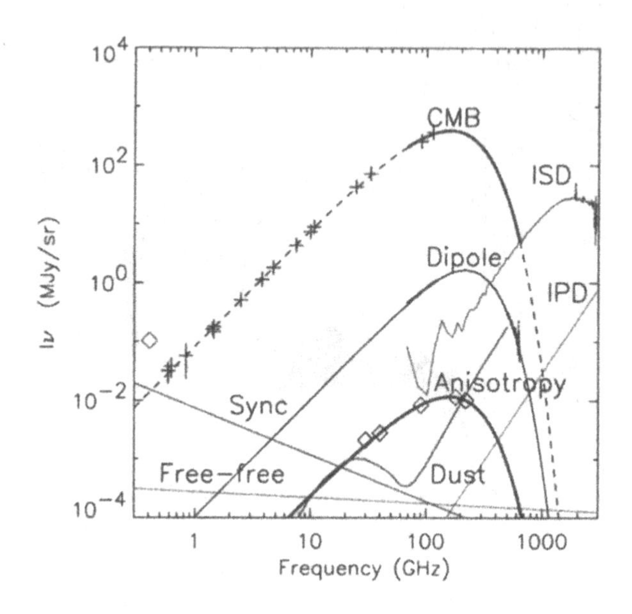

Figure 3. Plot of the CMB and the foreground emission at approximately the galactic latitude of the North Celestial Pole. For dust and synchrotron spectra, the fluctuating component has been plotted (de Oliveira-Costa *et al.* 1998) The flux levels of the free-free and dust emission have also been plotted. The corresponding l is about 20. At higher l, all these foregrounds have less fluctuation power. The FIRAS data (dust, dipole, and CMB spectra) are courtesy of Bill Reach.

5. Types of measurements

The scientific payoff from the CMB has motivated a large number of experiments; over twenty groups are trying to measure the anisotropy. The frequency coverage is large. The experiments and detector technologies are summarized in [3], [32], & [44].

Here we just note that there are three general classes of measurements. They are (1)beam switching or beam synthesis experiments, (2)direct mapping experiments, and (3)interferometers. By far the most data have come from the beam switching/synthesis method but this is certain to change soon.

The detectors of choice are high electron mobility transistor amplifiers (HEMTs [48], [49]) for frequencies below 100 GHz and bolometers for higher frequencies. The primary advantage of HEMTs is their ease of use and speed. A typical HEMT sensitivity is 0.5 $\mathrm{mKs}^{1/2}$. The advantage of bolometers is their tremendous sensitivity, e.g. < 0.1 $\mathrm{mKs}^{1/2}$. The CMB anisotropy has also been detected with SIS mixers[36].

Over the past year, five new results from experiments have come out of which I am aware. They are the IAC[18], CAT, QMAP, TOCO, and PYTHON 5. These data span from 30 to 150 GHz and from $l = 50$ to 400. They support the picture given in [9]. As an example of a beam switching experiment, I'll use TOCO; and as an example of a mapping experiment I'll use QMAP.

6. QMAP

The QMAP experiment is described in a trio of papers ([24], [33], & [20]). The purpose of QMAP, which was proposed before MAP, was to make a "true" map of the sky at 30 and 40 GHz. By true map we mean a map that is simply described by a temperature and temperature uncertainty per pixel. Ideally, the pixel to pixel covariance matrix is diagonal. For QMAP, this was not the case and the full covariance matrix was required. Maps that are reconstructed from beam switching measurements, for instance the SK[56], MSAM[37], MAX[62], and PYTHON 5[13] maps, are not true maps and cannot be analyzed as maps.

QMAP is a direct mapping experiment. The data stream is converted directly into a map and the covariance matrix is computed from the data. So far, only four experiments have directly mapped the CMB anisotropy (COBE/FIRAS, COBE/DMR, FIRS, and QMAP). QMAP, so far, has the highest S/N per pixel. We can look forward to the BOOMERanG [10] and MAXIMA [41] data which, with their tremendous detectors, should produce much higher sensitivity maps.

The key to making a map is to "connect each pixel with all the ones around it." [55] [65]. In simplest terms, one wants to sit at a pixel and know the derivatives in each direction. QMAP accomplished this connectedness by observing above the North Celestial Pole and letting the sky rotate through the beam. The scan lines thus intersect at a variety of angles [24]. Maps that are reconstructed from temperature differences (e.g. from beam switching experiments) do not generally have this property because the differences are all done at constant elevation (SK is an exception; again the rotation around the North Celestial Pole was used.)

The QMAP power spectrum is given in Table 1. With the two flights, and six channels per flight a number of cross checks can be made. The data common to both flights and between channels within one flight are consistent. In a chi-by-eye, the QMAP map looks very similar to the reconstructed SK map in areas where they overlap [22].

The QMAP data are extremely clean. There is essentially no editing of spurious points etc. One simply takes the data, calibrates it, removes a slowly varying offset (this is done self-consistently in the map solution) and produces the map. It is the type of data set for which the analysis pipeline

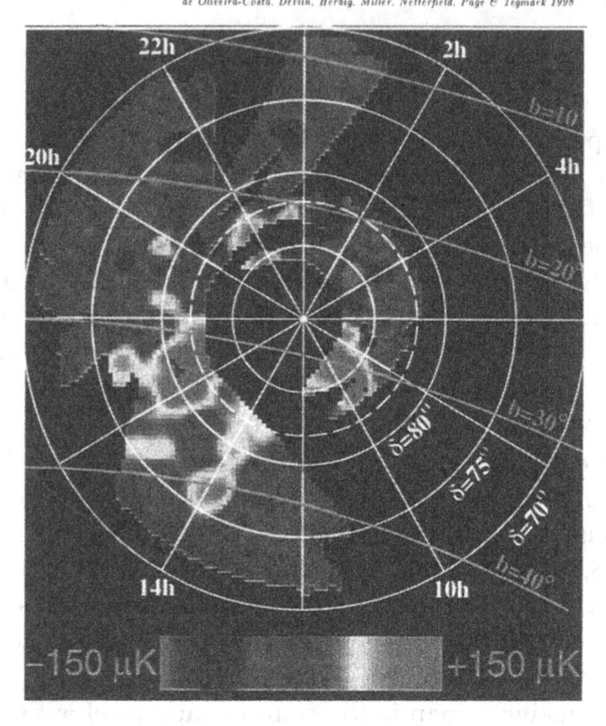

Figure 4. Results of two QMAP flights. There are multiple spots in the map with a signal to noise of 10 to 20. SK fills the region within $\delta = 82°$ shown by the dashed line. This plot shows a Weiner filtered version of the raw map. This is what the anisotropy looks like at degree angular scales.

could have been written before the experiment.

7. MAT/TOCO

The MAT/TOCO experiment is a collaboration between Mark Devlin's group at Penn and the Princeton group. We took the QMAP gondola and optics, changed the cooling from liquid helium to a mechanical refrigerator, and mounted the telescope on a Nike Ajax radar trailer. For two seasons (Oct.-Dec. 1997 and Jun.-Dec. 1998) we observed from Cerro Toco near the

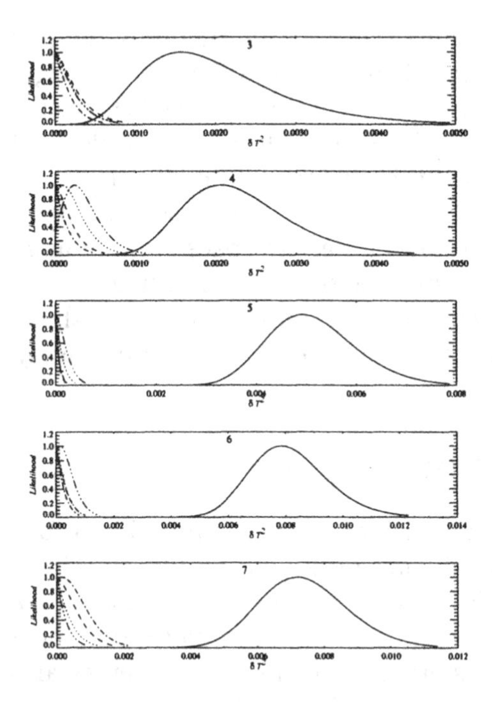

Figure 5. The likelihood of the data and the likelihood of the null combinations of the TOCO97 data set. The four null signals are data taken with the chopper scanning one way minus data with the chopper scanning the other, differences between subsequent 0.25 *s* segments, differences between subsequent 5 *s* segments, and the first half minus the second half of the campaign.From top to bottom, the panels correspond to $l = 63, 86, 114, 158, \& 199$ as given in Table 1.

ALMA site in the northern Chilean Andes.[2]

In the first season (TOCO97), all the HEMT channels worked but the two SIS channels did not. The problem with the SISs was fixed for the second season (TOCO98). Thus we cover from $l = 60$ to $l = 400$ in multiple frequency bands. In the field we were plagued by refrigerator problems. This resulted in a higher SIS temperature and thus lower sensitivity than we expected from the laboratory measurements.

In this sort of experiment, one must deal with the variable atmospheric temperature and variable local temperature. The data must be edited and one must go to great measures to ensure that the editing does not bias the answer. Of central importance is the correct assessment of the instrument noise. As δT_l^2 is proportional to the measured variance minus the instrument variance, an incorrect assessment of the noise will bias the result (see, for example, [43] and [59] for discussions).

[2]The Cerro Toco site of the Universidad Católica de Chile was made available through the generosity of Professor Hernán Quintana, Dept. of Astronomy and Astrophysics.

The straight forward way to make sure that the noise is understood is to make combinations of the data in which the sky signal is cancelled out. The analysis of such a null signal should yield the instrument noise. These null combinations should cover multiple time scales and spatial scales. As an example, we show the null tests from the TOCO97 data in Figure 5 (for TOCO98 see [42]). Note that in all cases, the signal is well above the instrument noise and that for each l-space bin, the null tests, regardless of time scale, give consistent noise levels. Additionally, the ratio of the noise between l-space bins can be computed; it agrees with the data. The results from both campaigns are in Table 1. There are roughly 100 more days of 30 and 40 GHz data to analyze.

8. SK/QMAP/TOCO (SQT)

I've tried to come up with an easy-to-state criteria for selecting data sets for a compendium of solid results above $l = 50$. Qualifications for entry onto this list would include confirmation of results, maturity of analyses (of order 30% of the reported data has undergone some sort of reanalysis after publication resulting in significantly different answers), some check for foreground contamination, internally consistent data, and measurements that include internal consistency checks of the data quality and noise levels. I have not been successful. As any caveated selection runs the hazard of being biased toward results that agree with our results, I will use instead a completely subjective criterion. Namely, experiments with which I have been involved over the past few years. The list is given in Table 1.

The SQT data span from 30 to 150 GHz, use different calibrators, involve different analysis packages, different radiometers, different platforms, and different observing strategies.

Except for the QMAP point at $l = 126$ which is somewhat correlated with the other QMAP points, these data can be considered uncorrelated. At low l, all data sets are sample variance limited. The sky coverage varies quite a bit. For SK it is 200 deg^2, for QMAP 530 deg^2, for TOCO97 600 deg^2, and for TOCO98 500 deg^2.

What is the significance of the down turn from the peak near $l = 200$? The first thing to note is that the D-band data points (TOCO98 in Figure 6) are essentially uncorrelated. There are two data points below the maximum. The net effect is a 5σ detection of a fall from just the TOCO98 data alone. In addition, including the last (low) SK point enhances the probability of a downturn. In other words, from just these data, the down turn is indisputable. and of course there are many other experiments as shown in Figure 1.

In Table 1 we also give the recent analysis of the MSAM data [63]. The

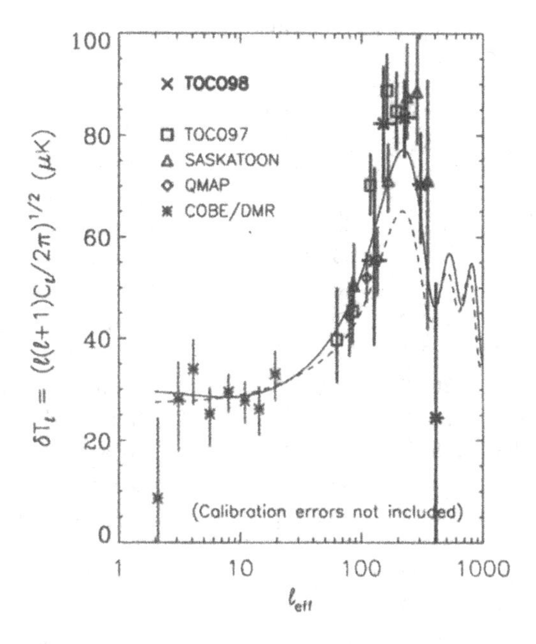

Figure 6. The SK/QMAP/TOCO and COBE/DMR data. For the highest l point, we show the data point using the convention of Knox. The models are the same as in Figure 1. Note that the peak is clearly not at $l = 400$.

MSAM point at $l = 200$ is about 3σ below the mean of the SQT points. Interestingly, the data comprising this point come from a section of sky examined in two MSAM flights [37] and on the ground in the SK experiment [43]. My interpretation is that this is a statistical fluke. MSAM covers only of order 10 $\deg^2$ of sky. The experiments are well enough documented to check this against the MAP data.

9. Finding the peak

Once we have the power spectrum we can either fit the data to models or we can look for nearly model independent parametrizations. Directly fitting to models has been done by a number of groups (e.g. [1], [4], [8], [25], [38], [50], & [58]) In broad brushstrokes, the data are consistent with spatially flat models with a cosmological constant and inconsistent with spatially open models (though this conclusion depends on the selection of data sets [50]). In the following, we give a model independent assessment of the position of the peak.

The richness of the CMB is the very thing that makes it so difficult to fit. In the current stage, one cannot look at the spectrum and get simple

TABLE 1. Selected data. Calibrations errors ($\approx$ 10%) are not included.

Name	l_{eff}	δT_l (μK)	Comments & Reference
TOCO97	63^{+18}_{-18}	$39.7^{+10.3}_{-6.5}$	Torbet et al.
QMAP	80^{+41}_{-41}	$44.3^{+6.7}_{-7.9}$	Foreground subtracted
MSAM	84^{+46}_{-45}	35^{+15}_{-11}	Foreground subtracted
TOCO97	86^{+16}_{-22}	$45.3^{+7.0}_{-6.4}$	Torbet et al.
SK	87^{+39}_{-27}	$50.5^{+8.4}_{-5.2}$	Foreground subtracted
QMAP	111^{+64}_{-64}	$52.0^{+5.0}_{-5.0}$	Foreground subtracted
TOCO97	114^{+20}_{-24}	$70.1^{+6.3}_{-5.8}$	Torbet et al.
QMAP	126^{+54}_{-54}	$55.6^{+6.4}_{-7.2}$	Foreground subtracted
TOCO98	128^{+26}_{-33}	$54.6^{+18.4}_{-16.6}$	Miller et al.
TOCO98	155^{+28}_{-38}	$82.0^{+11.0}_{-11.0}$	Miller et al.
TOCO97	158^{+22}_{-23}	$88.7^{+7.3}_{-7.2}$	Torbet et al.
SK	166^{+30}_{-43}	$71.1^{+7.3}_{-6.3}$	Foreground subtracted
TOCO97	199^{+38}_{-29}	$84.7^{+7.7}_{-7.6}$	Torbet et al.
MSAM	201^{+82}_{-70}	49^{+10}_{-8}	Foreground subtracted
TOCO98	226^{+37}_{-56}	$83.0^{+7.0}_{-8.0}$	Miller et al.
SK	237^{+29}_{-41}	$87.6^{+10.5}_{-8.4}$	Foreground subtracted
SK	286^{+24}_{-36}	$88.6^{+12.6}_{-10.5}$	Foreground subtracted
TOCO98	306^{+44}_{-59}	$70.0^{+10.0}_{-11.0}$	Miller et al.
SK	349^{+44}_{-41}	$71.1^{+19.9}_{-29.4}$	Foreground subtracted
MSAM	407^{+46}_{-123}	47^{+7}_{-6}	Foreground subtracted
TOCO98	409	< 67 (95%*conf*)	Miller et al

answers. There are multiple sets of parameters that give rise to the same power spectrum if only a certain region of l-space is covered [6]. This degeneracy is broken by including other data sets into the analysis; for instance, one may assume prior knowledge of the Hubble constant or the baryon density.

In an effort to say where the peak is, we have parametrized a generic spectrum by taking the lambda model in Figure 1[61], normalizing it to $\delta T_l = 32$ at $l = 25$ and stretching it in l and changing the amplitude while maintaining the normalization. (To my knowledge, this was done first by Barth Netterfield to the SK data. At that time, we found that the SK peak preferred h=0.35, a cosmological constant, or lots of baryons. Basically, anything to make the peak higher than the sCDM peak.) We then compute the likelihood as a function of l and amplitude of the peak. The results are shown in Figure 7 for both the SQT and all the data in Figure 1.

To show what the data prefer, we have done a simple χ^2 analysis assum-

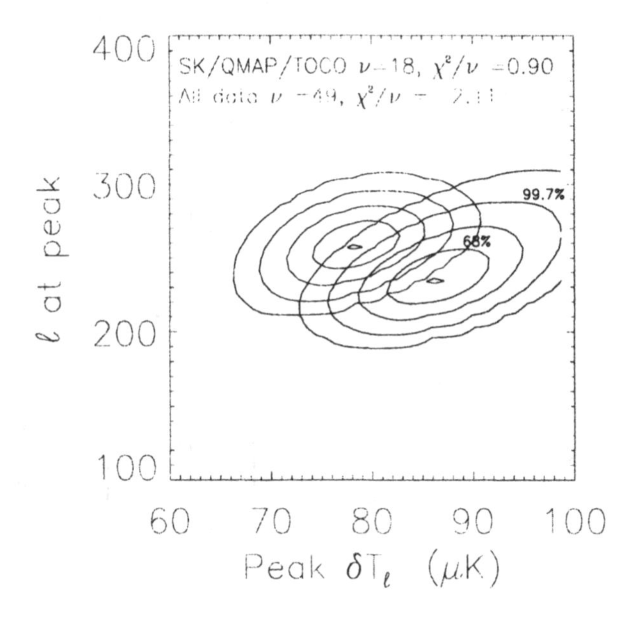

Figure 7. The peak position and amplitude for the SK/QMAP/TOCO data and all the data in Figure 1. DMR was not included. Note that the reduced chi-squared for SQT is completely consistent with statistical uncertainty. Note also that a peak near $l = 400$ is very unlikely. Calibration error shifts the contours left and right.

ing a spatially flat universe and limiting ourselves to variations of Ω_Λ, Ω_b and Ω_{cdm} with h=0.65. The SQT data are well described by $\Omega_b = 0.05$, $\Omega_{cdm} = 0.2$, & $\Omega_\Lambda = 0.75$. If one wants to explain the height of the peak without a cosmological constant, then one must have something like $\Omega_b = 0.12$, $\Omega_{cdm} = 0.88$, clearly at variance with a large body of cosmological data.

We have reached the point where there are well developed techniques for measuring and quantifying the anisotropy from balloons and the ground that give reliable and consistent results. This situation will continue to improve with the large interferometers and highly sensitive balloon data coming on line now. The current limitations to the experiments are 1)calibration, 2)sky coverage and 3)knowledge of the beam. The dominant error for the combined SQT data is the calibration. We find that the amplitude of the peak between $l = 150$ and $l = 250$ is $\delta T_l = 82 \pm 3.3 \pm 5.5$ μK, the first error is statistical and the second error is from the calibration. A typical intrinsic calibration source accuracy is 5%. Secondly, without full sky coverage, the ultimate error bars cannot be achieved. For instance, if one maps only a quarter the full sky, the error bars will be twice as large as

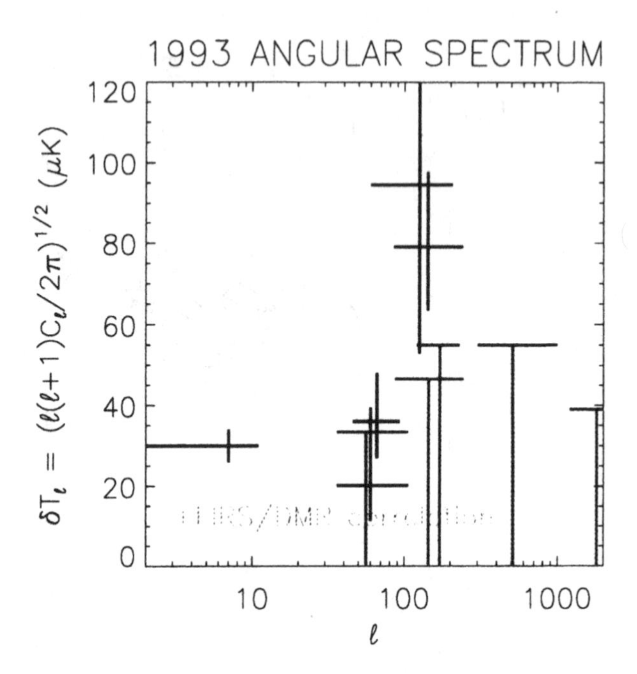

Figure 8. Plot of the CMB angular spectrum of data published in 1993 and before. The upper limits are all at 95% confidence. No measurement between $l = 100$ and $l = 1000$ has been confirmed.

potentially achievable. Finally, there is no reason that the beams cannot be known to high accuracy. The problem is that they must be measured *in situ* and these measurements can be difficult. To the accuracy needed, beams can no longer simply be modelled by two dimensional Gaussian profiles.

10. The past

It is amusing to look back and see how far we have come in the past six years. Figure 8, is a plot of all the data that were published in or before 1993 with sensitivities at an interesting level. I've converted the reported limits on the Gaussian autocorrelation function to the modern band powers using [5] and [30]. The full power spectrum of DMR had not been published at this time. Based on these data, one would be surprised by the models that currently fit so well.

There have been tremendous advances on the theoretical front as well. Model predictions have been brought to the masses through CMBFAST [52]; textures and similar mechanisms are generally believed to be inconsistent with the current data [45], and there are many classes of isocurvature models that are no longer viable. As the theories improve, some of these models may arise again or we may find that the anisotropy is an admixture of adiabatic, isocurvature, and texture perturbations.

11. The future

At $l < 1000$ the future is in multielement interferometers [11] [16], long duration balloon flights e.g ([10]), and satellite missions. At $l < 20$, the sky can only be mapped precisely from a satellite. For $l > 1000$, measurements can be made from the ground. Arrays of bolometers and interferometers seem ideal. The polarization has yet to be detected but experiments coming on line now should be able to do the job within a year or so.

There are now four space missions on the books. There is NASA's MAP satellite which just had its Launch-1 year review, there is ESA's Planck satellite (with collaborative NASA support) which is scheduled for a 2007 launch, there is the SPORT mission which plans to measure the polarization of the CMB at HEMT frequencies from the space station [14], and, in NASA's technological road map, there is a mission, CMBPOL, to measure the polarization of the CMB in ≈ 2015. The later is in the talking phase [46]. I shall focus on MAP and Anthony Lasenby will focus on Planck.

12. MAP

The primary goal of MAP[3] is to produce a high fidelity, polarization sensitive, full sky map of the cosmic microwave background anisotropy. From its inception, the focus has been on how one makes a full sky map with negligible systematic error. MAP is a MIDEX mission which means there is minimal redundancy as well as firm cost and schedule caps. MAP was proposed in June 1995, selected in April 1996, and is planned for a November 2000 launch. It was also proposed with the notion of getting the data to the community as fast as possible. We plan to make maps public nine months after we scan the whole sky (roughly one year after getting to L2).

Sources of systematic error in mapmaking include "1/f noise" (any variations from tenths of seconds to minutes) in the detectors and instrument, magnetic fields, and sidelobe contamination. Our systematic error budget allows for a total of 5 μK of extraneous signal *before* any modelling of the source of the contamination. The 5 μK applies to all angular scales and time scales though the most troublesome sources are the ones that are synchronous with the spin period.

[3]MAP is a collaboration between NASA (Chuck Bennett [PI], Gary Hinshaw, Al Kogut, & Ed Wollack), Princeton (Norm Jarosik, Michele Limon, Lyman Page, David Spergel & David Wilkinson), Chicago (Steve Meyer), UCLA (Ned Wright), UBC (Mark Halpern), and Brown (Greg Tucker)

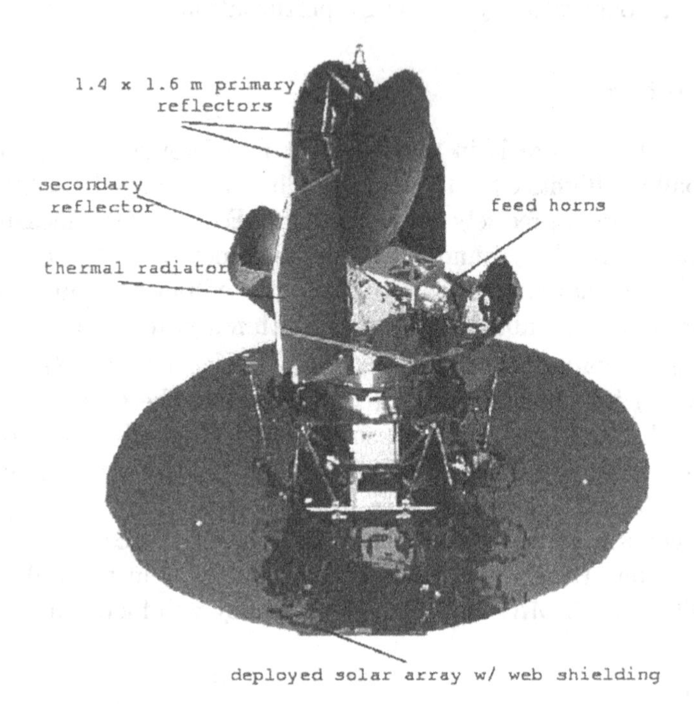

Figure 9. Picture of MAP with the solar arrays deployed. The receivers are located directly beneath the primary mirrors and are cooled by the thermal radiators. The Earth, Moon, and Sun are beneath the solar arrays.

MAP is being documented on the web as the instrument is built and tested. Official information may be obtained at http://map.gsfc.nasa.gov/; technical papers are in preparation.

12.1. MISSION OUTLINE

The design guidelines for MAP were:

- **Simplicity** Other than the thruster valves and attitude control reaction wheels, there are no moving parts when the satellite is at L2. For taking science data, there is only one mode of operation. MAP is passively cooled to ≈ 95 K.
- **Stability** The orbit at the Earth-Sun Lagrange point, L2, is thermally stable and has a negligible magnetic field. L2 is roughly 1.5×10^6 km from Earth and the Sun, Earth, and Moon are always "under" the spacecraft. (It will take approximately 3 months to get to L2.)

- **Heritage** No major new components were required except the NRAO W-band amplifiers. These were designed for MAP.
- **Ease of Integration** The mission relies on the complete understanding of the systematic noise levels. The magnitude of many of the systematic effects are easily determined when the instrument is warm.

with spacecraft and mission parameters as follows:

- **Mission Duration** Two years at L2. It can last longer.
- **Mass** 800 kg
- **Power** 400 W
- **Launch Vehicle** Delta 7425 (4 strap on motors).

12.2. RECEIVERS

MAP uses pseudo-correlation radiometers to continuously measure the difference in power from two input feeds on opposite sides of the spacecraft. The detecting elements are HEMTs designed by Marian Pospieszalski at the National Radio Astronomy Observatory (NRAO). The wide bandwidth and high sensitivity of the HEMTs, even while operating near 95 K, are what make MAP possible. There are ten feeds on each side of the spacecraft and each feed supports two polarizations, thus there are a total of twenty differential chains. Each receiver chain uses four amplifiers (two at $\approx$ 95 K and two at $\approx$ 290 K) for a total of eighty amplifiers. The radiometers are configured so that they difference two polarizations whose electric fields are parallel.

The key characteristic of the receivers is their low 1/f noise. The noise level at the spin rate, 8 mHz, is virtually the same as at the 2.5 kHz switching rate. This means that the receivers will not correlate noise from one pixel to the next. To put this receiver performance in perspective, we note that the 1/f knee of the W-band HEMTs alone is near 1 kHz. There is a complete model of the receivers from the feeds to the detector outputs, including all the support electronics. With the model, and measurements that correlate the model to reality, we determine the sensitivity of the output to temperature variations in each component. The temperatures of the most sensitive components are monitored during flight to roughly 1 mK accuracy.

Due to the wide HEMT bandwidth, the central effective frequency of the band depends on the source one is observing. Between dust and synchrotron sources, the shift is greater than 1 GHz in W-band. This will have to be taken into account when observing different sources. Below is a table of the representative central frequencies and noise bandwidths of the receivers. The values in flight will be different and are channel dependent.

TABLE 2. MAP band centers and noise bandwidth

Band	f_c (GHz)	Bandwidth (GHz)	# Channels
K	23	5.3	4
K_a	33	7	4
Q	41	8.4	8
V	61	12	8
W	95	17	16

12.3. OPTICS

The optics comprises two back-to-back telescopes. The secondary of each telescope is illuminated by a cooled corrugated feed which is the input to the receiver chain. The reflector surfaces are shaped to optimize the beam profiles but to a good approximation, the telescopes are Gregorian. The Gregorian design was chosen because it is more compact than the Cassegrain and because it could accommodate the back-to-back feed geometry required for the radiometers. We considered a wide variety of designs, including single and triple reflector systems, but the Gregorian suited our needs best.

The feeds and receivers occupy a large space. As a consequence the beam profiles are not symmetric (see Table 3). The scan strategy, to a first approximation, symmetrizes the beam profile. Thus even if the beams were 2-D Gaussians to start with, the symmetrized profile would not be Gaussian. Likewise, the window functions are not simply Gaussian. Nonetheless, the beam profiles and windows can be parametrized by Gaussians for most work. The data in the table are representative, the final beams will be measured in flight with sub-percent accuracy.

TABLE 3. Approximate E & H plane beam θ_{FWHM}

Band	θ_E (deg)	θ_H (deg)
K	0.95	0.75
K_a	0.7	0.6
Q	0.45	0.5
V	0.3	0.35
W	0.21	0.21

Outside of making the optics fit into the MIDEX fairing, one wants to ensure that one measures power from only the main beam. The Sun, Earth, and Moon, which are always at least 100° away from the main beam, are blocked by the solar arrays. The contribution from these sources is computed to be much less than 1μK. The more difficult source to block is the Galaxy which illuminates the feeds from just over the top of the secondary. To block it, we substantially oversized the secondary, so that at its edge the illumination from the feed is less that 10^{-5} the illumination at the center (< -50 dB edge taper).

The optics are modelled using a program from YRS Associates[66] that solves for the currents on the reflectors as waves propagate through the system. The measured beam profiles (all have been measured at ten frequencies across the band) are in excellent agreement with the predictions. Chris Barnes modified the code to run on a supercomputer so that we can also predict the sidelobes. The sidelobe measurements agree well with the predictions.

Using the models, we can estimate the contribution to MAP from the Galaxy. Our model includes a spinning dust component so that the frequency spectrum resembles something like that shown in [26]. We then fly MAP over the Galaxy and record two signals. One signal is the rms difference between the two telescopes with the response integrated over the whole sky; the second signal is the same except with the contribution from the main beams subtracted. In other words, the second method tells how much Galactic signal comes through the sidelobes or from angles greater than $\approx 4°$ from the main beam. We do this for $|b| > 15$. The model is only approximate and does not yet include a contribution from extragalactic sources. Of course, the model will be updated after we measure what the Galaxy is really like.

TABLE 4. Approximate Galactic contributions for $|b| > 15°$

Band	Galactic contribution (μK)	Sidelobe contribution (μK)
K	120	16
K_a	60	2
Q	40	4
V	20	0.2
W	...	...

To put these numbers in perspective, the *rms* magnitude of the CMB is about 120 μK. In K band, the *rms* Galaxy signal is roughly the same. To first order, these add in quadrature to produce a signal with an *rms* of

170 μK. In V-band, the galactic contribution is far less; it will change the power spectrum by $\approx 2\%$ if uncorrected. These rough numbers show that the power spectrum estimates are quite robust to the Galactic contamination.

As we have emphasized, the goal of MAP is to make maps; the power spectrum is just one way to quantify them. In producing a map of the CMB, we will clearly have to model and subtract the Galaxy. The rightmost column in Table 4 gives an indication of the contribution to the map if the sidelobe contribution is not accounted for.

12.4. SCAN STRATEGY

The scan strategy is at the core of MAP and can only be realized at a place like L2. To make maps that are equally sensitive to large and small scale structure, large angular separations must be measured with small beams. To guard against variations in the instrument, as many angular scales as possible should be covered in as short a time as possible. MAP spins and precesses like a top. There are four time scales. The beams are differenced at 2.5 kHz; the satellite spins at 0.45 rpm; the spin axis precesses around a 22.5° half angle cone every hour; and the sky is fully covered in six months. In one hour, roughly 30% of the sky is mapped.

As of this writing, the core of a pipeline exists to go from the time ordered data to maps to a power spectrum.

12.5. SCIENCE

Of primary interest, initially, will be the angular spectrum. It will be calibrated to percent accuracy and there will be numerous independent internal consistency checks. Of central importance will be the accompanying systematic error budget. The power spectrum will be sample variance limited up to $l \approx 700$. In other words, if the systematic errors and foreground contributions are under control, it will not be possible to determine the angular spectrum any better. MAP will be sensitive up to $l \approx 1000$.

The angular spectrum is not the best metric for assessing MAP. The primary goal is a map with negligible correlations between pixels. With such a map, analyses are simplified and the map is indeed a true picture of the sky. MAP should be able to measure the temperature-polarization cross correlation [15] and will be sensitive to the polarization signal itself. Correlations with X-ray maps will shed light on the extended Sunyaev Zel'dovich effect (not to mention the dozen or so discrete sources[51]). Correlations with the Sloan Digital Sky Survey will inform us about large scale structure. MAP will be calibrated on the CMB dipole and thus will be able to calibrate radio sources and planets to a universal system. It will also help elucidate the emission properties of the intergalactic medium.

13. Conclusions

This is a truly amazing era for cosmology. Our theoretical knowledge has advanced to the point at which definite and testable predictions of cosmological models can be made. For the CMB anisotropy, results from the angular power spectrum, frequency spectrum, statistical distribution, and polarization must all be consistent. In addition, these results must be consistent with the distribution, velocity flows, and masses of galaxies and clusters of galaxies as well as the age of the universe. There is a fantastic interconnecting web of constraints.

Using the CMB to probe cosmology is still in its early phase. After all, the anisotropy was discovered less than a decade ago. As cosmological models and measurements improve, the CMB and other measures will become a tool for probing high energy physics. For example, should the cosmological constant survive, we will have a handle on new physics in the early universe that we could never have obtained from accelerators.

14. Acknowledgements

I thank Chuck Bennett, Mark Devlin, Amber Miller, Suzanne Staggs, and Ned Wright for comments that improved this article. In the course of writing, a paper describing the VIPER experiment[47] appeared. These data are not included in the analysis nor do they change any conclusions.

References

1. Bahcall, N., Ostriker, J. P., Perlmutter, S., and Steinhardt, P. J., 1999, Science, 284, 1481-1488 (astro-ph/9906463)
2. Baker, J.C., Grainge, K., Hobson, M.P., Jones, M.E., Kneissl, R., Lasenby, A.N., O'Sullivan, C.M.M., Pooley, G. Rocha, G., Saunders, R., Scott, P.F., Waldram, E.M. 1999, Submitted to MNRAS, astro-ph/9904415
3. Barreiro, R. B., astro-ph/9907094
4. Bartlett, J. G., Blanchard, A. Douspis, M. Le Dour, M., Proc Evol of Large Scale Structure, garching, Aug 1998. astro-ph/9810318
5. Bond, J. R., Astro. Lett. & Comm. Vol 32, No. 1, 1995. Presented in 1994.
6. Bond, J. R., Crittenden, R., Davis, R. L, Efstathiou, G., & Steinhardt, P. J. 1994, Phys Rev Letters, 72, 1, pg 13
7. Bond, J. R., Efstathiou, G., Tegmark, M. 1998, MNRAS, 50, L33-41
8. Bond, J. R., Jaffe, A. H., Phil Trans R. Soc. Lond. astro-ph/9809043
9. Bond, J. R., Jaffe, A. H., Knox, L. 1999, Accepted in Ap.J., astro-ph/9808264
10. BOOMERanG at the web sites http://www.physics.ucsb.edu/ boomerang or http://oberon.roma1.infn.it/boomerang/
11. Cosmic Background Imager web site http://astro.caltech.edu/~tjp/CBI/
12. Coble, K., et al. 1999, astro-ph/9902195

13. Coble, K. Ph.D. Thesis, Univ. of Chicago, 1999.
14. Cortiglioni et al. astro-ph/9901362.
15. Crittenden, R. G., Coulson, D. & Turok, N. Phys Rev D, 1995, D52, 5402. See also astro-ph/9408001 and astro-ph/9406046
16. Degree Angular Scale Interferometer web site http://astro.uchicago.edu/dasi/
17. Davies, R.D., & Wilkinson, A. in "Microwave Foregrounds" APS Conference Series, pg. 77, Vol 181, 1999, A. de Oliveira-Costa & M. Tegmark eds.
18. Dicker et al. astro-ph/9907118. Submitted to MNRAS 1999.
19. de Oliveira-Costa, A., Kogut, A., Devlin, M. J., Netterfield, C. B., Page, L. A., Wollack, E. J. 1997 Ap.J. Letters, 482, L17-L20
20. de Oliveira-Costa, A., Devlin, M. J., Herbig, T. H., Miller, A.D., Netterfield, C. B., Page, L. A. & Tegmark, M. 1998 Ap.J. Letters, 509, L77
21. de Oliveira-Costa, A., Tegmark, M., Page, L.A., Boughn, S. 1998 Ap.J. Letters, 509, L9
22. de Oliveira-Costa, A. 1999. Private communication and work in progress. See related maps at http://www.sns.ias.edu/~angelica/skymap.html.
23. de Oliveira-Costa, A. 1999. Private communication. Work in progress on foreground subtraction for QMAP data.
24. Devlin, M. J., de Oliveira-Costa, A., Herbig, T., Miller, A. D., Netterfield, C. B., Page, L. A., & Tegmark, M. 1998, ApJ Letters, 509, L73
25. Dodelson, S. & Knox, L., 1999, astro-ph/9909454.
26. Drain, B. T. & Lazarian, A. 1999, astro-ph/9902356, in *Microwave Foregrounds*, ed A. de Oliveira-Costa & M. Tegmark APS Conference Series, Vol 181, pg 133, (ASP:San Francisco)
27. Fixsen, D. J., et al. 1997 (astro-ph/9704176)
28. Ganga, K. M. et al. 1993, Ap.J. 432:L15-L18.
29. Glanz, J. 1999. In the "News" section, Science, Vol 283. See also the VIPER web site at http://cmbr.phys.cmu.edu
30. Gundersen, 1999, Ph.D. Thesis from U.C. Santa Barbara, 1995
31. Gush, H., Halpern, M., & Wishnow, E. H., 1990, Phys. Rev. Letters, 65, 537
32. Halpern, M., & Scott, D. in *Microwave Foregrounds*, ed A. de Oliveira-Costa & M. Tegmark APS Conference Series, Vol 181, pg 283, (ASP:San Francisco)
33. Herbig, T., Devlin, M. J., de Oliveira-Costa, A., Miller, A. D., Page, L. A., & Tegmark, M. 1998, Ap.J. Letters, 509, L73
34. Hu, W. and White, M. 1996, Ap.J. 471:30 This is also available through http://www.sns.ias.edu/~whu
35. Huey, G. & Steinhardt, P., 1999, The updated cosmic triangle available through http://feynman.princeton.edu/~steinh/
36. Kerr, A. R., Pan, S. -K., Lichtenberger, A. W., and Lloyd, F. L. 1993, Proceedings of the Fourth International Symposium on Space Terahertz Technology, pp 1-10
37. Knox, L., Bond, J. R., Jaffe A. H., Segal, M. & Charbonneau, D. Phys.Rev. D58 (1998) 083004, astro-ph/9803272.
38. Lineweaver, C. H., Science, 284, 1503-1507, 1999. See also astro-ph/9909301.
39. Mather, J. C., et al. 1990, Ap.J. Letters, 354:L37.
40. Mather, J. C., et al. 1999, Ap.J., 512. See also astro-ph/9810373 and Fixsen, D. J., et al. 1996, Ap.J., 473:576. (astro-ph/9605054)
41. MAXIMA's web site http://cfpa.berkeley.edu/group/cmb/gen.html contains more information.
42. Miller, A. D., Devlin, M. J., Dorwart, W. Herbig, T., Nolta, Page, L., Puchalla, J., Torbet, E., & Tran, H. 1999, Ap.J. Letters, 524, L1-4.
43. Netterfield, C. B., Devlin, M. J., Jarosik, N., Page, L., & Wollack, E. J. 1997, Ap.J. 474, 47
44. Generation of Large Scale Structure ed D.N. Schramm and P. Galeotti (Kluwer, Netherlands), p75.
45. Pen, U., Seljak, U., & Turok, N. 1997, Phys Rev Letters, 79:1611

46. Peterson, J.B. et al. astro-ph/9907276
47. Peterson, J.B. et al. astro-ph/9910503
48. Pospieszalski, M. W. 1992, Proc. IEEE Microwave Theory Tech., MTT-3 1369; and Pospieszalski, M. W. 1997, Microwave Background Anisotropies, ed F. R. Bouchet (Gif-sur-Yvette: Editions Frontièrs): 23-30
49. Pospieszalski, M. W. et al. 1994 IEEE MTT-S Digest 1345
50. Ratra, B., *et al.* 1999 Ap.J. 517:549
51. Refregier, A., Spergel, D. & Herbig, T. Ap.J, 1998, astro-ph/9806349
52. Seljak, U. and Zaldarriaga, M. 1998. The CMBFAST code is available through http://www.sns.ias.edu/matiasz/CMBFAST/cmbfast.html. See also 1996 Ap.J. 469, 437-444
53. Smoot, G.F., *et al.* 1992, Ap.J. Letters 396, L1
54. Staggs, S. T., et al, 1996. Ap.J. Letters, 473, L1
55. Tegmark, M. 1997, PRD, 55, 5895
56. Tegmark, M., de Oliveira-Costa, A., Devlin, M., Netterfield, C.B., Page, L. & Wollack, E. 1997, Ap.J. Letters, 474:L77
57. Tegmark, M., Eisenstein, D., Hu, W., & de Oliveira-Costa, A. 1999, Submitted to Ap.J., astro-ph/9905257
58. Tegmark, M., 1999, Ap.J. 514, L69-L72. See also astro-ph/9809201
59. Torbet, E., Devlin, M. J., Dorwart, W. Herbig, T., Nolta, Miller, A. D., Page, L., Puchalla, J., & Tran, H. 1999, ApJ Letters, astro-ph/9905100, 521, L79
60. Turner, M.S., 1999, astro-ph/9904051
61. Wang, L., Caldwell, R.R., Ostriker, J. P. & Steinhardt, P.J. 1999, astro-ph/9901388
62. White, M., & Bunn, E. 1995, Ap.J. Letters 443:L53
63. Wilson, G. W., et al. 1999, Submitted to Ap.J., astro-ph/9902047
64. Wollack, E. J., Devlin, M. J., Jarosik, N.J., Netterfield, C. B., Page, L., Wilkinson, D. 1997, Ap.J., 476, 440-447
65. Wright E. IAS CMB Data Analysis Workshop. astro-ph/9612006.
66. Rahmat-Samii, Y., Imbriale, W., & Galindo, V. YRS Associates, 4509 Tobias Ave. Sherman Oaks, CA 91403

OBSERVATIONS OF THE COSMIC MICROWAVE BACKGROUND

A.N. LASENBY
Astrophysics Group
Cavendish Laboratory
Madingley Road
Cambridge CB3 OHE, U.K.

Abstract

This contribution aims to provide a brief introduction to the observational side of Cosmic Microwave Background (CMB) astronomy. Experimental techniques and contaminants are discussed, and then some recent experiments which are starting to provide the first accurate measures of the power spectrum are considered in detail. The implications for cosmological models are discussed, and then a survey given of some ongoing and future experiments, including balloons and satellite missions, with an emphasis on CMB interferometers. The role of CMB data in deriving fundamental parameters, and its fusion with other data sets in testing cosmological models, is emphasized.

1. Introduction — expected properties of the anisotropies

The origin of the CMB anisotropies has been described in detail elsewhere in this volume, so here we just give the minimum description of what is required theoretically in order to understand the experimental approaches and results. We do this first for inflationary predictions, and then mention briefly what happens in theories involving topological defects.

The key point about inflation is that it produces fluctuations in a single mode — the growing mode [1]. This has the eventual result that all the fluctuations at a given wavenumber k are coherent (in phase) and thus we see a series of oscillations in the power spectrum as a function of k — see Fig. 1.

Doing this properly over full sky implies decomposing into spherical

R.G. Crittenden and N.G. Turok (eds.), Structure Formation in the Universe, 215–239.

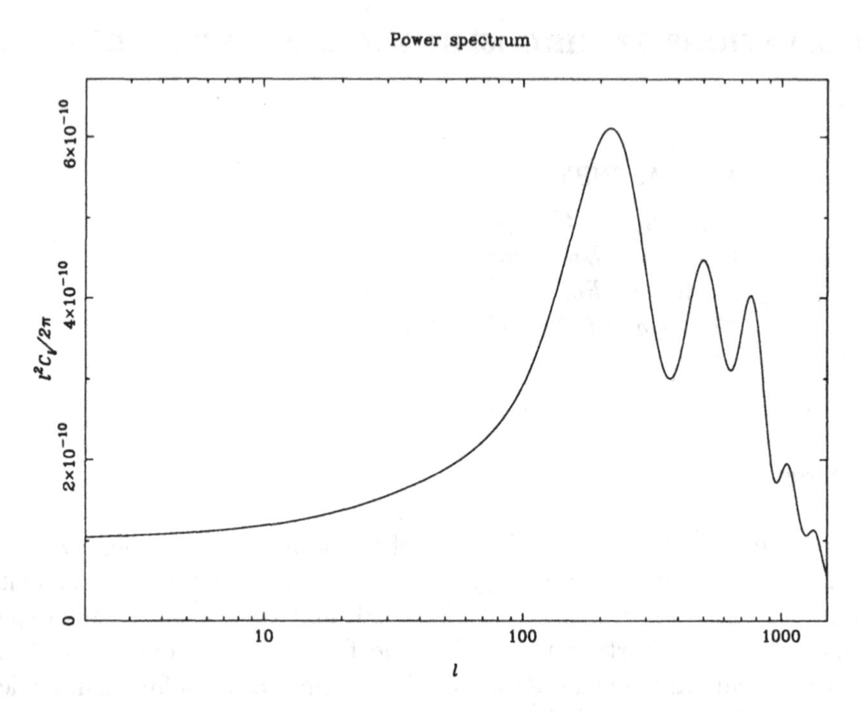

Figure 1. Power spectrum for standard CDM. Parameters assumed are $\Omega = 1$, $n = 1$, $H_0 = 50\,\mathrm{km\,s^{-1}\,Mpc^{-1}}$ and a baryon fraction of $\Omega_b = 0.04$.

harmonics ($k \mapsto \ell$):

$$\frac{\delta T}{T}(\theta, \phi) = \sum_{\ell,m} a_{\ell m} Y_{\ell m}(\theta, \phi),$$

from which the rotationally invariant power spectrum is defined by

$$C_\ell = \langle |a_{\ell m}|^2 \rangle.$$

Generally one plots $\ell(\ell + 1)C_\ell$ which is the power per unit log interval in ℓ.

Physical lengths at recombination get translated to an angle on the sky via the angular diameter distance formula. This is mainly a function of $\Omega_{\rm total}$ so the position of the peaks in the power spectrum as a function of ℓ is a sensitive indicator of the total energy density of the universe (baryonic, dark and an effective cosmological constant component).

There are two other important parameters: n, the slope of the primordial power-law power spectrum is defined via

$$\langle |\delta_k|^2 \rangle \propto k^n$$

Inflation (in its most common forms) predicts $n \approx 1$. (This is what would give roughly flat slope in the CMB power spectrum on large scales.)

The second is ratio of tensor (gravitational wave) to scalar (density) perturbations. We assume this is small below, but in various versions of inflation could be significant. (Note inflation predicts very small vector (vorticity) perturbations).

So, relative to a particular model, one can do detailed fitting of the major parameters (Ω_m, Ω_b, H_0, n etc.). The CMB will provide way to measure these with unprecedented accuracy, and one can simultaneously check the details of inflationary predictions and the particle physics potential presumed to give rise to inflation. As discussed elsewhere in this volume, it may even be possible to check theories such as the Hartle-Hawking no boundary proposal and details of instanton models for the very earliest stages of the universe [2].

Inflation predicts Gaussian fluctuations. If topological defects seeded structure formation we would expect non-Gaussian fluctuations, plus a washing out of the peaks in the power spectrum. Recently Magueijo *et al.* [3] have proposed a hybrid theory in which inflation occurs, and is primarily responsible for structure formation today, but an admixture of cosmic strings is also present. By varying the ratio of the contribution of strings versus inflation to the C_ℓ spectrum (at an arbitrarily chosen point $\ell = 5$) they demonstrate that one can produce spectra varying all the way from standard CDM form (as in Fig. 1), to a spectrum having a single peak, occurring near $\ell = 400$ with no secondary peaks at all. As we shall see below, a model of the latter kind is now definitely ruled out.

2. Experimental Problems and Solutions

2.1. THE CONTAMINANTS

The detection of CMB anisotropy at the level $\Delta T/T \sim 10^{-5}$ is a challenging problem and a wide range of experimental difficulties occur when conceiving and building an experiment. We will focus here particularly on the problem caused by contamination by foregrounds and the solutions that have been adopted to fight against them. The anisotropic components that are of essential interest are (i) The Galactic dust emission which becomes significant at high frequencies (typically > 100 GHz); (ii) The Galactic thermal (free-free) emission and non-thermal (synchrotron) radiation which are significant at frequencies lower than typically ~ 30 GHz; (iii) The presence of point-like discrete sources; (iv) The dominating source of contamination for ground- and balloon-based experiments is the atmospheric emission, in particular at frequencies higher than ~ 10 GHz; (v) Finally a possibility has emerged recently for a 'spinning dust' contribution in the range 10–100 GHz. We now discuss more fully point sources and spinning dust.

2.2. SPINNING DUST

Some existing CMB measurements give strong evidence for an anomalous component in the Galactic foregrounds. Evidence has come at high significance from the COBE [4], OVRO [5], Saskatoon [6] and Tenerife [7, 8] experiments that there is a component in the lower-frequency data (tens of GHz) which is correlated with dust emission, at a level much higher than likely for free-free emission alone. The constraints on free-free come from H_α limits (see e.g. [9]). To evade these limits very high temperatures (5×10^5 to 10^6 K) would be necessary. There is no indication from the X-ray results for such a component, however [5].

One explanation for this anomalous emission is as the rotational emission from 'spinning dust'. Draine & Lazarian [10] have pioneered this approach, which starts from the premise (borne out by some higher-frequency data) for a population of ultra-small dust grains, which could rotate rapidly enough to be a source in the tens of GHz region. An alternative, also being explored, is magnetic dipole emission from thermally excited grains, e.g. [11]. Initial evidence from the Tenerife data [7] suggested that the characteristic 'bump feature' in the predicted microwave spectrum of spinning dust (see Fig. 2) was indeed present in the Tenerife data, which showed a rise in the dust-correlated component of emission between 10 and 15 GHz. However, more recent work [8], with updated data, shows that while an anomalous component is indeed confirmed, the data is fully compatible as regards slope with a free-free spectrum (i.e. declining from 10 to 15 GHz) with only marginal evidence for a spinning dust component. Future CMB experiments with good 10–100 GHZ coverage such as MAP will doubtless shed much light on this, as well as the availability of improved H_α surveys (see e.g. [13]).

2.3. RADIO SOURCES

The properties of the radio source population at frequencies between 8 GHz and about 300 GHz are very poorly known [14]. At higher frequencies sub-mm surveys (such as with SCUBA and other instruments) can provide several frequency points and at lower frequencies good surveys exist from the VLA and Greenbank at 1.4, 5 and 8 GHz. Not very much work has been possible at intermediate frequencies however. This lack of knowledge is particularly unfortunate in the frequency range 15 – 100 GHz, since several ground-based CMB experiments work in this range, and radio source effects on these experiments tend to be larger than in the mm/sub-mm range

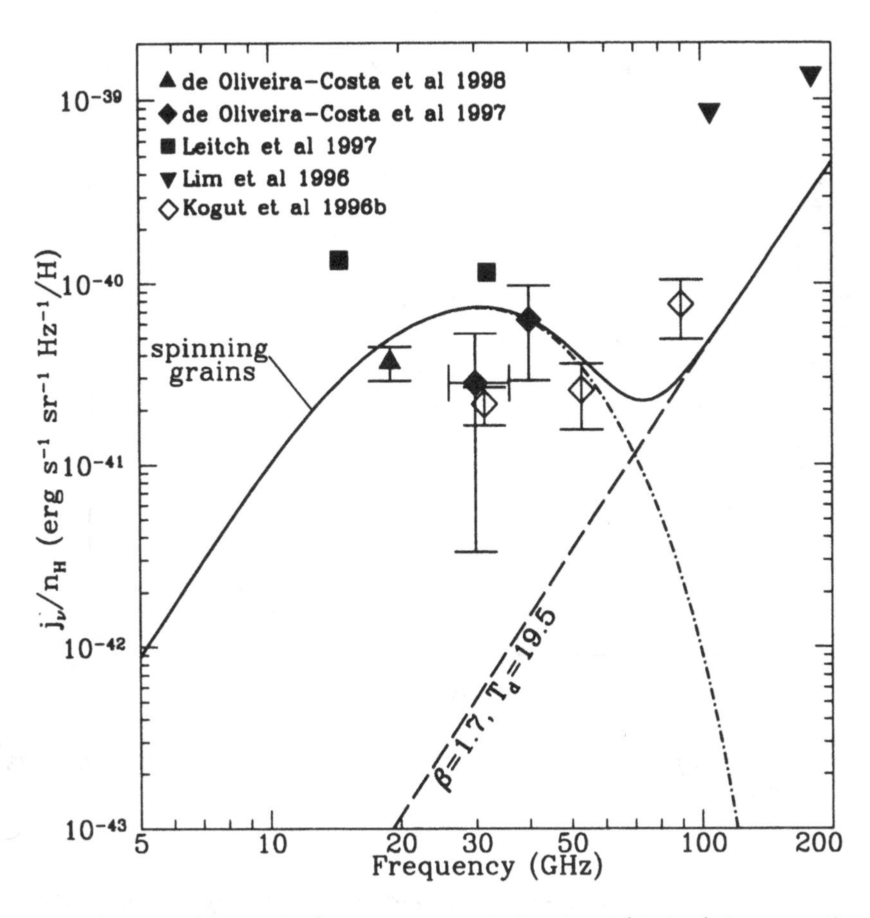

Figure 2. Spinning (dot-dashed) versus expected vibrational (dashed) dust contribution with experimental points marked (taken from [12]).

both from the flux behaviour of the sources and because the ground-based experiments can work at higher angular resolution. The larger antenna sizes this implies gives a corresponding increase in flux sensitivity to radio sources.

Another reason the 10-100 GHz range is so important to understand, is because of the existence of so-called GigaHertz Peaked Spectrum sources (or GPS). These sources, as their name implies, have a flux spectrum which rises roughly as ν^2 in the GHz range, reaches a peak then turns over again. The majority have spectra peaking around 1–10 GHz [15]. However, the most extreme example known (see Fig. 3) has its turnover at about 30 GHz, coinciding with where several current and future ground-based CMB experiments operate. The point about a ν^2 flux spectrum of course, is that it precisely mimics a CMB spectrum in this range. We note that Planck

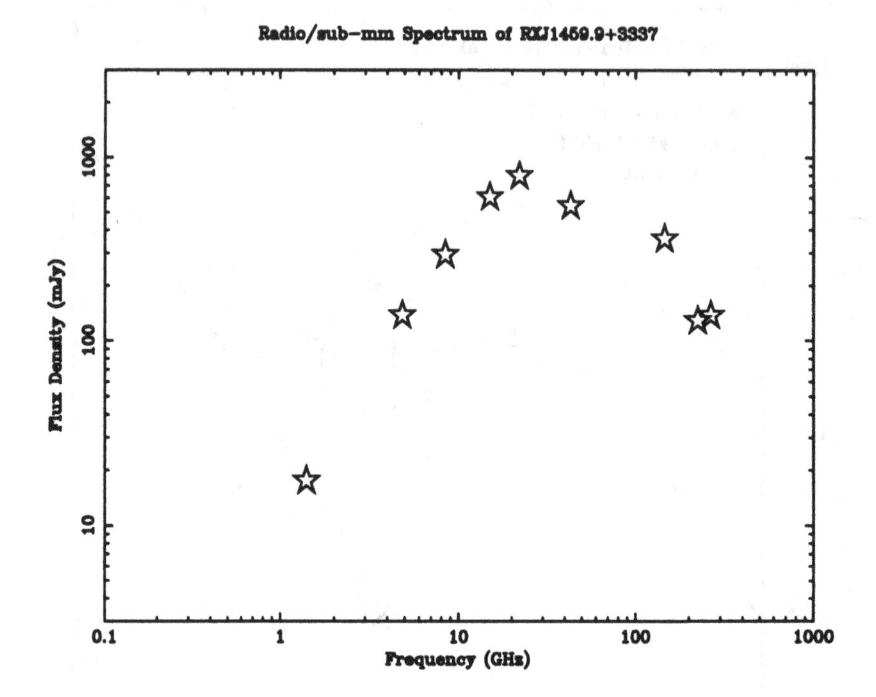

Figure 3. The radio/sub-mm spectrum for the X-ray selected quasar RXJ1459.9+3337. (Figure courtesy of A. Edge.)

satellite data will provide a unique opportunity to study the properties of radio sources over the range 30 – 350 GHz. The combination of 4 LFI and 4 HFI channels covering this range mean that the number counts and spectral properties of many thousands of radio sources between 30 to 300 GHZ will be available for the first time.

2.4. THE SOLUTIONS TO FOREGROUND PROBLEMS

A natural solution is to run the experiment at a suitable frequency so that the contaminants are kept low. There exists a window between $\sim$ 10 and $\sim$ 40 GHz where both atmospheric and Galactic emissions should be lower than the typical CMB anisotropies. For example, the Tenerife experiments are running at 10, 15 and 33 GHz and the Cambridge Cosmic Anisotropy Telescope (CAT) at 15 GHz. However, in order to reach the level of accuracy needed, spectral discrimination of foregrounds using multi-frequency data has now become necessary for all experiments. This takes the form of either widely spaced frequencies giving a good 'lever-arm' in spectral discrimination (e.g. Tenerife, COBE, most balloon experiments), or a closely spaced set of frequencies which allows good accuracy in subtraction of a particular known component (e.g. CAT, the forthcoming VSA, and the Saskatoon experiment).

Concerning point (iv) above, three basic techniques, which are all still being used, have been developed in order to fight against the atmospheric emission problem:

The Tenerife experiments (e.g. [16]) are using the *switched beam* method. In this case the telescope switches rapidly between two or more beams so that a differential measurement can be made between two different patches of the sky, allowing one to filter out the atmospheric variations.

A more modern and flexible version of the switched beam method is the *scanned beam* method (e.g. Saskatoon [17] and Python telescopes). These systems have a single receiver in front of which a continuously moving mirror allows scanning of different patches of the sky. The motion pattern of the mirror can be re-synthesised by software. This technique provides a great flexibility regarding the angular-scale of the observations and the Saskatoon telescope has been very successful in using this system to provide results on a range of angular scales.

Finally, an alternative to differential measurement is the use of interferometric techniques. Here, the output signals from each of the baseline horns are cross-correlated so that the Fourier coefficients of the sky are measured. In this fashion one can very efficiently remove the atmospheric component in order to reconstruct a cleaned temperature map of the CMB. The advantages of this method are discussed further below.

3. Updates and Results on Various Experiments

3.1. MOBILE ANISOTROPY TELESCOPE

The Mobile Anisotropy Telescope (MAT) is using the same optics and technology as the Saskatoon experiment at a high-altitude site in Chile (Atacama plateau at 5200m). This site is believed to be one of the best sites in the world for millimetre measurements and is now becoming popular for other experiments (e.g. the Cosmic Background Interferometer, see below) because of its dry weather. The experiment is mounted on a mobile trailer which is towed up to the plateau for observations and maintenance. The relevant point where MAT differs from the Saskatoon experiment is the presence of an extra channel operating at 140 GHz. This improves the resolution and the experiment probably had the honour of being the first able to establish the existence of the first Doppler peak from a single experiment. (Strong evidence for a Doppler peak was already available from the combination of the Saskatoon (e.g. [17]) and CAT [18] experiments.) See [19] for further details of the MAT observations. The data sets from

Frequency	90 GHz	150 GHz
Resolution	25′	16.5′

TABLE 1. Frequency channels and resolution for the North American test flight of Boomerang

Frequency	90 GHz	150 GHz	220 GHz	400 GHz
Resolution	20′	12′	12′	18′

TABLE 2. Frequency channels and approximate resolution for the Boomerang Antarctica flight.

this experiment have become known as the 'TOCO' data after the name of a local peak in the Andes.

3.2. BOOMERANG

The aim here was to use a scanning balloon experiment to cover the ℓ range $50 \lesssim \ell \lesssim 800$. The first long duration balloon flight (LDB) was in 1993 (a solar physics flight). They have now carried out one of these (December 1998/January 1999), now called the Antarctica or LDB flight. There was also a 'test flight' on North America in 1997, from which the first power spectrum results were released [20]. Each flight had different configuration for the instrument, with a quite different amount of data taken.

The focal plane arrangement for the North American test flight had two frequencies with 6 bolometers in total. 4.5 hours of data on the CMB were achieved, with approximately 600 μK sensitivity in 4,000 16 arcminute pixels. The 150 GHz channel was the one mainly used for power spectrum estimation. A brute-force likelihood approach was carried out in 23,561 one-third beam-sized 6.9′ pixels, with the results given in [20, 21].

For the Antarctica flight [22, 23], coverage of 4 frequencies with 16 bolometers in total were available. The flight was basically perfect first time, and the balloon gondola (the scientific package) was picked up just 20 miles from the original launch point after circling the South Pole. 190 hours of primordial CMB data on an approximately $30° \times 40°$ region were obtained and about 18 hours of data on three SZ clusters.

It is fair to say that the maps obtained [22] represent the start of a qualitatively new age in CMB astronomy. For the first time, primordial fluctuations are clearly visible over a large area and at several (here 3) frequencies, in a way which removes any lingering doubt as to their CMB origin. Quite a bit still remains to be done in analysis of this data, some of which suffers from pointing problems which limit the effective resolution.

The initial analysis has been limited to just 8000 14-arcminute pixels in just one of the 150 GHz detectors. This has provided results up to an ℓ of 600, although the instrument intrinsically has the sensitivity to reach $\ell \approx 800$.

The initial power spectrum results are shown in Fig. 4, in comparison with the MAXIMA-1 results (see below). They have been quite surprising, in that although a flat (perhaps even closed!) universe is confirmed, the second Doppler peak appears somewhat suppressed, if it is present at all. Much effort is currently going into seeing the implications of both this and the MAXIMA data (e.g. [23, 24]), with the currently most likely outcomes being the need for a higher $\Omega_{\rm baryon}h^2$ than usually derived from nucleosynthesis constraints, and perhaps an overall slope in the primordial power spectrum from inflation (n different from 1). Since a higher $\Omega_{\rm baryon}h^2$ tends to suppress the second peak while raising the third peak, this means the interferometer experiments (see below) are very well poised to make a decisive contribution here.

3.3. MAXIMA-1

Shortly after the announcement of the BOOMERANG Antarctica results, results were also announced from another balloon experiment, MAXIMA-1 [25]. This experiment is not long-duration, but has a very sensitive array receiver. The result is coverage over a field somewhat smaller than that observed by BOOMERANG (about 10 by 10 degrees), resulting in less resolution in ℓ space, but with enough sensitivity to make a highly significant determination of the power spectrum, over a similar range. The power spectrum from MAXIMA-1 is shown Fig. 4 in comparison with that of BOOMERANG. It tends to give a slightly higher first peak, at slightly higher ℓ, again strongly favouring a flat universe, but with a similar depression of the second peak. First cosmological constraints from MAXIMA-1 are given in [26].

3.4. FUTURE BALLOON FLIGHTS

As well as data already taken by MAXIMA and BOOMERANG which is yet to be analysed, new flights of both instruments are planned and funded which will feature enhanced capabilities. In particular B2K, a version of BOOMERANG scheduled for flight in December 2000, will be the first to have a polarisation capability. In addition, other long duration balloon experiments are expected to be flying shortly, including TOPHAT, which has the payload *on top of* the balloon instead of a gondola beneath, to minimise ground pickup. A northern hemisphere version of BOOMERANG (though with a different scanning strategy) called ARCHEOPS, has already

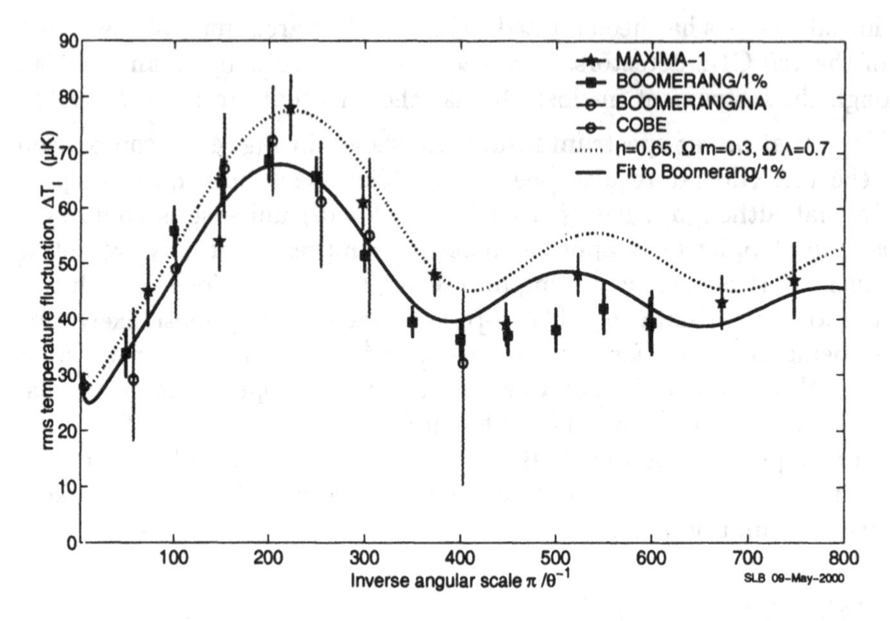

Figure 4. A linear scale comparison of the BOOMERANG North America (NA) and Antarctica (1%) results with MAXIMA-1. Also shown in a 'standard' flat model and a fit to the BOOMERANG data alone. Note the 'Inverse angular scale' can be read directly as ℓ.

had a successful test flight, and a long duration flight is planned for winter 2000.

4. Interferometers

Interferometers have become increasingly important in CMB research. This is due to their special properties in dramatically reducing the effects of atmospheric emission, and because their beams are electronically synthesized, giving lower levels of sidelobe pickup and better rejection of systematics. A summary of existing and recently operational interferometer experiments is given in Table 3.

The way in which an interferometer can reduce the effects of atmospheric emission is illustrated in Fig. 5 (taken from Lay & Halverson [27]). The top row (a,b,c) shows the response on the sky for each of a chopped beam system, swept beam system and interferometer, while the bottom row (d,e,f) shows the response in what can be considered as 'spatial frequency space' for each of them. The final row is a side profile of this response. For the chopped beam, the spatial frequency response is a sinc type function; for a swept beam system, one of the advantages of it is that the response can be re-synthesized in software to lie in any desired bin, as indicated by the

Experiment	*Name*	*Site*	*Status*
RYLE	Upgraded 5km	Cambridge	Current
ATCA	Compact array of AT	Australia	Current
CAT	Cosmic Anisotropy Telescope	Cambridge	Finishes soon
Int33	33GHz interferometer	Tenerife	Current
VSA	Very Small Array	Tenerife	Operational
DASI	Degree Angular Scale Interferometer	South Pole	Operational
CBI	Cosmic Background Imager	Chile	Operational

TABLE 3. Some current CMB Interferometers

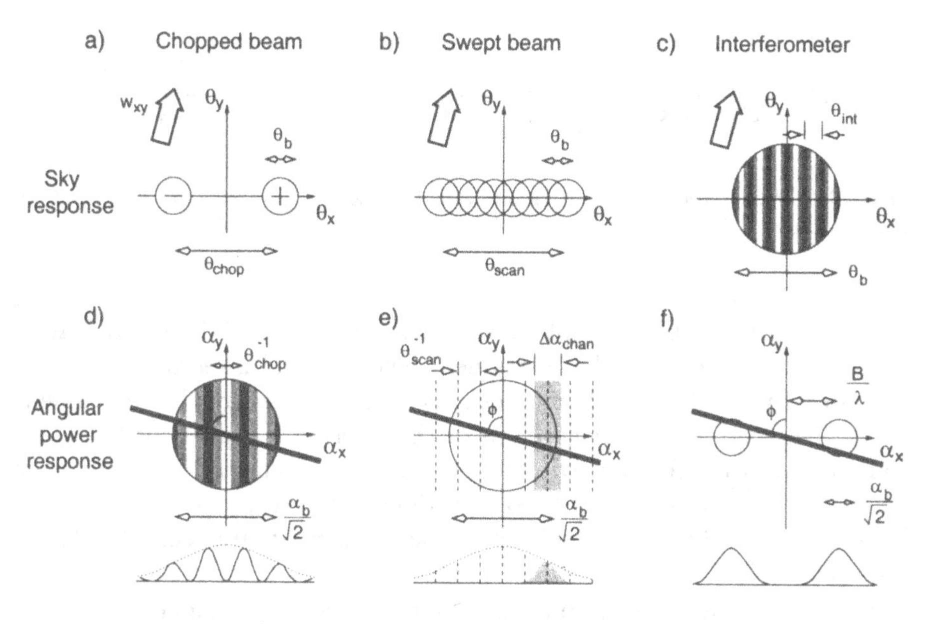

Figure 5. Effects of the atmosphere (taken from Lay & Halverson [27]).

vertical grey bar (a particular choice of bin), while for the interferometer, since the samples it takes are already in Fourier space, their transform (the spatial response) is just a picture of where the antennae lie in real space (the two circles in sub-figure (f)). Now how does this help with the atmosphere? Also shown in the figure is the wind direction (the arrow labelled w_{xy} in the top row), together with a black band (at an angle shown as ϕ) in the bottom row indicating the area in the spatial frequency response plane at

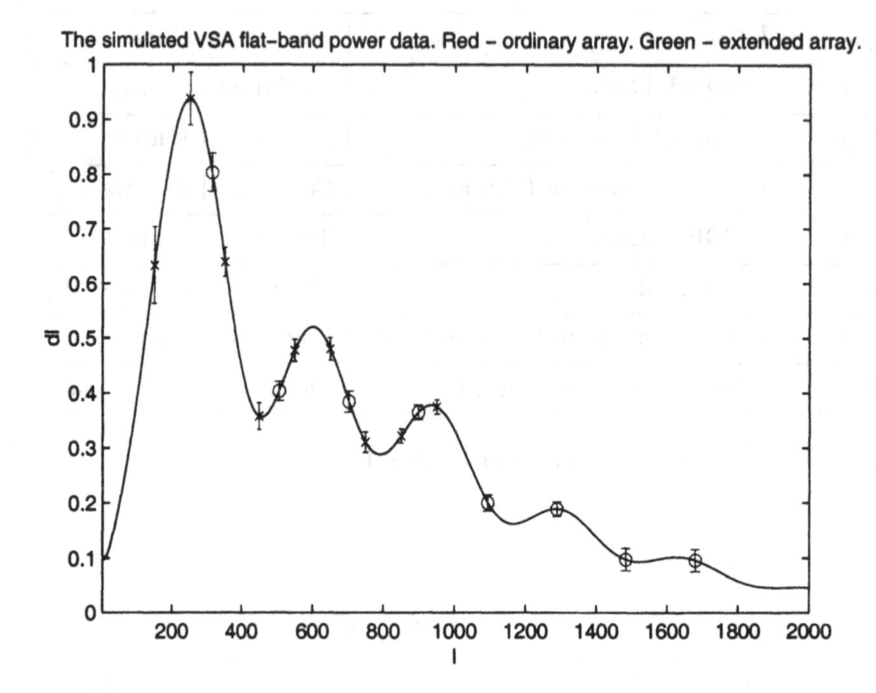

Figure 6. VSA power spectrum recovery in both arrays (normal and extended), in arbitrary units.

which effects from the atmosphere will lie given this wind direction. Taking the product of this with the instrument response and integrating gives an estimate of the effects that will actually be registered by the instrument. It can be seen that the effectiveness of the filtering is higher for the swept beam system than for the switched beam, but is higher still for the interferometer. In particular, for wind directions not directly perpendicular to the antenna baseline, there will be virtually no residual effects due to the atmosphere.

We now give a description of some of the newer experiments in Table 3.

4.1. THE VERY SMALL ARRAY

The Very Small Array (VSA) was first built and tested in Cambridge and is now sited in Tenerife. Active observations are beginning June 2000. The 14 elements of the interferometer operate from 26 to 36 GHz and cover angular scales well-matched to the secondary Doppler peaks. The results will consist of 9 independent bins in each of two array configurations (compact (= normal) and extended) regularly spaced from $l \sim 150$ to $l \sim 1700$ on the power spectrum diagram [28] (see Fig. 6). The information from the

Figure 7. The VSA with 11 antennas mounted and tracking.

VSA on this secondary peak region (as well as from the two other current primordial CMB interferometers — see below) is keenly awaited, given the apparent low height of the second Doppler peak in the BOOMERANG and MAXIMA-1 observations. A recent picture of the VSA on site in Tenerife is shown in Fig. 7.

4.2. THE COSMIC BACKGROUND IMAGER AND DASI

The two further interferometer projects which will complement the work of the VSA are the The Cosmic Background Imager (CBI), operating from Chile [29] and the Degree Angular Scale Interferometer [30, 31] (DASI) - formerly Very Compact Array (VCA) - which is operating at the South Pole (University of Chicago & CARA). They both share the same design (13 element interferometers) and the same correlator operating from 26 to 36 GHz. However the size of the baselines differ between CBI and DASI so that CBI will cover angular scales from 4 to 20 arcmin while DASI will cover the range between 15 arcmin and 1.4° (similar to the compact array of the VSA). A prediction for the ℓ-space sensitivity of both instruments is shown in Fig. 8. DASI has now been installed at the South Pole and a recent picture of it is shown in Fig. 9. The CBI has been installed in Chile since November 1999, and has is taking data of both blank fields and SZ clusters. A view of it prior to its transport to Chile is shown in Fig. 10. All three of these interferometric experiments (VSA, CBI and DASI) should have

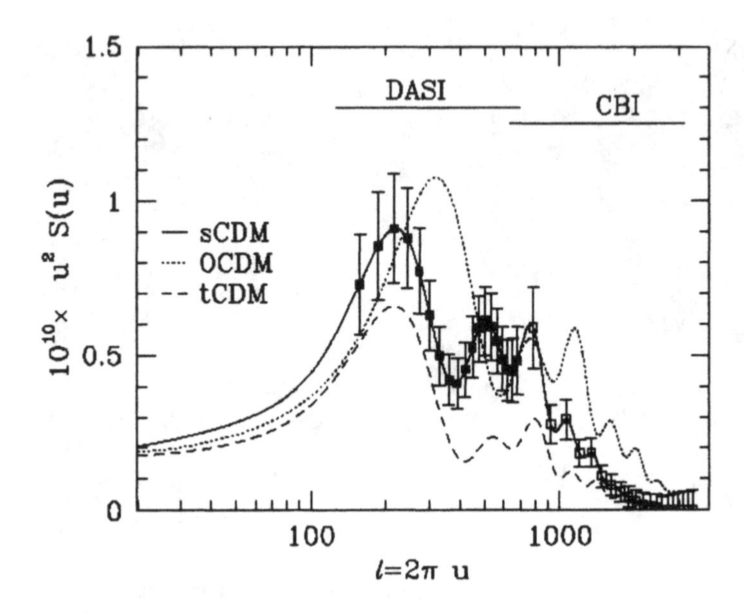

Figure 8. Joint DASI/CBI predicted sensitivity.

produced data with potentially very significant cosmological consequences by the end of 2000.

4.3. INSTRUMENTS FOR SECONDARY CMB EFFECTS AND BLANK S-Z SURVEYS

In addition to primordial CMB fluctuations, there is expected to be a spectrum of secondary fluctuations stretching to high ℓ. After the primordials and polarization, this is expected to be the next key area for CMB investigation, since it promises to reveal many details of what happened during the 'dark ages' of the universe. Among the effects expected are

Ostriker-Vishniac effect This is a second order effect arising from a cross term between electron density and bulk velocity in an ionized medium. It is fairly precisely calculable for a given thermal history [32, 33].

Inhomogeneous Reionization Here energy sources turning on at intermediate/high redshift ionize the IGM in their vicinity, while other areas remain neutral. This affects the CMB via Ostriker-Vishniac and SZ-type effects, but is in general hard to calculate due to the assumptions that have to be made about the turn-on and power of the initial sources [34].

Figure 9. DASI at the South Pole.

Figure 10. The CBI before installation in Chile.

SZ effect In addition to the SZ effect from individual resolved clusters, there will be a 'field' SZ effect from the set of all clusters along a given line of sight. (Typically about 1/3 of all lines of sight are expected to intersect a cluster at some redshift.) The magnitude of this effect is a sensitive indicator of both cosmology and cluster gas evolution [35].

Topological Defects Cosmic strings can exert a late-time ISW effect, which would give rise to sharp non-Gaussian features at high-resolution.

Winds from black holes Here outflows in black holed-seeded protogalaxies shock and heat the surrounding medium, resulting in an SZ effect of galactic scales [36].

Each of these effects has consequences for the power spectrum which peak in the ℓ range 2000 to 10,000 (except the last which is expected to peak in the region $\ell \approx 10^4$–10^5). The presence of any of them could upset parameter determination from other lower ℓ experiments based on an assumed 'primordial only' CMB power spectrum. Several groups are proposing, and in one case have funded, interferometer arrays which can survey this region of ℓ space. These include:

AMI: The Arcminute Microkelvin Imager (Cambridge) (see Figs. 11 and 12) This would be sited in Cambridge U.K. and operate at 15 GHz

AMIBA: The advanced microwave background imaging array. This is now funded to be built in Taiwan at 90 GHz, and includes a polarization capability [37]

The JCA: An array proposed by John Carlstrom (Chicago) to operate probably at 30 GHz.

5. Summary of current experimental results and constraints

Much work has gone into setting constraints on cosmological parameters using the CMB power spectrum, in the form in which it has gradually begun to be revealed by the observations. In addition, different groups have embarked on setting *joint* constraints using not just CMB, but other cosmological indicators such as Type Ia supernovae, peculiar velocities, cluster abundance and large scale structure results. The point here, is that especially with the current levels of accuracy, the CMB results alone suffer degeneracies with regard to particular combinations of parameters (e.g. cosmological constant versus tensor mode contribution), and in any case do not provide direct constraints on other quantities of interest today, such as σ_8, the *rms* in 8 Mpc spheres, or b, the assumed linear bias variable. By combining with other data sets, each of which has its own degeneracies and possible systematic effects of course, we can hope to recover a wider range of parameters, as well as check the underlying assumptions of the joint models used and for the presence of systematics.

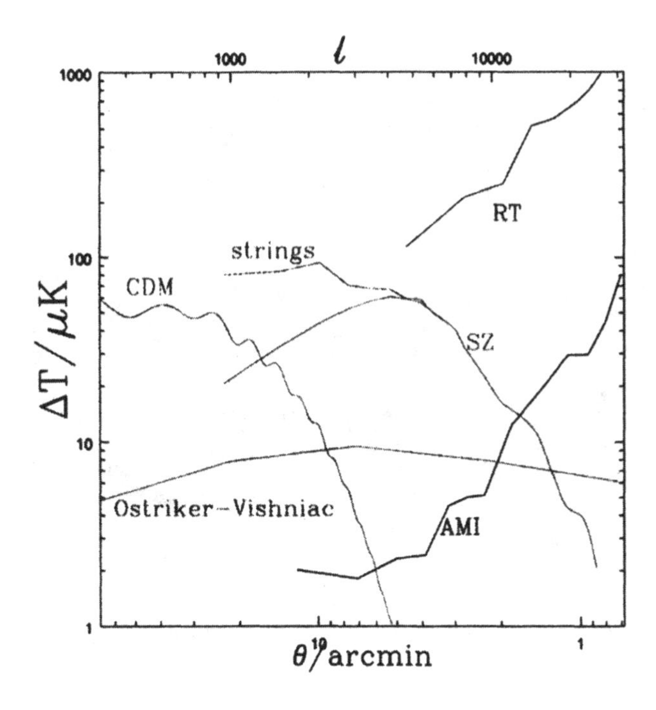

Figure 11. Comparison of expected primordial CMB power spectrum (CDM) with a particular model of the Ostriker-Vishniac effect and with the instrumental sensitivities of the Ryle Telescope (RT) and AMI. Also shown are effects due to a single cosmic string or S-Z cluster in the field of view.

Our own group in Cambridge has carried out several such studies already, starting with combining CMB data with IRAS 1.2 Jy galaxy redshift survey data [38]. This has been followed by combination with cluster abundance data [39], Type Ia supernovae [40, 41] and most recently peculiar velocity data [42]. Currently the results including cluster abundance are somewhat discrepant from those obtained including peculiar velocities, indicating a perhaps deeper problem. Thus the most recent analysis [42] includes as a case study just CMB (the latest data set including both the BOOMERANG Antarctica results [22] and MAXIMA-1 [25]) peculiar velocities (the SFI catalogue of Haynes *et al.* [43, 44]) and Supernovae (the results of Perlmutter *et al.* [45]). These three data sets are quite interesting to combine since all three are effectively clear of assumptions about the biasing of mass versus the galaxy distribution. The CMB data used is shown in Fig. 13, which also serves as a summary of the current experimental position. The joint study is carried out assuming a flat universe, which seems justified given the current CMB data. As can be seen, the CMB data indicate the position of the first acoustic peak, near $\ell \sim 200$ which corre-

Figure 12. Simulation of final appearance of AMI.

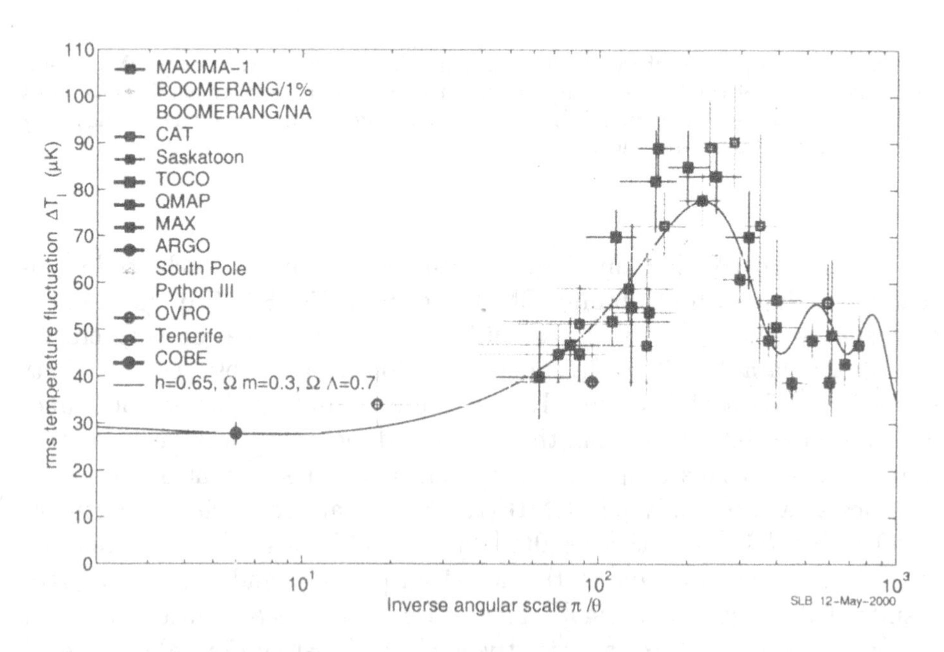

Figure 13. Current CMB power spectrum results including BOOMERANG and MAXIMA. (The 'Inverse angular scale' can be read directly as spherical harmonic coefficient ℓ.)

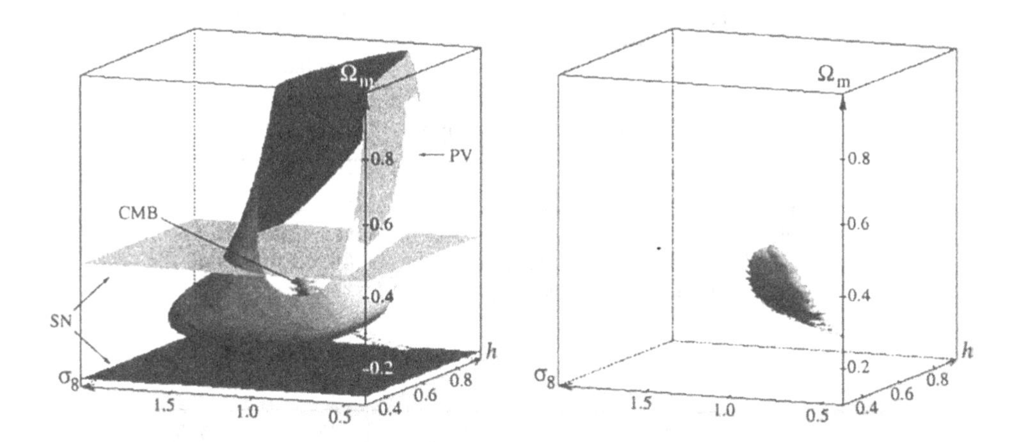

Figure 14. Left: peculiar velocities (PV), CMB (whole compilation including Maxima-1 and Boomerang Antarctica results) and supernovae (SN) 2σ iso-probability surfaces. For PV and CMB the surfaces are at Δlog(Likelihood)=4.01, and for the SN the surfaces are at Δlog(Likelihood)=2.00, corresponding to the 95 per cent limits for 3 and 1 dimensional Gaussian distributions respectively. The SN surfaces are two horizontal planes. Right: The 2-sigma surface for the joint PV, CMB and SN likelihood function. Taken from [42].

sponds to a wavenumber of $k \sim 0.03\ h\ \mathrm{Mpc}^{-1}$. This constrains the combination $\Omega_\mathrm{m} + \Omega_\Lambda$ to be roughly around unity (e.g. Efstathiou et al. 1999 [40], Lange et al. 2000 [23], Balbi et al. 2000 [26]), consistent with the flat universe assumed in our current analysis. In fact using just BOOMERANG and COBE, Lange et al. (2000) find $\Omega_\mathrm{m} + \Omega_\Lambda \sim 1.1$, whereas using just MAXIMA-1 and COBE, Balbi et al. (2000) find $\Omega_\mathrm{m} + \Omega_\Lambda \sim 0.9$.

Within this framework of a flat model, we can see some of the results of our joint study in Fig. 14. This shows that the CMB, peculiar velocity (PV) and supernovae surfaces (SN) in the 3-d space of σ_8, Hubble constant h and matter density Ω_m are indeed usefully complementary, and appear to intersect within their allowed 2-sigma regions (see [42] for further details). The final best fit parameters and allowed (1-d marginalised) 95 per cent confidence regions are shown in Table 4. We find here a somewhat higher Hubble constant here than in our previous studies, which is probably mainly due to the new BOOMERANG and MAXIMA-1 data indicating a smaller first peak than previously thought, and we also find a higher σ_8. In particular, the result for σ_8 is higher than the Bridle et al. (1999) [39] constraint, $\sigma_8 = 0.74 \pm 0.1$ (95% confidence) obtained by combining the CMB with cluster abundance and IRAS and allowing for linear biasing. This may reflect the preference of the peculiar velocities for a slightly higher value of $\sigma_8\Omega_\mathrm{m}^{0.6}$ than favoured by the cluster abundance analysis. Clearly there is work to be done in reconciling these results, but overall the methods appear

Parameter	Best fit point	95 per cent confidence limits
h	0.74	$0.64 < h < 0.86$
$\Omega_{\rm m}$	0.28	$0.17 < \Omega_{\rm m} < 0.39$
σ_8	1.17	$0.98 < \sigma_8 < 1.37$

TABLE 4. Parameter values at the joint PV, CMB, SN optimum. The 95% confidence limits are given, calculated for each parameter by marginalising the likelihood function over the other parameters.

	LFI (HEMT)				HFI (Bolometers)					
ν GHz	30	44	70	100	100	143	217	353	545	857
No. of detectors	4	6	12	34	4	12	12	6	8	6
$\theta_{\rm FWHM}$	$33'$	$23'$	$14'$	$10'$	$10.7'$	$8'$	$5.5'$	$5'$	$5'$	$5'$
$\Delta T/T \times 10^{-6}$	1.6	2.4	3.6	4.3	1.7	2.0	4.3	14.4	147	6670
Polarization	yes	yes	yes	yes	no	yes	yes	no	yes	no

TABLE 5. Experimental parameters of the Planck satellite. HFI refers to the high frequency component of the instrument package, and LFI is the low frequency instrument. The $\Delta T/T$ sensitivity is per beam area in one year (thermodynamic temperature).

very promising.

6. Satellite experiments

Two new satellite experiments to study the CMB are currently under construction.These are MAP, or Microwave Anisotropy Probe, which has been selected by NASA as a Midex mission, and will launch in early 2001, and the Planck Surveyor, which has been selected by ESA as an M3 mission, and will be launched by 2007. The MAP satellite, has five frequency channels from 30 GHz to 100 GHz, with best resolution 12 arcmin, and is described in detail elsewhere in this volume.

The Planck Surveyor satellite combines both HEMT and bolometer technology in 10 frequency channels covering the range 30 GHz to 850 GHz, with best resolution 5 arcmin. An artists impression of this satellite is shown in Figure 15. Both these missions are of course of huge importance for CMB research and cosmology, and even well ahead of launch have sparked off intense theoretical interest, and many new research programmes in theoretical CMB astronomy and data analysis.

The experimental parameters of the Planck mission are shown in Table 5. The mission is designed to give high sensitivity to CMB structures,

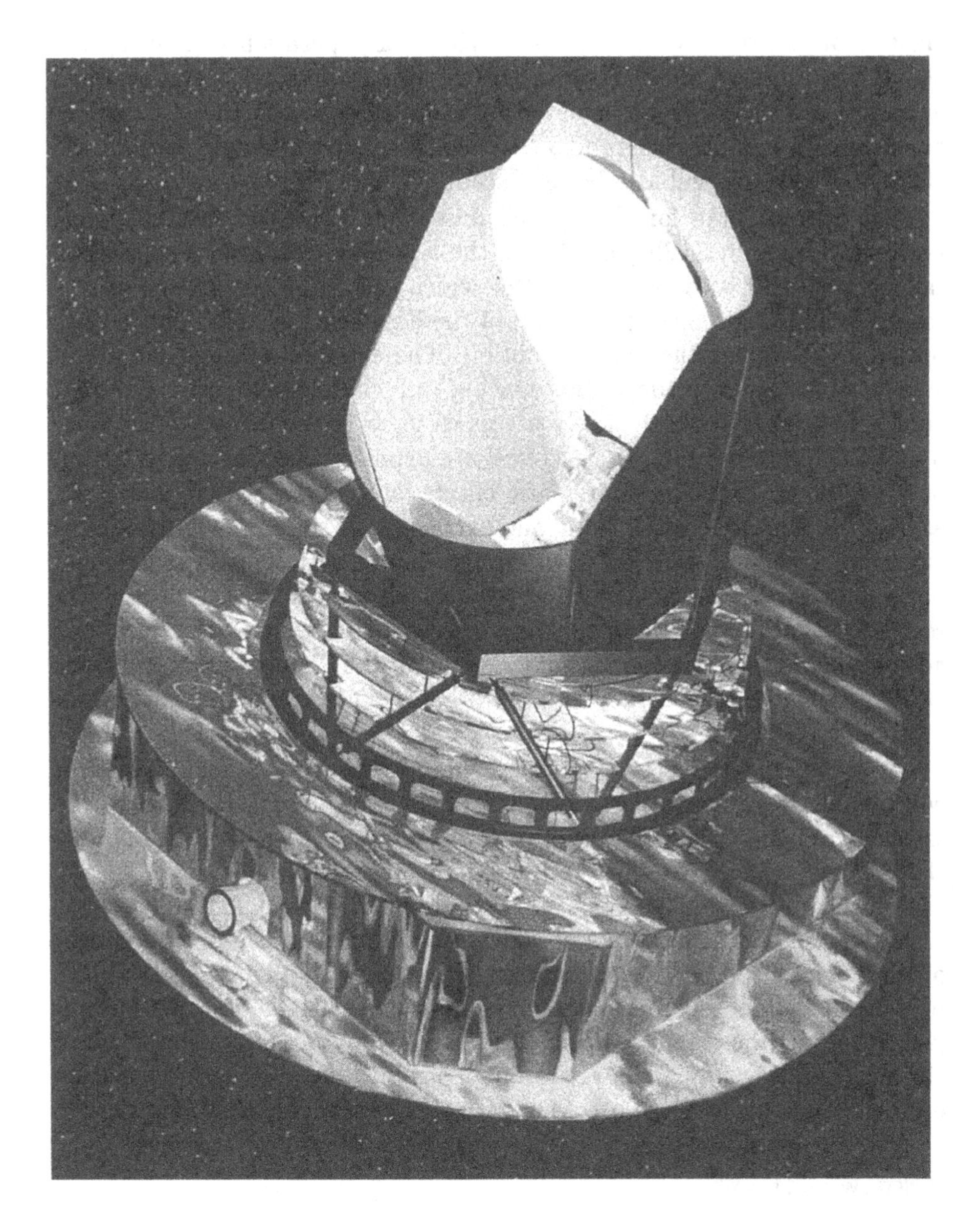

Figure 15. Artists impression of the Planck Satellite.

together with sufficient frequency coverage to enable accurate separation of the non-CMB physical components. These will typically be Galactic dust, synchrotron and free-free emission, together with extragalatic radio and sub-mm/FIR sources. Also present will be the effects of Sunyaev-Zeldovich distortion of the CMB as it passes through the hot intracluster gas of clusters of galaxies. Schemes for achieving this separation (e.g. using the Maximum Entropy Method, as in [46]), typically show that the required accuracy can probably be achieved over the majority of the sky, and as an extremely significant byproduct, catalogues of extragalactic sources and clusters of galaxies will also be produced.

The current plan for the Planck mission involves the satellite being launched while physically joined with the FIRST satellite, with separation taking place in low Earth orbit, after which each proceeds independently to the second Lagrangian point (L2) of the Earth-Sun system. This launch option is known as the 'carrier' option. The satellite will spin at about 1 rpm with the detector beams tracing out circles quite close to great circles, in a pattern which slowly precesses to cover the full sky once every 6 months. Thus two full sky coverages are expected over the 1 year course of the mission. The main systematic effects expected for the Planck data, aside from those due to Galactic and extragalactic sources, will arise from 'scanning effects' around the scan circles, due to variations in instrument noise and d.c. levels. Schemes for removal of these scan effects and the reconstruction of the sky using the total set of scan circles are now well advanced, and there is good reason to hope that the Planck results will reach close to the levels of accuracy indicated by the noise sensitivity projections in Table 5.

A crucial aspect of the Planck mission, will be the extent to which it will be able to recover *polarization* information on the CMB. Polarization sensitive bolometers will be used in 3 of the High Frequency Instrument (HFI) channels, and all 4 of the Low Frequency Instrument (LFI) HEMT channels can discriminate polarized radiation. The projected polarization sensitivity should be sufficient to make measurement of the B component of the polarization power spectrum measurable, with a consequent ability to unambiguously separate a tensor mode from scalar mode contributions [47]. However, the extent to which foreground contamination compromises the achievable accuracy has yet to be determined. For further information on the Planck project, see e.g. `http://astro.estec.esa.nl/Planck/`.

Acknowledgments

I would like to thank many colleagues in the Cavendish Astrophysics Group for assistance and in particular S. Bridle for help with figures. I would also like to thank S. Bridle, M. Hobson, O. Lahav, A. Dekel and I. Zehavi for permission to quote from joint work before publication.

References

1. D. Polarski and A.A. Starobinsky. Semi-classicality and decoherence of cosmological

perturbations. *Class.Quant.Grav.*, 13:377, 1996.
2. S. Gratton and N. Turok. Cosmological perturbations from the no boundary Euclidean path integral. *Phys.Rev.D*, 6012:3507, 1999.
3. J. Magueijo, C. Contaldi, and M. Hindmarsh. Structure formation with strings plus inflation: a new paradigm. Proceedings of the 3K conference, Rome98, astro-ph/9903050, 1999.
4. A. Kogut, A.J. Banday, C.L. Bennett, K.M. Gorski, G. Hinshaw, G.F. Smoot, and E.L. Wright. Microwave emission at high Galactic latitudes in the four-year DMR sky maps. *Astrophys.J.*, 464:L5, 1996.
5. E.M. Leitch, A.C.S. Readhead, T.J. Pearson, and S.T Myers. Anomalous component of galactic emission. *Astrophys. J.*, 486:L23, 1997.
6. A. Oliveira-Costa, A. Kogut, M.J. Devlin, C.B. Netterfield, L. Page, and E.J. Wollack. Galactic microwave emission at degree angular scales. *Astrophys. J.*, 482:L17, 1997.
7. A. de Olivera-Costa, M. Tegmark, C.M. Gutierrez, A.W. Jones, R.D. Davies, A.N. Lasenby, R. Rebolo, and R.A. Watson. Cross-correlation of Tenerife data with Galactic templates — evidence for spinning dust? *Astrophys.J.Lett.*, 527:L9, 1999.
8. P. Mukherjee, A.W. Jones, R. Kneissl, and A.N. Lasenby. On dust-correlated Galactic emission in the Tenerife data. Submitted to *Mon.Not.R.astr.Soc.*, astro-ph/0002305, 2000.
9. P.R. McCullough, J.E. Gaustad, and W. Rosing. Implications of H_α observations for studies of the CMB. In A. de Oliveira-Costa and M. Tegmark, editors, *Microwave Foregrounds*, page 253. Astronomy Society of the Pacific Conference Series, Vol. 181, 1999.
10. B.T. Draine and A. Lazarian. Diffuse galactic emission from spinning dust grains. *Astrophys.J.*, 494:L19, 1998.
11. B.T. Draine and A. Lazarian. Magnetic dipole microwave emission from dust grains. *Astrophys.J.*, 512:740, 1999.
12. B.T. Draine and A. Lazarian. Microwave emission from galactic dust grains. In A. de Oliveira-Costa and M. Tegmark, editors, *Microwave Foregrounds*, page 133. Astronomy Society of the Pacific Conference Series, Vol. 181, 1999.
13. G. Lagache, L. M. Haffner, R. J. Reynolds, and S. L. Tufte. Evidence for dust emission in the Warm Ionised Medium using WHAM data. *Astron.Astrophys.*, 354:247, 2000.
14. L. Toffolatti, G. De Zotti, F. Argüeso, and C. Burigana. Extragalactic radio sources and CMB anisotropies. In A. de Oliveira-Costa and M. Tegmark, editors, *Microwave Foregrounds*, page 153. Astronomy Society of the Pacific Conference Series, Vol. 181, 1999.
15. C. Stanghellini, C.P. O'Dea, D. Dallacasa, S.A. Baum, R. Fanti, and C. Fanti. A complete sample of GHz-peaked-spectrum radio sources and its radio properties. *Astro.Astrophys.Supp.*, 131(2):303, 1998.
16. C.M. Gutierrez, R. Rebolo, R.A. Watson, R.D. Davies, A.W. Jones, and A.N. Lasenby. The Tenerife Cosmic Microwave Background maps: observations and first analysis. *Astrophys.J.*, 529:47, 2000.
17. C.B. Netterfield, M.J. Devlin, N. Jarosik, L. Page, and E.J. Wollack. A measurement of the angular power spectrum of the anisotropy in the cosmic microwave background. *Astrophys. J.*, 474:47, 1997.
18. J.C. Baker, K. Grainge, M.P. Hobson, M.E. Jones, R. Kneissl, A.N. Lasenby, C. O'Sullivan, G. Pooley, G. Rocha, R.D.E. Saunders, P.F. Scott, and E.M. Waldram. Detection of cosmic microwave background structure in a second field with the Cosmic Anisotropy Telescope. *Mon.Not.R.astr.Soc.*, 308:1173, 1999.
19. A.D. Miller, R. Caldwell, M.J. Devlin, W.B. Dorwart, T. Herbig, M.R. Nolta, L.A. Page, J. Puchalla, E. Torbet, and H.T. Tran. A measurement of the angular power spectrum of the CMB from $\ell = 100$ to 400. *Astrophys.J.*, 524:L1, 1999.
20. P.D. Mauskopf, P.A.R. Ade, P. de Bernardis, J.J. Bock, J. Borrill, A. Boscaleri,

B.P. Crill, G. DeGasperis, G. De Troia, P. Farese, P.G. Ferreira, K. Ganga, M. Giacometti, S. Hanany, V.V. Hristov, A. Iacoangeli, A.H. Jaffe, A.E. Lange, A.T. Lee, S. Masi, A. Melchiorri, F. Melchiorri, L. Miglio, T. Montroy, C.B. Netterfield, E. Pascale, F. Piacentini, P.L. Richards, G. Romeo, J.E. Ruhl, E. Scannapieco, F.Scaramuzzi, R. Stompor, and N. Vittorio. Measurement of a peak in the cosmic microwave background power spectrum from the North American test flight of BOOMERANG. Submitted to *Astrophys.J.*, astro-ph/9911444, 1999.

21. A. Melchiorri, P.A.R. Ade, P. de Bernardis, J.J. Bock, J. Borrill, A. Boscaleri, B.P. Crill, G. De Troia, P. Farese, P.G. Ferreira, K. Ganga, G. de Gasperis, M. Giacometti, V.V. Hristov, A.H. Jaffe, A.E. Lange, S. Masi, P.D. Mauskopf, L. Miglio, C.B. Netterfield, E. Pascale, F. Piacentini, G. Romeo, J.E. Ruhl, and N. Vittorio. A measurement of Omega from the North American test flight of BOOMERANG. Submitted to *Astrophys.J.*, astro-ph/9911445, 1999.
22. P. de Bernardis, P.A.R. Ade, J.J. Bock, J.R. Bond, J. Borrill, A. Boscaleri, K. Coble, B.P. Crill, G. De Gasperis, P.C. Farese, P.G. Ferreira, K. Ganga, M. Giacometti, E. Hivon, V.V. Hristov, A. Iacoangeli, A.H. Jaffe, A.E. Lange, L. Martinis, S. Masi, P.V. Mason, P.D. Mauskopf, A. Melchiorri, L. Miglio, T. Montroy, C.B. Netterfield, E. Pascale, F. Piacentini, D. Pogosyan, S. Prunet, S. Rao, G. Romeo, J.E. Ruhl, F. Scaramuzzi, D. Sforna, and N. Vittorio. A flat universe from high-resolution maps of the cosmic microwave background radiation. *Nature*, 404:955, 2000.
23. A.E. Lange, P.A.R. Ade, J.J. Bock, J.R. Bond, J. Borrill, A. Boscaleri, K. Coble, B.P. Crill, P. de Benardis, P. Farese, P. Ferreira, K. Ganga, M. Giacometti, E. Hivon, V. V. Hristov, A. Iacoangeli, A.H. Jaffe, L. Martinis, S. Masi, P.D. Mauskopf, A. Melchiorri, T. Montroy, C.B. Netterfield, E. Pascale, F. Piacentini, D. Pogosyan, S. Prunet, S. Rao, G. Romeo, J.E. Ruhl, F. Scaramuzzi, and D. Sforna. First estimations of cosmological parameters from BOOMERANG. Submitted, astro-ph/0005004, 2000.
24. M. Tegmark and M. Zaldarriaga. New CMB constraints on the cosmic matter budget: trouble for nucleosynthesis? Submitted to *Astrophys.J.*, astro-ph/0004393, 2000.
25. S. Hanay, P. Ade, A. Balbi, J. Bock, J. Borrill, A. Boscaleri, P. de Bernardis, P.G. Ferreira, V.V. Hristov, A.H. Jaffe, A.E. Lange, A.T. Lee, P.D. Mauskopf, C.B. Netterfield, S. Oh, E. Pascale, B. Rabii, , P.L. Richards, G.F. Smoot, R. Stompor, C.D. Winant, and J.H.P. Wu. MAXIMA-1: a measurement of the cosmic microwave background anisotropy on angular scales of $10'$ to $5°$. Submitted to *Astrophys.J.*, astro-ph/0005123, 2000.
26. A. Balbi, P. Ade, J. Bock, J. Borrill, A. Boscaleri, P. de Bernardis, P.G. Ferreira, S. Hanany, V.V. Hristov, A.H. Jaffe, A.T. Lee, S. Oh, E. Pascale, B. Rabii, P.L. Richards, G.F. Smoot, R. Stompor, C.D. Winant, and J.H.P. Wu. Constraints on cosmological parameters from MAXIMA-1. Submitted to *Astrophys.J.*, astro-ph/0005124, 2000.
27. O.P. Lay and N.W. Halverson. The impact of atmospheric fluctuations on degree-scale imaging of the Cosmic Microwave Background. Submitted to *Astrophys.J.*, astro-ph/9905369, 1999.
28. M.E. Jones. In F.R. Bouchet, R. Gispert, B. Guiderdoni, and J. Tran Thanh Van, editors, *'Microwave Background Anisotropies', Proceedings of the XVIth Moriond Astrophysics Meeting*, page 161. Editions Frontières, 1997.
29. T.J. Pearson. http://astro.caltech.edu/~tjp/CBI/, 1998.
30. M. White, J.E. Carlstrom, and M. Dragovan. Submitted to Astrophys.J., astro-ph/9712195, 1997.
31. A.A. Stark, J.E. Carlstrom, F.P. Israel, K.M. Menten, J.B. Peterson, T.G. Phillips, G. Sironi, and C.K. Walker. pre-print, astro-ph/9802326, 1998.
32. J.P. Ostriker and E.T. Vishniac. Generation of microwave background fluctuations from non-linear perturbations at the era of galaxy formation. *Astrophys. J.*, 306:L51, 1986.

33. A.H. Jaffe and M. Kamionkowski. Calculation of the Ostriker-Vishniac effect in cold dark matter models. *Phys.Rev.D*, 5804:3001, 1998.
34. L. Knox, R. Scoccimarro, and S. Dodelson. Impact of inhomogeneous reionization on cosmic microwave background anisotropy. *Phys.Rev.Lett.*, 81:2004, 1998.
35. G.P. Holder, J.J. Mohr, J.E. Carlstrom, A.E. Evrard, and E.M. Leitch. Expectations for an interferometric Sunyaev-Zel'dovich effect survey for galaxy clusters. Submitted to *Astrophys.J.*, astro-ph/9912364, 2000.
36. N. Aghanim, C. Balland, and J. Silk. Sunyaev-Zel'dovich constraints from black hole-seeded proto-galaxies. To appear in Astron.Astrophys., astro-ph/0003254, 2000.
37. Web page for AMIBA telescope. http://www.asiaa.sinica.edu.tw/amiba/.
38. M. Webster, S.L. Bridle, M.P. Hobson, A.N. Lasenby, O. Lahav, and G. Rocha. Joint estimation of cosmological parameters from CMB and IRAS data. *Astrophys.J.Lett.*, 509:L65, 1998.
39. S.L. Bridle, V.R. Eke, O. Lahav, A.N. Lasenby, M.P. Hobson, S. Cole, C.S. Frenk, and J.P. Henry. Cosmological parameters from cluster abundances, cosmic microwave background and IRAS. *Mon.Not.R.astr.Soc.*, 310:565, 1999.
40. G.P. Efstathiou, S.L. Bridle, A.N. Lasenby, M.P. Hobson, and R.S. Ellis. Constraints on Ω_m and Ω_Λ from distant Type Ia supernovae and Cosmic Microwave Background anisotropies. *Mon.Not.R.astr.Soc.*, 303:L47, 1999.
41. A.N. Lasenby, S.L. Bridle, M.P. Hobson, and G.P. Efstathiou. Constraints on H_0 from combining CMB, LSS, Supernovae and Clusters. To appear in *Proceedings of the 19th Texas Symposium on Relativistic Astrophysics, Paris, 1998*, 1999.
42. S.L. Bridle, I. Zehavi, A. Dekel, O. Lahav, M.P. Hobson, and A.N. Lasenby. Cosmological parameters from velocities, CMB and supernovae. Submitted to *Mon.Not.R.astr.Soc.*, astro-ph/0006170, 2000.
43. M.P. Haynes, R. Giovanelli, J.J. Salzer, G. Wegner, W. Freudling, L.N. da Costa, T. Herter, and N.P. Vogt. The I-band Tully-Fisher relation for Sc galaxies: Optical imaging data. *Astronom.J.*, 117:1668, 2000.
44. M.P. Haynes, R. Giovanelli, P. Chamaraux, L.N. da Costa, W. Freudling, J.J. Salzer, and G. Wegner. The I-band Tully-Fisher relation for Sc galaxies: 21 centimeter HI line data. *Astronom.J.*, 117:2039, 2000.
45. S. Perlmutter, G. Aldering, G. Goldhaber, R.A. Knop, P. Nugent, P.G. Castro, S. Deustua, S. Fabbro, A. Goobar, D.E. Groom, I.M. Hook, A.G. Kim, M.Y. Kim, J.C. Lee, R. Pain N.J. Nunes, C.R. Pennypacker, R. Quimby, C. Lidman, R.S. Ellis, M. Irwin, R.G. McMahon, P. Ruiz-Lapuente, N. Walton, B.J. Boyle B. Schaefer, A.V. Filippenko, T. Matheson, A.S. Fruchter, N. Panagia, H.J.M. Newberg, and W.J. Couch. Measurements of Ω and Λ from 42 high-redshift supernovae. *Astrophys.J.*, 517:565, 1999.
46. M.P. Hobson, A.W. Jones, A.N. Lasenby, and F.R. Bouchet. Component separation methods for satellite observations of the cosmic microwave background. *Mon.Not.R.astr.Soc.*, 300:1, 1998.
47. M. Zaldarriaga and U. Seljak. An all-sky analysis of polarization in the microwave background. *Phys.Rev.D.*, 55:1830, 1997.

CMB ANALYSIS

J. RICHARD BOND
Canadian Institute for Theoretical Astrophysics
Toronto, ON M5S 3H8, CANADA

AND

ROBERT G. CRITTENDEN
DAMTP, Centre for Mathematical Sciences,
University of Cambridge, Cambridge CB3 0WA, United Kingdom.

Abstract

We describe the subject of Cosmic Microwave Background (CMB) analysis — its past, present and future. The theory of Gaussian primary anisotropies, those arising from linear physics operating in the early Universe, is in reasonably good shape so the focus has shifted to the statistical pipeline which confronts the data with the theory: mapping, filtering, comparing, cleaning, compressing, forecasting, estimating. There have been many algorithmic advances in the analysis pipeline in recent years, but still more are needed for the forecasts of high precision cosmic parameter estimation to be realized. For secondary anisotropies, those arising once nonlinearity develops, the computational state of the art currently needs effort in all the areas: the Sunyaev-Zeldovich effect, inhomogeneous reionization, gravitational lensing, the Rees-Sciama effect, dusty galaxies. We use the Sunyaev-Zeldovich example to illustrate the issues. The direct interface with observations for these non-Gaussian signals is much more complex than for Gaussian primary anisotropies, and even more so for the statistically inhomogeneous Galactic foregrounds. Because all the signals are superimposed, the separation of components inevitably complicates primary CMB analyses as well.

1. Introduction

1.1 What is CMB Analysis? The subject we call "CMB analysis" is a blend of basic theory, simulation and statistical data analysis. The goal of CMB analyzers is to extract the physics of the various signals that contribute to the data. CMB analyzers may therefore be theorists or ex-

R.G. Crittenden and N.G. Turok (eds.), Structure Formation in the Universe, 241–280.

perimentalists. Depending upon how advanced the state-of-the-art is, the relevant analysis topic may lean more towards the theory/simulation side or more towards the statistical analysis and Monte Carlo simulation side. In the CMB field, we adopt a theorist's distinction between *primary*, *secondary* and *foreground* anisotropies, though of course the observations do not know of this distinction. The primary ones are those we can calculate using linear perturbation theory (or, in the case of cosmic defects, with linear response theory). This covers the crucially important epoch of photon decoupling near redshift 1100, and even until the current time on very large scales, and to redshifts of a few on intermediate scales. Secondary anisotropies are those associated with nonlinear phenomena, either calculable via weakly nonlinear perturbation theory, semi-analytic methods or by more direct simulation of nonlinear patterns. Gravitational lensing effects on the CMB, quadratic nonlinearities and the kinetic Sunyaev-Zeldovich effect associated with Thomson scattering from flowing matter, reionization inhomogeneities, the thermal Sunyaev-Zeldovich effect associated with Compton upscattering from hot gas, the Rees-Sciama effect associated with nonlinear potential wells, all come under this heading. We also traditionally call emission by dust in high redshift galaxies a secondary process, and even emission from extragalactic radio sources. On top of this, there are various foreground emissions from dust and gas in our Milky Way galaxy — signals which are nuisances to the primary CMBologist but of passionate interest to interstellar medium astronomers. Fortunately, most of the secondary and foreground signals have very different dependencies on frequency (Fig. 1), and rather statistically distinct sky patterns (Fig. 2). Collateral information from non-CMB observations can also be brought to bear to unravel the various components.

1.2 Primary Theory and Current Data: We think we know how to calculate the primary signal in exquisite detail. The fluctuations are so small at the epoch of photon decoupling that linear perturbation theory is a superb approximation to the exact non-linear evolution equations. Intense theoretical work over three decades has put accurate calculations of this linear cosmological radiative transfer problem on a firm footing, and there are speedy, publicly available and widely used codes for evaluation of anisotropies in a wide variety of cosmological scenarios, e.g. "CMBfast" [1]. These have been further modified by many different groups to attack even more structure formation models.

The simplest and thus least baroque versions of inflation theory predict that the fluctuations from the quantum noise that give rise to structure form a Gaussian random field. Linearity implies that this translates into temperature anisotropy patterns that are drawn from a Gaussian random

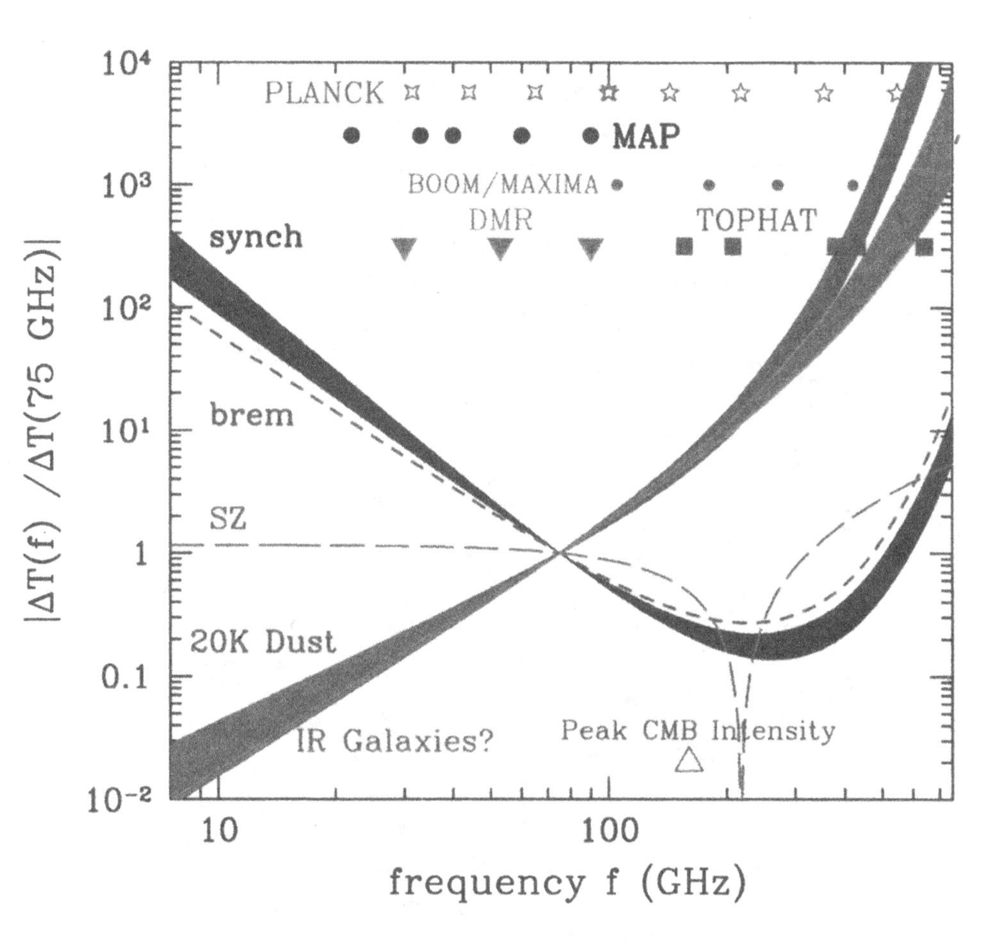

Figure 1. Foregrounds and secondary anisotropies depend differently on photon frequency, the key property for separating the components. Plotted is the frequency dependence of the effective thermodynamic temperature fluctuations, normalized to their values at 75 GHz. Thus the primary CMB fluctuations correspond to a horizontal line in this figure, synchrotron and bremsstrahlung dominate at low frequencies, dust at high. The bands represent a measure of our uncertainty in the appropriate foreground spectra. The highly distinctive shape for the Sunyaev-Zeldovich (SZ) effect is negative at low frequencies, positive at high. The actual level of contamination depends on the angular scale. The foreground emission is minimal around 90 GHz, not far from where the CMB intensity peaks. Detector frequencies for some notable experiments are denoted by the symbols at the top. Currently, detectors below 100 GHz are HEMTs, above are bolometers. Note the wide coverage planned for the Planck satellite which uses both.

process and which can be characterized solely by their power spectrum. The emphasis is therefore on confronting the theory with the data in power spectrum space, as in Fig. 3. The panels show how the power spectrum C_ℓ

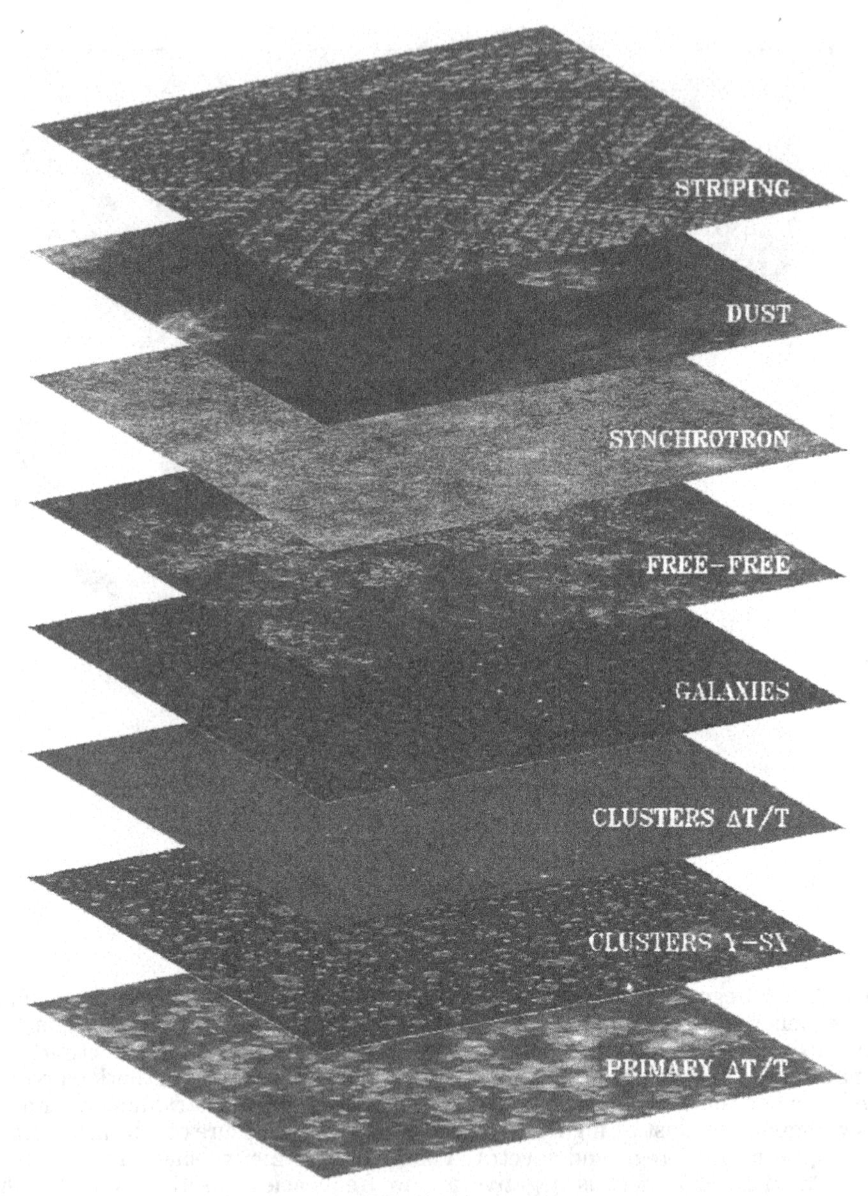

Figure 2. A schematic view by Bouchet and Gispert of many of the possible microwave foregrounds that need to be separated from the primary CMB anisotropy pattern shown in the 10° map at the bottom. These include noise effects (striping), galactic foregrounds (dust, synchrotron and bremsstrahlung or free-free emission) as well as secondary anisotropies (extragalactic radio and infrared galaxies, CMB upscattering by hot gas in clusters – "clusters Y-SX," and the Doppler effect arising from moving clusters – the kinetic SZ effect or "clusters $\Delta T/T$"). Each of these has a unique temperature pattern on the sky, and all but the kinetic SZ effect have a different spectral signature (Fig. 1).

responds as various cosmological parameters are individually changed.

Until a few years ago, the emphasis was on fairly restricted parameter spaces, but now we consider spaces of much larger dimension. These are used for forecasts of how well proposed experiments can do, and for the first round of data from a sequence of high-precision experiments covering large areas of sky: ground-based single dishes and interferometers; balloons of short and long duration (LDBs, $\sim$ 10 days) and eventually of ultra-long duration (ULDBs, $\sim$ 100 days); NASA's MAP satellite in 2001 and ESA's Planck Surveyor in 2007. The lower right panel of Fig. 3 gives a forecast of how well we think that the two satellite experiments can do in determining C_ℓ if everything goes right. This rosy prognosis should be compared with the compressed bandpowers shown in the other panels estimated from all of the current data: DMR and some 19 other pre-conference "prior" experiments, TOCO [2], unveiled about the time of the conference, and four that appeared after the conference, the Boomerang flights (the LDB flight [3, 4] and its North America test flight [5]) the Maxima-I [6, 7] flight, the first results from CBI, the Cosmic Background Imager [8], and from DASI, the Degree Angular Scale Interferometer [9]. Although new methods have been applied to these data sets, much was in place and some algorithms were well tested on previous data, in particular DMR, SK95, MSAM and QMAP.

Although we sketch the current state of the art in CMB analysis in this paper, note that this is a very fast moving subject. A large number of researchers in a handful of groups are working on a broad range of issues, often as members of the experimental teams associated with the increasingly complex experiments. A snapshot of the state two years ago and references to the relevant papers was given in [33], but many advancements have been made since then. For example, the MAP plan for analysis is described in [40]. One issue that still needs to be addressed is that the strong simplifying assumptions of signal and noise Gaussianity cannot be correct in detail, and could yield misleading results.

1.3 Primary Parameter Sets for Gaussian Theories: A "minimal" inflation-motivated parameter set involves about a dozen parameters, e.g., $\{\omega_b, \omega_{cdm}, \omega_{hdm}, \omega_{er}, \Omega_k, \Omega_\Lambda, \tau_C, \sigma_8\, n_s, \tilde{r}_t, n_t\}$. For many species we use $\omega_j \equiv \Omega_j \mathrm{h}^2$ rather than Ω_j since it directly gives the physical density of the particles rather than as a ratio to the critical density: thus, ω_b for baryons, ω_{cdm} for cold dark matter, ω_{hdm} for hot dark matter (massive but light neutrinos), and ω_{er} for relativistic particles present at decoupling (photons, very light neutrinos, and possibly weakly interacting products of late time particle decays). The total non-relativistic matter density is $\omega_m \equiv \omega_{hdm} + \omega_{cdm} + \omega_b$.

The curvature energy relative to the closure density is $\Omega_k \equiv 1-\Omega_{tot}$ and Ω_Λ is the energy relative to closure in the cosmological constant. Ω_Λ can also be interpreted as a vacuum energy, a sector which may be more complex and require more parameters. For example, a scalar field which dominates at late times, "quintessence", has potentials and other interactions which must be set. A simple phenomenology of quintessence has added to Ω_Q an average equation of state pressure-to-density ratio $w_Q \equiv \bar{p}_Q/\bar{\rho}_Q$.

Another factor crucial for predicting the CMB anisotropies is τ_C, the Compton depth to Thompson scattering from now back through the period when the universe was reionized by early starlight.

To characterize the initial fluctuations, amplitudes and tilts are required for the various fluctuation modes present. For adiabatic scalar perturbations, the tilt is n_s and the amplitude is most appropriately the overall power in curvature fluctuations, but is often replaced by σ_8^2, a density bandpower on the scale of clusters of galaxies, $8\,\mathrm{h}^{-1}$ Mpc. For tensor modes induced by gravity waves, the amplitude parameter is some measure of the ratio of gravity wave to scalar curvature power, $\tilde{r}_t$; usually a ratio of $\mathcal{C}_\ell$'s is used, e.g., at ℓ=2 or 10. Inflation theories often give relationships between the tensor tilt n_t and $\tilde{r}_t$ which can be used to reduce the parameter space e.g., [10]. Parameters for the amplitude and tilt of scalar isocurvature modes could also be added to the mix. The characterization of the modes present in the early universe by amplitude and tilt could be expanded to include parameters describing the change of the tilts with scale $(dn_s/d\ln k)$, the change of that change, and so on. Relatively full functional freedom in $n_s(k)$ is possible in inflation models, and substantial freedom also exists for $n_t(k)$.

1.4 Well and Poorly Determined Parameter Combinations: In Fig. 3, we compare the compressed bandpowers with $\mathcal{C}_\ell$ sequences as ω_b, Ω_{tot}, Ω_Λ, ω_{hdm} and n_s are individually varied. These show the discriminatory power of the current data. Fixing the age at 13 Gyr as we do here defines a specific relation between ω_k, ω_Λ and ω_m. From these figures, we expect that four can be determined reasonably well ($\omega_b, \Omega_{tot}, n_s$ and the overall amplitude). This is borne out in the full study with the real data, modulo parameter correlations which mean the best determined quantities are linear combinations of parameters, or parameter eigenmodes [12, 13, 10]. For example, it is actually a combination of Ω_{tot} and Ω_Λ which is well determined, but it happens to be Ω_{tot}–dominated. Another orthogonal combination of the two will remain very poorly determined no matter how good the CMB data gets [12], an example of a near-degeneracy among cosmological parameters. Other near-degeneracy examples are less severe: e.g., $\Omega_{hdm}/(\Omega_{cdm}+\Omega_{hdm})$ is not well determined by current data, but could be

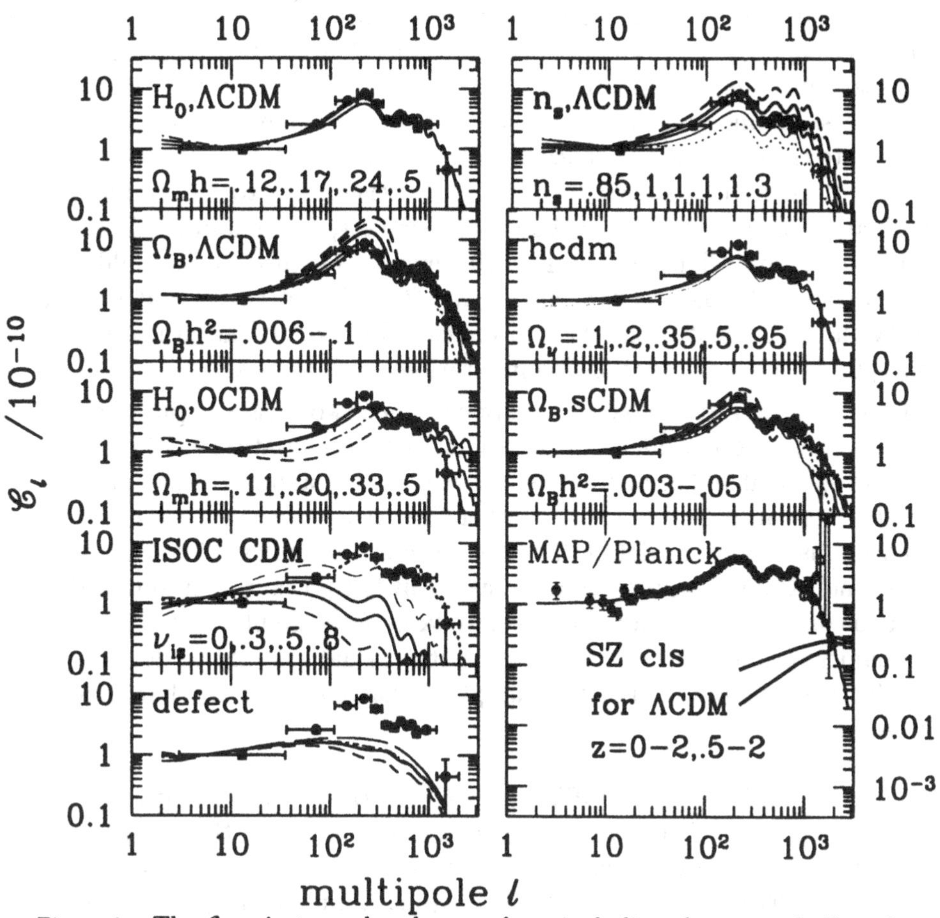

Figure 3. The $\mathcal{C}_\ell$ anisotropy bandpower data, including the recent balloon-borne and interferometry data, compressed to 13 bands using the methods of [14] are compared with various $\mathcal{C}_\ell$ model sequences, each for universes with age 13 Gyr (left to right): (1) untilted flat ΛCDM sequence with H_0 varying from 50 to 90, Ω_Λ from 0 to 0.87; (2) n_s varying from 0.85 to 1.3 for an $H_0 = 70$ (Ω_Λ=.66) ΛCDM model – dotted is 0.85 with gravity waves, next without, upper dashed is 1.25, showing visually why n_s is found to be nearly unity; (3) $\Omega_B\mathrm{h}^2$ varying from 0.0063 to 0.10 for the H_0=70 Ω_Λ=.66 ΛCDM model; (4) neutrino fractions Ω_{hdm}/Ω_m varying from 0.1 to 0.95 for an H_0=50 Ω_Λ=0 sequence with Ω_m=1; (5) H_0 from 50 to 65, Ω_k from 0 to 0.84 for the untilted oCDM sequence, showing the strong ℓ-shift of the acoustic peaks with Ω_k; (6) $\Omega_B\mathrm{h}^2$ varying from 0.003 to 0.05 for the $H_0 = 50$ Ω_Λ=0 sCDM model; (7) an isocurvature CDM sequence with positive isocurvature tilts ranging from 0 to 0.8; (9) sample defect $\mathcal{C}_\ell$'s for textures, etc. from [15] – cosmic string $\mathcal{C}_\ell$'s from [16] are similar and also do not fare well compared with the current data. The bottom right panel is extended to low values to show the magnitude of secondary fluctuations from the thermal SZ effect for the ΛCDM model. The kinematic SZ $\mathcal{C}_\ell$ is significantly lower. Dusty emission from early galaxies may lead to high signals, but the power is concentrated at higher ℓ, with a weak tail because galaxies are correlated extending into the $\ell \lesssim 2000$ regime. Forecasts of how accurate $\mathcal{C}_\ell$ will be determined for an sCDM model from MAP (error bars growing above $\ell \sim 700$) and Planck (small errors down the $\mathcal{C}_\ell$ damping tail) are also shown.

by Planck.

Accompanying the idealized bandpower forecasts are predictions for how well the cosmological parameters in inflation models can be determined after integrating, i.e., marginalizing, the likelihood functions over all other parameters. In one exercise that allowed a mix of nine cosmological parameters to characterize the space of inflation-based theories (the 11 above with ω_{er} fixed and n_t slaved to $\tilde{r}_t$), COBE was shown to determine one combination of them to better than 10% accuracy, LDBs and MAP could determine six, and Planck seven [13]. All currently planned LDBs could also get two combinations to 1% accuracy, MAP could get three, and Planck five. With the current data, one can forecast four to 10% accuracy, which is actually borne out in detailed computations with the data.

Two panels in Fig. 3 show what happens when two classes of pure isocurvature models are considered, one a Gaussian isocurvature CDM model with the spectral index allowed to tilt arbitrarily, another a set of non-Gaussian cosmic defect models. Neither fare well with the current data. Primary anisotropies in defect theories are more complicated to calculate because non-Gaussian patterns are created in the phase transitions which evolve in complex ways and which require large scale simulations. The non-Gaussianity means that the comparison with the data should be done more carefully, without the Gaussian assumption that goes into the bandpower estimations.

1.5 Current CMB and CMB+LSS Constraints: Although the purpose of this paper is to describe CMB analysis techniques, the comparison of the radically-compressed current bandpowers with the models in Fig. 3 invite a brief description of the results. Explicit numbers are those derived in [4] for the "minimal" inflation-motivated 7-parameter set for the combination of DMR and Boomerang, with a weak prior probability assumption on the Hubble parameter ($0.45 < h < 0.90$) and age of the Universe (> 10 Gyr.) Including all other current CMB data as well gives very similar estimates and errors, reflecting the consistency between the older, less statistically significant data and the new experiments. (DMR+DASI numbers [9] are quite close to DMR+Boomerang numbers, since the spectra are similar. For a more complete description, see [32, 56, 65, 4, 9, 7].)

1.5.1 CMB-only Estimates: As the upper right panel of Fig. 3 suggests, the primordial spectral tilt n_s is well determined using CMB data alone: $n_s = 0.97^{+.10}_{-.08}$. This is rather encouraging for the nearly scale invariant models preferred by inflation theory. The figure also shows that constraints on Ω_Λ will not be very good, but the strong dependence of the position of the acoustic peaks on Ω_k means that it is better restricted: $\Omega_{tot} \approx 1.02 \pm 0.06$. The baryon abundance is also well determined, $\Omega_b h^2 = 0.022 \pm 0.004$, near the 0.019±0.002 estimate from deuterium observations in quasar absorption line spectra combined with Big Bang Nucleosynthesis theory.

The isocurvature CDM models with tilt $n_{is} > 0$ and the isocurvature defect models (*e.g.* strings and textures) [15, 16] shown in Fig. 3 clearly have more difficulty with the CMB bandpowers.

There is a region of the spectrum that nominally seems to be out of reach of CMB analysis, but is actually partly accessible because ultralong waves contribute gentle gradients to our CMB observables. For example, the size of compact spatial manifolds has been constrained; [66] find for flat equal-sided 3-tori, the inscribed radius must exceed $1.1\chi_{lss}$ from DMR at the 95% confidence limit, where χ_{lss} is the distance to the last scattering surface. For asymmetric 1-tori, the constraint is weaker, $> 0.7\chi_{lss}$, but still reasonably restrictive. It is also not as strong if the platonic-solid-like manifolds of compact hyperbolic topologies are considered, though the overall size of the manifold should as well be of order the last scattering surface radius to avoid conflict with the large scale power seen in the COBE maps [66].

1.5.2 CMB+LSS Estimates: We have always combined CMB and LSS data in our quest for viable models. For example, DMR normalization determines σ_8 to within 7%, and comparing with the $\sigma_8 \sim 0.55\Omega_m^{-0.56}$ target value derived from cluster abundance observations severely constrains the cosmological parameters defining the models. This is further restricted when the possible variations in the spectral tilt allowed by the COBE data are constrained by the higher ℓ Boomerang data. More constrictions arise from galaxy-galaxy and cluster-cluster clustering observations: the shape of the linear density power spectrum must match the shape reconstructed from the data. In [32, 56, 65, 4], the LSS data were characterized by a simple shape parameter and σ_8, with distributions broad enough so that these prior probability choices would not be controversial, but encompass the ranges that almost all cosmologists would believe.

Using all of the current CMB data and the LSS priors, [4] got $H_0 = 56\pm9$, $\Omega_\Lambda = 0.55\pm0.09$, with n_s, Ω_{tot} and $\Omega_b h^2$ virtually unchanged from the CMB-only result. Apart from the significant Ω_Λ detection, the dark matter is also strongly constrained, $\Omega_{cdm}h^2 = 0.13^{+.03}_{-.02}$. Restricting Ω_{tot} to be unity as in the usual inflation models, with CMB+LSS, $\Omega_b h^2 = 0.021 \pm 0.003$, $\Omega_{cdm}h^2 = 0.13 \pm 0.01$, $\Omega_\Lambda = 0.62 \pm 0.07$ and $n_s = 0.98^{+.10}_{-.07}$. If the equation of state of the dark energy is allowed to vary, $w_Q < -0.3$ is obtained, becoming substantially more restrictive when supernova information is folded in, $w_Q < -0.7$ at 2 σ [65].

Fig. 4 gives a visual perspective from [65] on how the parameter estimations for the adiabatic models evolved as more CMB data were added (but with always the LSS prior included for this figure). With just the COBE-

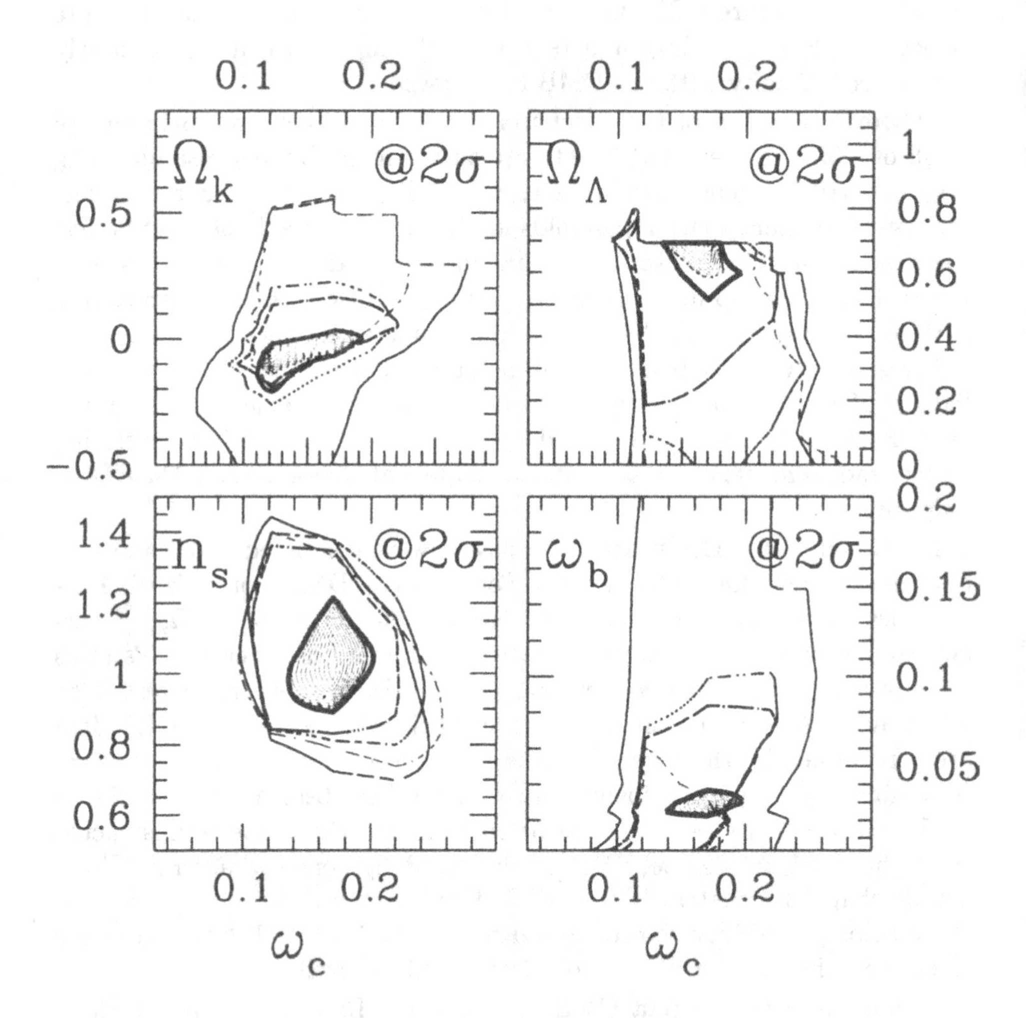

Figure 4. This figure from [65] shows 2-σ likelihood contours for the dark matter density $\omega_c = \Omega_{cdm}h^2$ and $\{\Omega_k, \Omega_\Lambda, n_s, \omega_b\}$ for the LSS prior combined with weak limits on H_0 (45-90) and cosmological age (> 10 Gyr), and the following CMB experimental combinations: DMR (short-dash); the "April 99" data (short-dash long-dash); TOCO+4.99 data (dot short-dash); Boomerang-NA+TOCO+4.99 data (dot long-dash, termed "prior-CMB"); Boomerang-LDB + Maxima-1 + Boomerang-NA + TOCO + 4.99 data (heavy solid, all-CMB). These 2σ lines tend to go from outside to inside as more CMB experiments are added. The smallest 2-σ region (dotted and interior) shows SN1+LSS+all-CMB, when SNI data is added. For the Ω_Λ, n_s and ω_b plots, Ω_{tot}=1 has been assumed, but the values do not change that much if Ω_{tot} floats. The main movement with the most recent data [4] is that ω_c localizes more around 0.13 in all panels, and the ω_b contour in the lower right panel migrates downward a bit to be in better agreement with the Big Bang nucleosynthesis result.

DMR+LSS data, the 2-σ contours were already localized in $\Omega_{cdm}h^2$. Without LSS, it took the addition of Maxima-1 before it began to localize. Ω_k localized near zero when TOCO was added to the April 99 data, more so when Boomerang-NA was added, and much more so when Boomerang-LDB and Maxima-1 were added. Some n_s localization occurred with just "prior-CMB" data. $\Omega_b h^2$ really focussed in with Boomerang-LDB and Maxima-1, as did Ω_Λ. If only DMR plus the most recent Boomerang results [4] to $\ell = 600$ are used (the limit in [3]), the plot is rather similar. However, when all recent Boomerang [4] results to $\ell = 1000$ are used, the inner contour in the $\Omega_{cdm}h^2$ direction sharpens up in all panels, and the $\Omega_b h^2$ contour lowers to be in the good agreement with the big bang nucleosynthesis result indicated above.

2. Computing Non-Gaussian Secondary Signals: the SZ Example

The secondary fluctuations involve nonlinear processes, and the full panoply of N-body and gas-dynamical cosmological simulation techniques are being used to study them. It is realistic to hope that the thermal and kinetic Sunyaev-Zeldovich effects can be understood statistically this way with sufficient accuracy for CMB analysis, and this is the main example used in this section. On the other hand, star-bursting galaxies will be quite difficult to understand from simulations alone, but CMB analysis will be greatly aided by their point-source character for most experimental resolutions. Galactic foregrounds, however, cannot be "solved" by hydro calculations.

2.1 Hydrodynamical Calculations: The Santa Barbara test [17] compared simulations of an individual cluster of $10^{15}\ M_\odot$ done with a variety of hydrodynamical and N-body techniques, in particular grid-based Eulerian methods and both grid-based and smooth particle (SPH) Lagrangian methods. (Eulerian grids are fixed in comoving space, Lagrangian grids adapt to the density of the flow.) Fig 5 shows this cluster, computed with the treePM-SPH code of Wadsley and Bond [19, 20], seen at the present in weak lensing, SZ and X-rays, at contour thresholds designed to represent potentially observable levels. For a review of the great strides made in SZ experimental methods and results on individual clusters up to 1998 see [18].

Another hydrodynamical example, the simulation of a supercluster [20, 21] using the treePM-SPH code, is used to illustrate the computational challenges once we extend beyond individual cluster simulations. Fig. 7 shows thermal SZ maps of the supercluster at redshift $z = 0.5$. The computation evolved from redshift $z = 30$ to the present a high resolution 104 Mpc diameter patch with $100^3/2$ gas and $100^3/2$ dark matter particles, surrounded by gas and dark particles with 8 times the mass to 166 Mpc, in turn surrounded by "tidal" particles with 64 times the mass to 266 Mpc.

There were a total of 1.6 million particles including the medium and low resolution regions, but much larger simulations are possible in this era of massive parallelization. For example, a parallel tree-SPH code ("gasoline"), with timesteps that can vary from particle to particle, can routinely do 256^3 gas and 256^3 dark matter simulations, and even 512^3 simulations on relatively modest (< 100 processor) BEOWULF systems [31].

The treePM technique is a fast, flexible method for solving gravity and can accurately treat free boundary conditions. The SPH part of the code includes photoionization as well as shock heating and cooling with abundances in chemical (but not thermal) equilibrium, incorporating all radiative and collisional processes. The species considered were H, H^+, He, He^+, He^{++} and e^-. The extra cooling associated with heavy elements injected into the intracluster media during galactic evolution was ignored, but so was the feedback on the medium from energy injected from galactic supernovae: the computation does not have high enough spatial resolution to calculate galaxy collapse.

We now describe some aspects of simulation design for the supercluster simulation [20, 21] which are also relevant for the larger parallel calculations. One first decides on the mass resolution to achieve, which is set by the lattice spacing of particles on an initial high resolution grid, a_L, chosen to be ~ 1 Mpc comoving to ensure that there will be adequate waves to form the target objects, in this case clusters. [1] Next one needs to determine the spatial resolution of the gas and of the gravitational forces, preferably highly linked. In Lagrangian codes like treePM-SPH, this varies considerably, being very high in cluster cores, moderate in filaments, and not that good in voids. Since accurate calculations of the cluster cores were a target, resolution better than ~ 40 kpc was desired. The best resolution obtained was about 15 kpc. Given a_L, the size of the high resolution part of the simulation volume is determined by CPU limitations on the number of particles that can be run in the desired time. This means the high resolution volume may distort considerably during the simulation. To combat this, progressively lower resolution layers were added to ensure accurate large scale tides/shearing fields operated on the high resolution patch. For

[1]The mass resolution limits the high k power of the waves that can be laid down in the initial conditions (Nyquist frequency, π/a_L), but for aperiodic patches there is no constraint at the low k end: the FFT was used for high k, but a power law sampling for medium k, and a log k sampling for low k, the latter two done using a slow Fourier transform, i.e., a direct sum over optimally-sampled k values, with the shift from one type of sampling to another determined by which gives the minimum volume per mode in k-space. By contrast, more standard periodic "big box" calculations are limited by the fundamental mode, although one can alleviate this by nesting the high resolution box in lower resolution ones as in the supercluster simulation. Even so, this simulated supercluster could not be done in a periodic code with this number of particles because of gross mishandling of the significant large scale tidal forces.

ITP95 Cluster
Comparison

seen in Lensing,
SZ & X at z=0

cosmology
τCDM:Ω=1,Ω_B=.1,
σ_8=.65,Γ=.25

final state
$M_{cl}=10^{15}\ M_\odot$
$L_X=2 \times 10^{45}$ erg/s
$T_{max}=10^8$ K
(adiabatic cl)

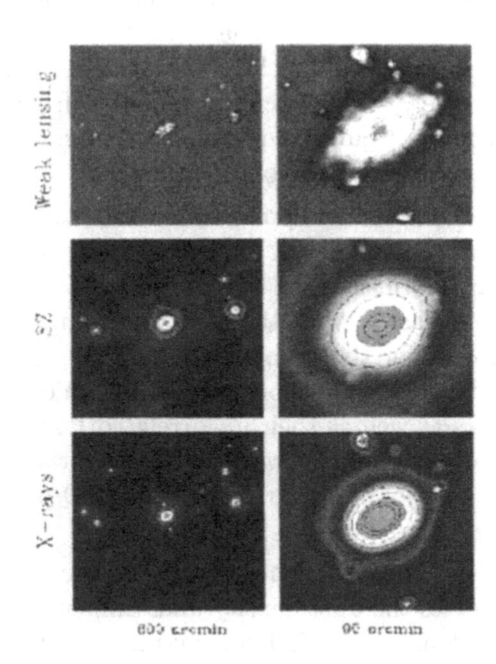

Figure 5. SZ maps for the ITP comparison cluster seen at $z = 0$ subtending the angles shown. The observing wavelength was taken to be in the Rayleigh-Jeans region of the spectrum, so $\Delta T/T = -2y$ here. The core dark regions interior to the white areas are above 32×10^{-6}; the dark contours surrounding the white are at 2×10^{-6}, levels now accessible to ground-based instrumentation. At higher redshift, gaseous filaments bridge the subclusters which merge to make the final state. Even the far-field outskirts of these subclusters is observable, but precision below 10^{-6} would be needed to probe well the filaments. (See also Fig. 7.)

the calculations shown, the high resolution region (grid spacing a_L, 100^3 sphere) sits within a medium resolution region ($2\,a_L$, 80^3), and in turn within a low resolution region ($4\,a_L$, 64^3). The mean external tide on the entire patch is linearly evolved and applied during the calculation.

2.2 Non-Gaussian Source Model: It is clear from Figs. 6 and 7 that the dominant signals are quite patchy if one is interested in amplitudes above a few μK. This has the disadvantage for analysis that non-Gaussian aspects of the predicted patterns are fundamental, and an infinite hierarchy of connected N-point spectra beyond the 2-point spectrum C_ℓ are required to specify them (as for defect models). However, it also seems reasonable that we could represent the results in terms of extended emission from localized sources. This concentration of power is characteristic of many types of secondary signals. Indeed some, such as radiation from dusty star-burst regions in galaxies, can be treated as point sources since they are much smaller than the observational resolution of most CMB experiments. However, highly accurate emission from dusty galaxies is too difficult to calculate from first principles, and statistical models of their distribution must be guided by observations.

Extended and point source models for the temperature fluctuations express the signal as a projection along the line-of-sight of a 3D random field, $\mathcal{G}(\mathbf{r},t)$, which is the convolution of profiles $g(\mathbf{r}|\mathcal{C},t)$ surrounding objects of some class $\mathcal{C}$, with the comoving number density $n_{\mathcal{C}*}(\mathbf{r},t)$ defining a point process for those objects [10, 28]. In [28, 29], the points were referred to as "shots", from shot-noise, although of course continuous clustering of the shots as well as their uncorrelated Poissonian self-correlation has to be included. Simple derivations for the form of $\mathcal{G}(\mathbf{r},t)$ and g and the relation of the 2D power spectrum C_ℓ to the 3D power spectrum of the $\mathcal{G}$ (and hence of $n_{\mathcal{C}*}$) are given in [10].

For dusty star-burst regions, the shots are galaxies and the profile g involves the dust density and temperature [28, 29], while for the SZ effect, the dominant emission comes from clusters and groups and the profile g is a constant times the pressure [28, 10, 30].

Various attempts have been made to use hydrodynamical codes to do an entire region up to $z = 2$ or so. However, it should be clear from the discussion of the numerical setup that approximations are always needed to do this. One approach was to calculate the full pressure power spectrum from a single simulation, $\langle|\widetilde{\mathcal{G}}(k,t)|^2\rangle$, and appropriately project it to determine the SZ C_ℓ [22, 23]. A more ambitious approach was to tile the space with boxes out to some redshift [24, 25, 26, 27, 31], but because of computer limitations and resolution requirements, the boxes were relatively small and computationally expensive. To make predictions, tricks were required, e.g., "observing' translated and rotated versions of the few simulated boxes at many different output redshifts.

2.3 The Shots and their Profiles: The shot model provides a way forward semi-analytically, by laying down pressure profiles around a catalogue of halos. The halos and their properties could be computed with N-body-only simulations using much larger boxes than those required for hydro. Another approach is to use "peak-patches", a method for identifying halos in the initial conditions of calculations which has been shown to be accurate for clusters [30]. For this approach, the entire space can be tiled with contiguous boxes that are smoothly joined, and have all the required long wavelength power to treat the really rare halo concentrations, or "super-

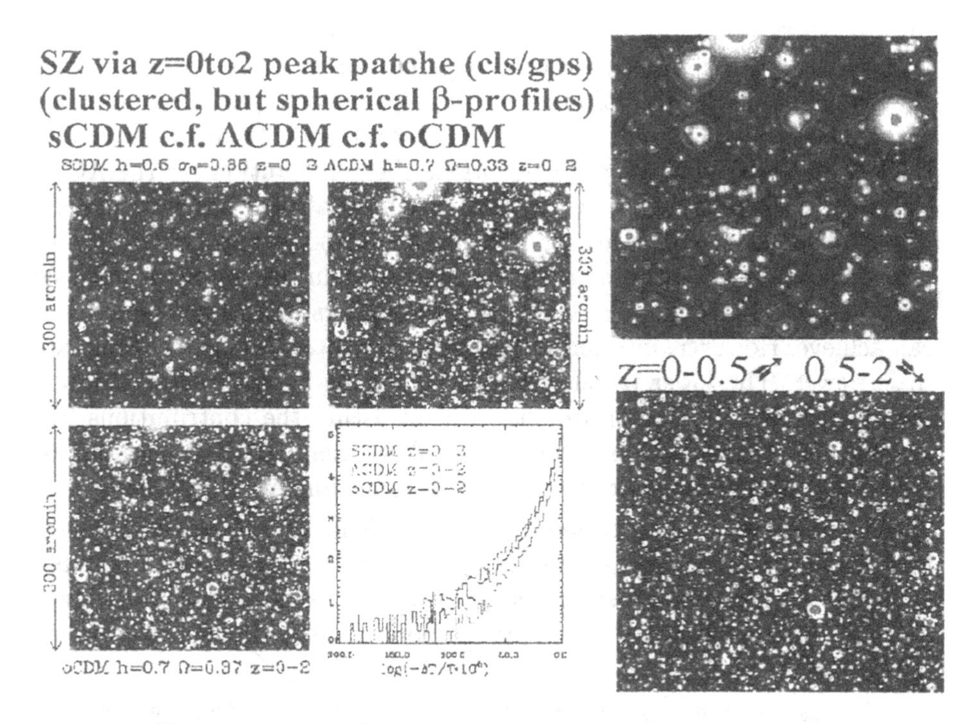

Figure 6. SZ maps derived from spherical pressure-profiles imposed upon halos identified up to z=2 with the peak-patch method, for three cluster-normalized cosmological models. The observing wavelength and contours are the same as in Fig. 5. The histograms show ΛCDM, oCDM and sCDM can be clearly differentiated. Blank field SZ surveys using interferometers and bolometer experiments promise to revolutionize our approach to the cluster system, especially at $z \gtrsim 0.5$. The contributions from clusters below and above this redshift are shown on the right for the ΛCDM model.

duper-clusters", that can appear.

What to do for the pressure profiles, $p(\mathbf{r}|\mathcal{C},t)$, is of course debatable. One strategy is to use profiles calculated from hydro simulations, but what has invariably been done is even simpler: pressure profiles scaled by bulk halo properties of the observed form derived from low redshift X-ray cluster data (and extrapolated to the more poorly known high redshift regime). For example, adopting a spherically symmetric isothermal "beta" profile is common: $p(\mathbf{r}|\mathcal{C},t) = p_c(1+r^2/r^2_{core*})^{-3\beta/2}\vartheta(R_{\mathcal{C}*}-r)$, where p_c is the central pressure, r_{core*} is a core radius, and $R_{\mathcal{C}*}$ is an outer truncation radius exterior to which the local pressure contribution is taken to be zero, and interior to which ϑ is unity. The radius r here is comoving and the $*$ subscript denotes comoving radii. $\beta \approx 2/3$ is a reasonable fit to the X-ray data.

This rapid semi-analytic peak-patch method for simulating clusters using isothermal beta-profiles was used in the construction of the two SZ layers (thermal and kinetic) of the sandwich in Fig. 2. The panels of Fig 6 show peak-patch-derived SZ maps for three cosmologies, and the lower right corner shows the distribution of $\Delta T/T$ in pixels, with its non-Gaussian distinct tails. The right two panels contrast the contribution for the ΛCDM example of clusters below $z = 0.5$ with those above. The top panel of Fig. 7 shows the ΛCDM example with a filter scale appropriate for the Planck satellite resolution and with an ideal resolution for the cluster system, below two arcminutes, which interferometers and large single dishes can achieve. However, interferometers typically filter the low ℓ, losing large scale power. The lower panel of Fig. 3 shows the SZ C_ℓ spectrum derived for the flat ΛCDM simulation shown, contrasting the contributions from clusters and groups above $z = 0.5$ with all of them. Both Poissonian and clustering contributions are included, since the simulation has them.

The power is not distributed in a democratic fashion as it is for Gaussian theories, but is concentrated in cold spots below 218 GHz, and in hot spots at higher frequencies. Thus, the naive visual comparison with bandpowers derived using a Gaussian assumption about the distribution of power is misleading.

The assumption that the SZ effect is dominated by the high pressure regions associated with clusters and groups of galaxies, with only weak modifications coming from the intercluster medium, is borne out visually in the supercluster simulation. The emission from the gas in filaments outside of the groups and poor clusters that reside in them is weak, below observable levels. What we can also see, however, is that an anisotropic pressure distribution is expected, with elongation along the filament directions, inviting improvements over spherical approximations.

To properly analyze the SZ effect in ambient fields, having fast Monte Carlo simulation methods along the lines developed here is clearly essential. That does not take away from the challenge of how best to do the analysis of such distinctly non-Gaussian signals, which consist of extended rather than point-like sources. Even with relatively small beams, these will be somewhat confused by overlapping sources. At least with the cluster system, we may hope to model the individual elements with hydrodynamical simulations. For other secondary sources such as dusty galaxies, a priori theoretical models are not really feasible, and the properties of the sources will have to be derived from the observations. At least in some of those cases, for typical beam sizes, the emission can be treated as from point sources.

2.4 Foreground Complications: The *foreground* signals from the interstellar medium are also non-Gaussian, but are not well-modelled by extended sources, since their emission power spectrum C_ℓ has been shown to

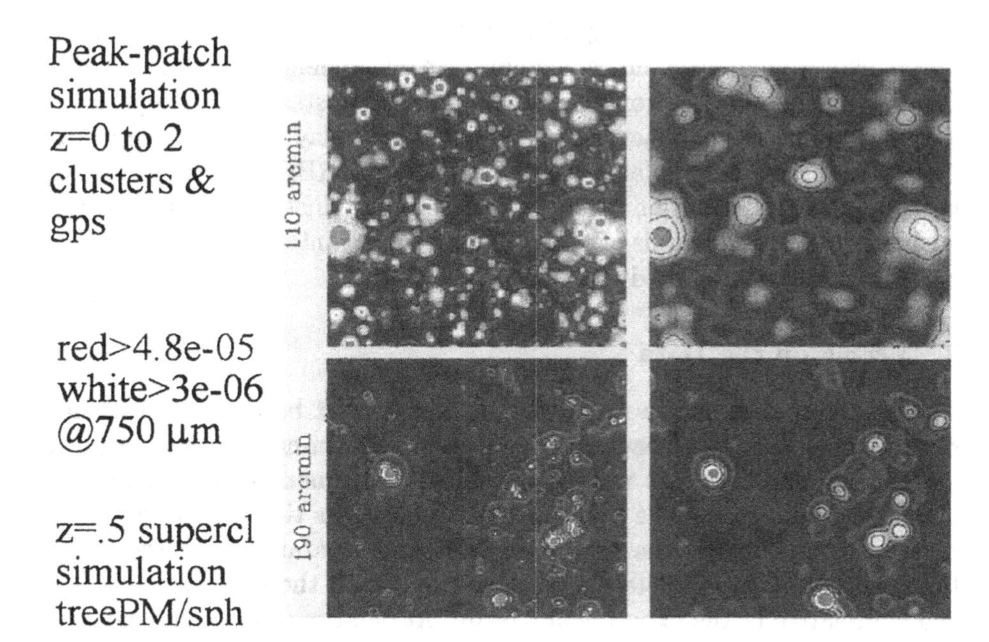

Simulation of 9 element array, 2.5m dishes, 30 Ghz, 10 GHz bandwidth, 10 hrs/pointing, 53 pointings, ~1 month through a BPKW treeP3M/SPH simulated dense supercluster seen at z=1, by Leitch, Mohr, Holder & Carlstrom

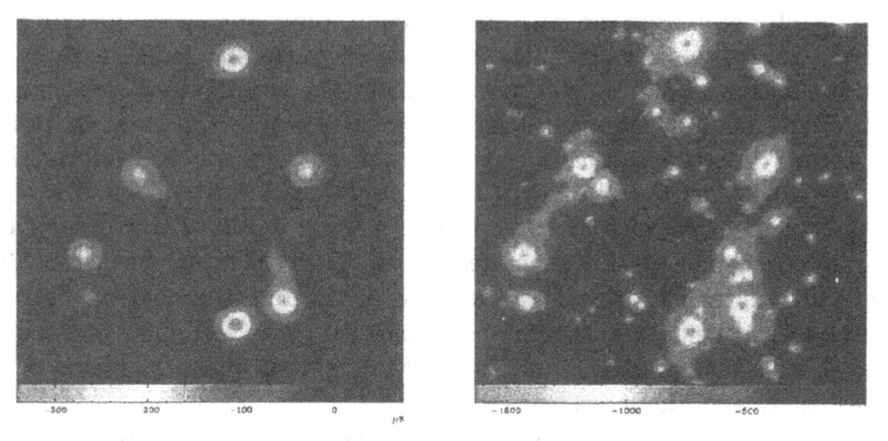

Figure 7. SZ maps subtending the angles shown for peak patches (above) and a single supercluster region seen at redshift 0.5. The core dark regions interior to the white areas are above 48×10^{-6}, and the dark contours surrounding the white are at 3×10^{-6} at $750\mu m$, levels that can be achieved with blank field SZ bolometer experiments. Filamentary bridges between the clusters are typically below 10^{-6}, but the far-field of clusters can be probed. The bottom left panel passes the supercluster region at $z = 1$ in the right panel through a simulated 30 GHz HEMT-based interferometer with the indicated characteristics.

rise to lower ℓ; further, no simplifications such as statistical isotropy apply. Direct hydrodynamical studies may increase our understanding of the ISM, but will not be able to provide strong guidance on statistical comparisons with data. The use of other data sets will clearly play a strong role in the final analysis; e.g., templates from the IRAS+DIRBE data is useful for modelling the distribution of shorter wavelength dust radiation. Exactly what to do for the analysis of foregrounds is under intense study and some current ideas are described in § 3.

3. Acting on the Data

The data comes in as raw timestreams, which must be cleaned of cosmic rays, obvious sectors of bad data, and other contaminants. Encoded in it are the sky and noise, as well as unwanted instrument, atmospheric and other residual signals. The pointing matrix identifies time-bits to a chosen pixel basis and allows the spatial signal to be separated from the noise; the resulting map completely represents the data if the pixelization is fine enough relative to the experiment's beam. At one time, the cosmology was drawn from the maps, perhaps using new bases beyond the spatial pixel ones, and possibly truncated to *compress* the data into more manageable chunks (e.g., using signal-to-noise eigenmodes to get rid of informationless noisy components of the map, [11]).

The statistical distribution of various operators on the pixelized data can be estimated from the map. In particular, the power spectrum C_ℓ represents a *radical compression* of the data, but of course is no longer a complete representation of the map. However, under the assumption that the sky maps only contain Gaussian signals characterized by an isotropic power spectrum, this is a complete representation of the statistical information in the map when the binning of the bandpowers is fine enough. For such Gaussian theories, such as those usually arising in inflation, cosmological parameters can be quickly estimated directly from the power spectrum. The radical compression has allowed millions of theoretical models to be confronted with the data in the large cosmological parameter spaces described above, e.g., [32], a situation infeasible if full statistical confrontation of all the models with the maps was required.

Basic aspects of the Boomerang, Maxima, DASI and CBI pipelines are described in [33], [35, 37, 4, 38], [36, 39], [9] and [8]. Pipelines were also developed for DMR and for QMAP. The MAP pipeline as currently envisaged is described in [40]. The Planck pipeline is still very much under discussion and development, complicated by the necessity of delivering high precision component maps and derived parameters from such a huge volume of data.

3.1 Bayesian Chains: The data analysis pipeline can be viewed as a

Bayesian likelihood chain:

$$\begin{aligned}
&\mathcal{P}(parameters|data,\ theoretical\ framework)\\
&= \mathcal{P}(data\ timestreams|maps,\ noise)\\
&\quad \otimes \mathcal{P}(noise|noise - obs)\\
&\quad\ \otimes \mathcal{P}(maps|signals)\\
&\qquad \otimes \mathcal{P}(signals|parameters)\\
&\qquad\ \otimes \mathcal{P}(parameters|prior\ knowledge)\\
&\qquad\quad \otimes [1/\mathcal{P}(data|theoretical\ framework)]
\end{aligned} \tag{1}$$

Each conditional probability in the unravelling of the chain is very complex in full generality. This means the exact statistical problem is probably not solvable, and only applying simplifying approximations to sections of the chain have made it tractable. Of particular widespread use has been the assumption that both noise and signals are Gaussian-distributed; others are being explored, especially by Monte Carlo means, but much remains to be done for us to attain the high precision levels that have been forecast.

We now describe the various terms, beginning with the last two. The priors $\mathcal{P}(parameters|prior\ knowledge)$ for the cosmological parameters are often taken to be uniform, but can represent information from other experiments, e.g., LSS observations, prior CMB experiments, age or H_0 constraints. The hope is that the new information from the rest of the probability chain will be so concentrated that the precise nature of the prior choice will not matter very much. As we have seen, prior information from non-CMB experiments is crucial, even given the most ideal CMB experiment, in order to break near-degeneracies among cosmological parameters.

The quantity $\mathcal{P}(data|theoretical\ framework)$ is an overall normalization. It can address whether the overall theory is crazy as an explanation of the data. It has some characteristics in common with goodness-of-fit concepts in "frequentist" probability analysis.

We often refer to entropies $\mathcal{S} \equiv \ln \mathcal{P}$ in the Gibbs sense so the Bayesian chain involves a sum of the individual conditional entropies that make up the whole.

3.2 Signal/noise separation: The information comes in the form of a discretized timestream, with d_{ct} as the data for the bit of time t in frequency channel c, and a pointing vector on the sky, $\hat{q}_c(t)$. The sky is gridded into pixels, and these are used to define signal maps which are functions of direction and frequency. We denote these by Δ_{cp}, for the signal at pixel p and frequency channel c. The pointing matrix, P_{ctp}, is an operator mapping time-bit t to pixel p with some weight. The difference $d - P\Delta$ is the noise, η, plus further residuals $\mathcal{R}$, which may be sky-based or experimental systematics. The probability of a given timestream can be written as

$$\mathcal{P}(d|\Delta,\eta,P_{ctp}) = \prod_{ct} \delta(d_{ct} - (\sum_{p} P_{ctp}\Delta_{cp} + \mathcal{R}_{ct} + \eta_{ct})). \qquad (2)$$

3.2.1 Map-making: Map-making for a given channel involves using just the information encoded in d and P to separate what is sky signal (and residuals) from what is noise, i.e., the construction of $\mathcal{P}(\Delta|d)$.

The pointing matrix is a projection operator, with many more time bits than pixels. The simplest example for P_{ctp} is the characteristic set function for the pixel $\chi_p(\hat{q}_c(t))$ (one inside the pixel, zero outside). For a chopping experiment such as MAP, the two pixels where the two beams are pointing (separated by 141°) are coupled at a given time bit. There are also choices to be made about scattering the information in the timestream among pixels. For example, one can include the beam in Δ or in P. At the moment the former is preferred, but there are good arguments for the latter if the beam changes in time (e.g., asymmetric beams with rotations). In that case, $\chi_p(\hat{q}_c(t))$ is replaced by $A_p\mathcal{B}(\hat{q}_p - \hat{q}_c(t))$, where $\mathcal{B}$ is the beam function and A_p is the pixel area, and the derived signal Δ_{cp} does not have the beam function convolved with it as it does in the "top hat" χ_p case. Of course, any other function normalized to the pixel area upon integration can be used, e.g., approximations to the beam which might not contain all sidelobe information. As long as the signals are appropriately convolved, and the pixelization choices are fine enough, the results should be identical. Pointing uncertainty further complicates the simplicity of P.

In map-making, there is an implicit assumption on the prior probability of the map, $\mathcal{P}(\Delta|prior)$, that it is uniform so any value is *a priori* possible. The map distribution should involve integration over all possible noise distributions; in practice, only the measured noise is used, which delivers the maximum likelihood map $\overline{\Delta}_{cp}$ and a pixel noise correlation $C_{N,cpc'p'}$, allowing for possible channel-channel correlations in the noise.

For many previous experiments, one could just compute the average $\overline{\Delta}$ and variance C_N of the measured d's contribution to each pixel, relying on the central limit theorem to ensure a Gaussian-distribution. However, these "naive maps", described below in more detail, also relied on an implicit assumption, namely that η was uncorrelated from time bit to time bit, which is not correct with the $1/f$ noise in detectors. The noise was also often so strong compared with the signal that the latter could be ignored in solving for Δ_{cp}; this is no longer the case.

So far it has been essentially universal to assume that η is Gaussian-distributed stationary noise with noise power $w^{-1}(\omega)$ a smooth function. Thus, the noise is completely specified by its covariance $c_{n,ct,c't'} \equiv \mathcal{S}[\langle\eta_{ct}\eta_{c't'}\rangle]$ which is assumed to be a function of $t - t'$, hence diagonal in the time frequencies ω. Here, $\mathcal{S}$ denotes the smoothing operation. Drifts and non-Gaussian elements can be included with timestream filtering and in the catch-all residual $\mathcal{R}$ using time templates (see below). Ignoring $\mathcal{R}$ for the moment, the probability of the time-ordered data given a map and noise-power w^{-1}, $\mathcal{P}(d|\Delta, w)$, is just a Gaussian:

$$\begin{aligned}
\mathcal{P}(d|\Delta, w) &= \exp[-(d-P\Delta)^\dagger w(d-P\Delta)/2]/\{(2\pi)^{N_{tbits}/2}\sqrt{\det(w^{-1})}\} \\
&= \exp[-(\overline{\Delta}-\Delta)^\dagger W_N(\overline{\Delta}-\Delta)/2]/\{(2\pi)^{N_{pix}/2}\sqrt{\det(W_N^{-1})}\} \\
&\quad \times \exp[-\bar{\eta}^\dagger \widetilde{w}\bar{\eta}/2]/\{(2\pi)^{(N_{tbits}-N_{pix})/2}\sqrt{\det(\widetilde{w}^{-1})}\}, \qquad (3)
\end{aligned}$$

$$\begin{aligned}
&W_N\overline{\Delta} = P^\dagger w d, && W_N \equiv P^\dagger w P \equiv C_N^{-1} = \langle nn^\dagger\rangle^{-1}, && (4)\\
&n \equiv \overline{\Delta} - \Delta = C_N P^\dagger w \eta, && \bar{\eta} \equiv \langle \eta | d\rangle = d - P\overline{\Delta}, &&\\
&\widetilde{w} = w(1 - P(P^\dagger w P)^{-1}P^\dagger w), && P^\dagger w \bar{\eta} = 0. && (5)
\end{aligned}$$

The maximum likelihood solution for $w(\omega)$ is $(\mathcal{S}[|\mathrm{FT}(d-P\overline{\Delta})|^2])^{-1}$, where FT denotes the Fourier transform operator. Note that the $\bar{\eta}$ noise term is multiplied by a projector which is perpendicular to the spatial sector; i.e., $C_N P^\dagger w\bar{\eta}=0$. The first Gaussian in eq.(3) is $\mathcal{P}(\overline{\Delta}|\Delta)$. We can interpret it as the integration over the spatial Gaussian random noise field n defined in eq.(4) of the product of $\mathcal{P}(\overline{\Delta}|\Delta, n) = \delta(\overline{\Delta} - (\Delta + n))$ and the distribution of n, $\exp[-n^\dagger C_N^{-1} n/2]/[(2\pi)^{N_{pix}/2}[\det C_N]^{1/2}]$.

3.2.2 General *cf.* Optimal Maps: Any linear operation from timestream to generalized pixels, $\overline{\Delta}^{(\mathcal{M})} = \mathcal{M}^\dagger d$ defines a map, albeit not an optimal one as for the maximum likelihood solution eq.(4). The map-noise $n^{(\mathcal{M})} \equiv \overline{\Delta}^{(M)} - \Delta^{(M)}$ has a correlation matrix $C_N^{(\mathcal{M})} = \mathcal{M}^\dagger w^{-1}\mathcal{M}$. Here the relation of the true sky signal seen in the $\mathcal{M}$-filter-space to the "optimal map" signal Δ is $\Delta^{(M)} = \mathcal{M}^\dagger P\Delta$. Of special interest are therefore classes of operators $\mathcal{M}$ for which $\mathcal{M}^\dagger P = P^\dagger \mathcal{M}$ is a projector, i.e., self-adjoint and equal to its square. For example, $\mathcal{M}^\dagger = (P^\dagger w_* P)^{-1}P^\dagger w_*$ satisfies $\mathcal{M}^\dagger P = I$, the identity, for any w_* provided $(P^\dagger w_* P)$ is invertible. The price one pays for such a non-optimal map is enhanced noise, with noise-weight matrix $W_N^{(\mathcal{M})} = (P^\dagger w_* P)(P^\dagger w_* w^{-1} w_* P)^{-1}(P^\dagger w_* P)$ which has maximum eigenvalues if $w_* = w$, but is otherwise "less weighty".

An interesting example is when w_* is taken to be a constant and P_{tp} is zero or one depending if the pointing at a time-bit t lies within the pixel p or not; in that case $P^\dagger P$ just counts the number of time-bits that fall into each pixel, and $\overline{\Delta}^{(\mathcal{M})}$ is just the average of the anisotropies over these. If we use $\eta_t\eta_{t'}$ in place of its smoothed version w^{-1}, the noise matrix just counts the variance in the amplitudes of the pixel hits about the mean. For this reason, $(P^\dagger P)^{-1}P^\dagger d$ is called a "naive map" [4]. For other cases, computing the error matrix may not be simple if w is not known.

For the full maximum likelihood map, even with stationarity, the convolution of w with d appears to be an $O(N_{tbits}^2)$ operation, but it is really $O(N_{tbits})$ because $w(t-t')$ generally goes nearly to zero for $t-t' \gg 0$. Similarly, the multiplication of the pointing matrix is also $O(N_{tbits})$ because of its sparseness. Thus, we can reduce the timestream data to an estimate of the map and its weight matrix in only $O(N_{pix}^2)$ operations.

3.2.3 Iterative Map-making Solutions: An iterative method [35] has proved quite effective to solve this: the map on iteration $j+1$ is estimated from $W_*\Delta_{(j+1)} = W_*\Delta_{(j)} + P^\dagger w_{(j)}\eta_{(j)}$, where the noise and noise power on iteration j are determined from $\eta_{(j)} = d - P\Delta_{(j)}$, $w_{(j)}^{-1}(\omega) = \mathcal{S}[|FT(\eta_{(j)})|^2(\omega)]$. Here $W_* \equiv P^\dagger w_* P$ is a matrix chosen to be easy to invert: e.g., a constant w_* was used for Boomerang [35]. As the maps converge, the noise orthogonality condition holds, so the solution is the correct one. What $w(\omega)$ looks like is white noise at high temporal ω, crashing to zero because of $1/f$ noise in the instruments and time-filtering. These low frequencies correspond to large spatial scales, and the iteration converges slowly there, inviting multigrid speedup of the algorithm, which is now being implemented [37]. The final $\overline{\Delta}$ and C_N are computed using d and the converged $w_{(j)}$ in eq.(4). In some instances, we can work directly with $W_N\overline{\Delta}$ and W_N rather than $\overline{\Delta}$ and C_N, avoiding the costly $O(N_{pix}^3)$ inversion. If sections of the map are cut out after W_N determination, through a projector χ_{cut} which is zero outside and one inside the cut, the cut weight matrix is $(\chi_{cut}W_N^{-1}\chi_{cut})^{-1}$. This corresponds to integration over all possible values of the now unobserved $\overline{\Delta}$'s outside of the cut, which increases the errors on the regions inside because of the correlation. Unfortunately, two matrix inversions are required, and the first one is potentially quite large even if the cut is severe.

Hinshaw *et al.* [40] describe the current pipeline for MAP. To a good approximation the MAP noise is white, so $w_{(j)} = w_*$ can drop out, and the critical element is computation of $P^\dagger P$, which is more complex than counting pixel-hits because MAP is a difference experiment. The MAP team choose W_* to be $diag(P^\dagger w_* P)$, but otherwise use the same iterative algorithm.

Since P is all that differentiates signal from noise in the timestream, it is essential that it be sufficiently complex; in practice this means many cross linkages among pixels in the scan strategy. This is because there are random long term drifts in the instruments that make it hard to measure the absolute value of temperature on a pixel, though temperature differences along the path of the beam can be measured quite well because the drifts are small on short time scales. If there are not sufficient cross-links, the maps are *striped* along the directions of the scan pattern, and the weight matrix W_N has to be called upon to demonstrate that these features are not part of the true sky signal we are searching for. The rapid on-board differencing for MAP effectively eliminates this.

3.3 Pixelization: Pixels are any discrete basis that give a complete representation of the data. Square top-hat functions tiling the space in a grid quite a bit smaller than the beam size were the norm in the past. As the size of the data increased, more elaborate hierarchical schemes were needed. For COBE, the Quadrilateralized Spherical Cube was used: the sky was broken into six base pixels corresponding to faces of a cube. Higher resolution pixels were created by dividing each pixel into four smaller pixels of approximately equal area, in a tree with pixels that are physically close having their data stored close to each other. There are (quite) small errors associated with the projections of the sphere onto the faces. At resolution r, there are $6 \times 2^{2(r-1)}$ pixels. The COBE data came with resolution $r = 6$, 6144 pixels of size 2.6°, and we sometimes use resolution 5, still safely smaller than the 7° beam.

For large sky coverage, it is beneficial to have a pixelization which is azimuthal, with many pixels sharing a common latitude, to facilitate fast spherical harmonic transforms (FSHTs) [41] between the pixel space, where the data and inverse noise matrix are simply defined, and multipole space, where the theories are often simple to describe. The FSHT operation count is a $\mathcal{O}(N_{pix}^{3/2})$ operations, with $\mathcal{O}(N_{pix} \ln N_{pix}^2)$ potentially possible to achieve. HEALPix[42] is an example which has been adopted for Boomerang, MAP and Planck: it has a rhombic dodecahedron as its fundamental base, which can be divided hierarchically while remaining azimuthal. There is also an extensive software package that accompanies it to allow manipulations. See also [43] for the alternative hierarchical "igloo" choice, with many HEALPix features, and some other advantages.

In some cases *generalized pixels* are used, involving direct projection of the data onto spatially-extended mode functions. The first extensive use of this was for the SK95 dataset. Signal-to-noise eigenfunctions used in data compression for COBE and SK95 are another example [11, 14]. Direct projection onto pixels in wavenumber space, with the mode functions being discrete Fourier transform waves, also form a fine alternative basis. Still there is nothing like a real-space visual representation to explore strange features in the data.

3.4 Filtering in Time and Space: Getting rid of unwanted systematic effects in the data has been fundamental throughout the history of CMB experiments. Sometimes this was done in hardware, e.g., rapid Dicke switching from one direction in the sky to another in "two-beam" chop experiments and "three-beam" chop-and-wobble experiments; sometimes it was done in editing, e.g., bad atmospheric contamination, cosmic rays; and now it is invariably done in software, e.g., $1/f$ noise in bolometers or HEMTs, spurious spin-synchronous signals associated with rotating platforms. In the linear operation $\sum_{t'} \mathcal{F}_{tt'} d_{t'}$ on the timestream, the filter matrix $\mathcal{F}_{tt'}$ is often time-translation invariant, hence diagonal in frequency. High pass filters are an example, translating through the time-space pointing matrix P_{tp} into primarily low ℓ spatial filtering, with the penalty the removal of part of the target sky signals as well as the unwanted ones.

3.4.1 Spatial Filters and Wavenumber Space: In direct filtering of the maps, the signal amplitude Δ_p in a pixel p can be written in terms of generalized pixel-mode functions $\mathcal{F}_{p,\ell m}$ and the spherical harmonic components of the anisotropy, $a_{\ell m}$: $\Delta_p = \sum_{\ell m} \mathcal{F}_{p,\ell m} a_{\ell m}$. $\mathcal{F}_{p,\ell m}$ encodes the basic spatial dependence (e.g., $Y_{\ell m}(\hat{q}_p)$) and filters, including the experimental beam $\mathcal{B}_{\ell m}$ and pixelization effects at high ℓ, and the switching strategy at low ℓ.

A Fourier transform representation of the signals in terms of wavenumber $\mathbf{Q}$ is often justified. There is a projection from the sphere with coordinates (θ, ϕ) onto a disk with radial coordinate $\varpi = 2\sin(\theta/2)$ and azimuthal angle ϕ which is an area-preserving map and one-to-one — except that one pole gets mapped into the outer circumference of the disk. It is highly distorted for angles beyond 180°, but even for the 140°-diameter COBE NGP and SGP cap maps of Fig. 8 it is an excellent representation. The associated discrete Fourier basis for the disk is close to a continuum Fourier basis $\mathbf{Q}$, with $|\mathbf{Q}| = \ell + 1/2$.

Instead of a cap, consider a rectangular patch of size $L \times L$. The amplitude $\Delta_p = \sum_{\mathbf{Q}} \tilde{\mathcal{F}}_p(\mathbf{Q}) a_{\mathbf{Q}}$ then involves a mode-sum with differential element $L^2 d^2\mathbf{Q}/(2\pi)^2 = f_{sky} 2Q dQ d\phi/(2\pi)$, where the fraction of the sky in the patch is $f_{sky} = L^2/4\pi$. The effective number of modes contributing to ℓ is therefore $g_\ell = f_{sky}(2\ell + 1)$, the usual $2\ell + 1$ in the f_{sky}=1 all-sky case. Of course the reduction by just f_{sky} is approximate, since modes with wavenumber below the fundamental one of $2\pi/L$ hardly contribute, inviting a more sophisticated relation for g_ℓ to characterize effects from filtering, discreteness and apodization (smooth weighting of the target region) [4]. The decomposition of the spatial-mode function in this "momentum space" now involves a Fourier phase factor associated with the pixel position instead of a $Y_{\ell m}$, the beam $\mathcal{B}(\mathbf{Q})$ and the rest, $U_p(\mathbf{Q})$, which includes discretization into time bins and pixelization as well as the switching-strategy and any long wavelength filtering done in software. (In the spherical harmonic decomposition, $U_{p,\ell m m'}$ is a function of m' as well as the ℓm and possibly the pixel position.)

Examples of $U_p(\mathbf{Q})$ are given in [10]. For switching experiments through a throw angle ϖ_{throw}, $U_p(\mathbf{Q})$ is a first power of $2\sin(\mathbf{Q}\cdot\varpi_{throw}/2)$ if a chop, a second power if a chop-and-wobble. A separable form $U_p(\mathbf{Q}) = \mu(\varpi_p|\varpi_C)u(\mathbf{Q})$, involving a spatial-mask function μ centered at some point ϖ_C and a position-independent filter $u(\mathbf{Q})$, is appropriate for the recent Boomerang analysis [4], with $u(\mathbf{Q})$ a combination of temporal and spatial filters and μ unity inside an ellipsoidal region, zero outside it.

Beams $\mathcal{B}_c(\mathbf{Q})$ for each channel c must be determined experimentally, typically through the patterns of point source on the sky, but also through detailed "optics" computations of the telescope setup. Most often there is a nice monotonic fall-off from the central point to relatively low levels of power, though there is invariably at least some beam-asymmetry and lurking side lobes. In the past, it was common practice to use circularly-symmetric monotonically-falling $\mathcal{B}_\ell$'s in CMB analysis, even "circularizing" pixelization effects, but now highly accurate treatments are needed to achieve our goal of great precision. The full width half maximum θ_{fwhm} is the usual way to quote beam scale. If a Gaussian approximation to the profile is appropriate, the Gaussian smoothing scale is $\ell_s \approx 810(10'/\theta_{fwhm})$ in multipole space.

3.4.2 Signal Correlation Matrices and Window Functions: For isotropic theories for which the spectra $\mathcal{C}_{T\ell}$ are only functions of ℓ, the pixel–pixel correlation function of the signals can be expressed in terms of $N_{pix} \times N_{pix}$ "$\mathcal{C}_\ell$-window" matrices $\mathcal{W}_{pp',\ell}$:

$$C_{Tpp'} \equiv \langle \Delta_p \Delta_{p'} \rangle = \mathcal{I}[\mathcal{W}_{pp',\ell}\mathcal{C}_{T\ell}], \quad \mathcal{W}_{pp',\ell} \equiv \frac{4\pi}{2\ell+1}\sum_m \mathcal{F}_{p,\ell m}\mathcal{F}^*_{p',\ell m}, \tag{6}$$

$$\mathcal{I}(f_\ell) \equiv \sum_\ell \frac{(\ell+\frac{1}{2})}{\ell(\ell+1)} f_\ell \, . \tag{7}$$

It is expressed in terms of $\mathcal{I}(f)$, the discrete "logarithmic integral" of a function f. The average $\overline{\mathcal{W}}_\ell \equiv \sum_{p=1}^{N_{pix}} \mathcal{W}_{pp,\ell}/N_{pix}$ is often used to characterize the ℓ-sensitivity of the experiment, since $\sqrt{\mathcal{I}[\overline{\mathcal{W}}_\ell \mathcal{C}_{T\ell}]}$ gives the *rms* anisotropy amplitude. For the simple case of the COBE map we have $\mathcal{W}_{pp',\ell} = \overline{\mathcal{W}}_\ell P_\ell(\cos\theta_{pp'})$ in terms of the Legendre polynomials for $\ell \geq 2$, in which both beam and pixelization effects are taken into account in $\overline{\mathcal{W}}_\ell$. In general, the implicit isotropization that makes this only a function of ℓ does not work at high ℓ because of the pixelization, and this can significantly complicate the treatment. Switching and other spatio-temporal filters applied in software or hardware further complicate the expressions.

3.5 Templates in Time and Space: Templates are specific temporal patterns that we probe for in the timestream, $\upsilon_{ct,A}$. Templates could have frequency structure (c) as well. They are often associated with specific spatial patterns $\Upsilon_{cp,A}$ that we can probe for in the map or directly in the timestream with $\upsilon_{ct,A}=\sum_p P_{ctp}\Upsilon_{cp,A}$. The unknowns are the amplitudes κ_A (and the errors on them), and possibly other parameters characterizing their position, orientation and scale. The residual signal associated with the templates is therefore $\mathcal{R}_{ct} = \sum_A \upsilon_{ct,A}\kappa_A$. Templates can model contamination by radiation from the ground, balloon or sun, components of atmospheric fluctuations and cosmic ray hits. Often they are synchronous with a periodic motion of the instrument.

In maps, spatial templates have been used in the removal of offsets, gradients, dipoles, quadrupoles and looking for point sources and extended sources (e.g., IRAS/DIRBE structures). Sample forms for Υ_{cp} are: a constant for an average offset, $\hat{q}_{pi}$ for the three dipole components, $\hat{q}_{pi}\hat{q}_{pj}-\delta_{ij}/2$ for the five quadrupole components, the pattern of dust emission in the IRAS/DIRBE maps, and the beam pattern for point sources, $\Upsilon_{cp} = \mathcal{B}_c(\hat{q}_p - \hat{q}_{source})$. Even the individual pixels themselves define spatial templates.

Since $\upsilon_{ct,A}$ is of the same mathematical form as P_{ctp} we can solve simultaneously for Δ_{cp} and κ_A, delivering an extended map $\overline{\Delta}$ and $\bar{\kappa}$, with appropriate error covariances for each, including a cross term. The assumption is that the *a priori* probability for the template amplitudes κ is uniform, as for the map amplitudes.

Often we just want to remove an undesired pattern from the map. We can do this by directly projecting it from the map, and modifying the map statistics. One can also marginalize the unobserved κ_A, i.e., integrate over them. A nice way to deal with this is to assume that the prior probability for κ is Gaussian with zero mean and correlation $K_{AA'} = \langle \kappa_A \kappa_{A'} \rangle$. Letting the dispersion become infinite reduces to the projection. The explicit result of marginalizing over the template amplitudes yields modified noise weights and a modified average map:

$$
\begin{aligned}
\widetilde{w} &\equiv w - w\upsilon(\upsilon^\dagger w \upsilon + K^{-1})^{-1}\upsilon^\dagger w = (w^{-1} + \upsilon K \upsilon^\dagger)^{-1}\,, \qquad (8)\\
\widetilde{W}_N \overline{\Delta} &= P^\dagger \widetilde{w} d\,, \qquad \widetilde{W}_N \equiv P^\dagger \widetilde{w} P\,, \qquad \widetilde{C}_N \equiv \widetilde{W}_N^{-1}.
\end{aligned}
$$

Letting the eigenvalues of $K \to \infty$ gives the uniform prior, and in that case the modified w has the template vectors projected out, i.e., $\upsilon\widetilde{w}\upsilon=0$.

Therefore the effect of marginalizing out κ is to enhance the noise. One price to pay is that time-translation invariance of the modified w is lost: i.e., $\widetilde{w}$ is not diagonal in temporal frequency ω, and cannot be computed directly by the FFT, followed by smoothing. However, the relationship eq.(8) allows $\widetilde{c}_n \equiv \widetilde{w}^{-1}$ to be computed with one $N_\kappa \times N_\kappa$ matrix inversion in addition to the FFT.

3.6 $\mathcal{P}$(signals|parameters): The signal component of the map,

$$\Delta_{cp} = \sum_j s_{(j)cp} + \sum_A \Upsilon_{cp,A}\kappa_A\,, \tag{9}$$

includes the frequency-independent primary, $s_{(1)cp}$ say, foreground and secondary anisotropy components $s_{(j)cp}$ and additional template-based signals $\Upsilon_{cpA}\kappa_A$.

3.6.1 Gaussian Signals and Wiener Filters: For primary anisotropies, the signal is often assumed to be Gaussian,

$$\ln \mathcal{P}(s|y) = -\tfrac{1}{2} s^\dagger C_T^{-1} s - \tfrac{1}{2}\mathrm{Tr}\ln C_T - N_{pix}\ln\sqrt{2\pi}\,, \tag{10}$$

where $C_T(y) \equiv W_T^{-1}$ is the theory pixel-pixel correlation matrix and we have denoted the set of cosmological parameters it depends on by y_α. (Tr denotes trace.) A great advantage of assuming all signals and noise to be Gaussian is that marginalizing all signal amplitudes $\sum s_{(j)cp}$ and template amplitudes κ_A yields another simple Gaussian:

$$\begin{aligned}
&\ln \mathcal{P}(\overline{\Delta}|y) = -\tfrac{1}{2}\overline{\Delta}^\dagger \widetilde{W}_t \overline{\Delta} + \tfrac{1}{2}\mathrm{Tr}\ln \widetilde{W}_t - N_{pix}\ln\sqrt{2\pi}\,, \\
&W_t \equiv (C_N + C_T)^{-1}\,,\ \widetilde{W}_t \equiv (\widetilde{C_N} + C_T)^{-1}\,,\ \widetilde{C_N} \equiv C_N + \Upsilon K \Upsilon^\dagger\,, \\
&\widetilde{W}_t \equiv W_t - [\Upsilon^\dagger W_t]^\dagger (K^{-1} + \Upsilon^\dagger W_t \Upsilon)^{-1} [\Upsilon^\dagger W_t]\,.
\end{aligned} \tag{11}$$

As for the time templates, the marginalization can be thought of as inducing enhanced noise in the template spatial structures.

The other part of the probability chain, $\ln \mathcal{P}(\Delta|\overline{\Delta})$, gives a Gaussian for the signal given the observations, with mean the Wiener filter of the map:

$$\begin{aligned}
\ln \mathcal{P}(\Delta|\overline{\Delta}) = \quad & -\tfrac{1}{2}(\Delta - \langle\Delta|\overline{\Delta}\rangle)^\dagger C_{\Delta\Delta}^{-1}(\Delta - \langle\Delta|\overline{\Delta}\rangle) \\
& -\tfrac{1}{2}\mathrm{Tr}\ln C_{\Delta\Delta} - N_{pix}\ln\sqrt{2\pi}, \\
\langle\Delta|\overline{\Delta}\rangle = C_T \widetilde{W}_t \overline{\Delta}\,, \qquad & \langle \delta\Delta \delta\Delta^\dagger|\overline{\Delta}\rangle = C_T - C_T \widetilde{W}_t C_T\,.
\end{aligned} \tag{12}$$

The brackets indicate an ensemble average. The lemma

$$B(A+B)^{-1} = (A^{-1} + B^{-1})^{-1} A^{-1} \tag{13}$$

between two invertible matrices A and B is convenient to remember. Thus the Wiener filter can also be written as $(\widetilde{W}_N + W_T)^{-1}\widetilde{W}_N\overline{\Delta}$.

We can also determine the distribution, given the observations, of the various component signals $s_{(j)}$ and of the noise; all of these are Gaussians with means and variances given by:

$$\begin{aligned}
&\langle s_{(j)}|\overline{\Delta}\rangle = C_{T(j)}\widetilde{W}_t\overline{\Delta}, && \langle \delta s_{(j)} \delta s^\dagger_{(j')}\rangle = C_{T(j)}\delta_{jj'} - C_{T(j)}\widetilde{W}_t C_{T(j')}, \\
&\langle n|\overline{\Delta}\rangle = C_N \widetilde{W}_t \overline{\Delta}, && \langle \delta n \delta n^\dagger\rangle = C_N - C_N \widetilde{W}_t C_N, \\
&\delta s_{(j)} \equiv s_{(j)} - \langle s_{(j)}|\overline{\Delta}\rangle, && \delta n \equiv n - \langle n|\overline{\Delta}\rangle.
\end{aligned} \tag{14}$$

3.6.2 Non-Gaussian Signals and Maximum Entropy Priors: For foregrounds and secondaries, all higher order correlations are needed since the distributions are non-Gaussian. One sample non-Gaussian prior distribution that has been analyzed for these signals and even for primaries is the so-called maximum entropy distribution, which has been widely used by radio astronomers in image construction from interferometry data. Although "max-ent" is often a catch-all phrase for finding the maximum likelihood solution, the implementation of the method involves a specific assumption for the nature of $\mathcal{P}(s_{(j)cp}|\text{theory})$. For positive signals, $s > 0$, it is derived as a limit of a Poisson distribution, $P_n = \mu^n e^{-\mu}/n!$, in the Stirling formula limit,

$$\begin{aligned} \ln n! &= \Gamma(1+\mu\zeta) \approx \mu\zeta\ln(\mu\zeta) - \mu\zeta + \ln(2\pi\mu\zeta)/2 : \\ \ln\mathcal{P}(s|y) &= (-\zeta\ln(\zeta) + \zeta - 1)\mu \;\; \text{where} \;\; \zeta = s/\mu . \end{aligned} \tag{15}$$

The subdominant $\ln(2\pi\mu\zeta)$ term is dropped and the Poisson origin is largely forgotten, replaced by an interpretation involving the classic Boltzmann entropy, $-\zeta\ln\zeta$, with the constraint on the average $\langle\zeta\rangle = 1$ and μ now interpreted as a measure.

Another example of positive signals is standard emitting sources. These often have power law distributions over some flux range, with leading term $\ln\mathcal{P} = -\gamma\ln\zeta$. These have to be regulated at small and/or large ζ to converge.

For symmetric positive and negative signals, a possible form is

$$\begin{aligned} \mathcal{S}_P(s_{(j)}) &= \ln\mathcal{P}(s_{(j)}|y) = -x\ln(x+\sqrt{1+x^2}) + \sqrt{1+x^2} - 1, \\ x^2 &\equiv s^{\dagger}_{(j)} C^{-1}_{T(j)} s_{(j)}, \end{aligned} \tag{16}$$

which reduces to $-x^2/2$ in the small fluctuation (small x) limit, but has non-Gaussian wings. This form has been used in CMB forecasting for Planck by [44, 47], but any form which retains the basic $-x^2/2$ limit for small x and a levelling off at high x can have a similar effect. For example, instead of the $-\text{arcsinh}(x)$ for $\partial\ln P/\partial x$, the more drastic deviation from Gaussian, $-\arctan(x)$, has been used in radio astronomy.

Of course, in spite of the analytic form for $\mathcal{S}_P$, the integration over the various $s_{(j)}$ to get $\mathcal{S}(\overline{\Delta}|y)$ does not have an analytic result. Nonetheless, iterative techniques can be used to solve for the maximum entropy solution, and errors can be estimated from the second derivative matrix of the total entropy.

3.7 Cleaning and Separating: Separating foregrounds and secondaries from the primary CMB into statistically accurate maps is a severe challenge.

It has been common to suppose $s_{(j)cp} = f_{(j)c}s_{(j)p}$ is separable into a product of a given function of frequency times a spatial function. But this is clearly not the case for foregrounds, and for most secondary signals, although for some, such as the Sunyaev-Zeldovich effect, it is nearly so (Fig. 1). Another approach which is crude but reasonably effective is to separate the signals using the multifrequency data on a pixel-by-pixel basis, but the accuracy is limited by the number of frequency channels.

3.7.1 Gaussian *cf.* Maximum Entropy Component Separation: It is clearly better to incorporate the knowledge we do have on the $s_{(j)cp}$ spatial patterns. This has been done so far by either assuming the prior probability for $\mathcal{P}(s_{(j)cp}|\text{theory})$ is Gaussian or of the max-ent variety, or is an amplitude times a template. Even the obviously incorrect Gaussian approximation for the foreground prior probabilities has been shown to be relatively effective at recovering the signals in simulations for Planck performed by Bouchet and Gispert [45]; see also [46].

The Poisson aspect of the maximum entropy prior makes it well-suited to find and reconstruct point sources, and more generally concentrated ones: for Planck-motivated simulations, it has been shown to be better at recovery than the Gaussian approximation [47]. Since errors are estimated from the second derivative matrix, non-Gaussian aspects of the errors are ignored. Further, foregrounds and secondary anisotropies have non-Gaussian distributions which are certainly not of the max-ent form, and can only be determined by Monte Carlo methods, simulating many maps — and then only if we know the theory well enough to construct such simulations, a tall task for the foregrounds. Fortunately, with current data these issues have not been critical to solve before useful cosmological conclusions could be drawn, but much more exploration is needed to see what should be done with separation for the very high quality Planck and MAP datasets.

3.7.2 Component Separation with Templates: For foreground removals, maps at higher or lower frequencies than the target one, where the foregrounds clearly dominate, can be used as the templates. The marginalized formulas given in §2.5, 2.6 get rid of the templates but their amplitudes and distribution may also be of great interest. The other part of the likelihood (before marginalization) is $\ln \mathcal{P}(\kappa|\overline{\Delta})$; with the Gaussian prior assumption, it is a Gaussian distribution with the mean a weighted template-filtering of the data, akin to a Wiener-filtering:

$$\langle\kappa|\overline{\Delta}\rangle = [K^{-1} + \Upsilon^\dagger W_t \Upsilon]^{-1}\Upsilon^\dagger W_t\overline{\Delta} \to [\Upsilon^\dagger W_t \Upsilon]^{-1}\Upsilon^\dagger W_t\overline{\Delta}\,,$$
$$C_{\kappa\kappa} = \langle\delta\kappa\delta\kappa^\dagger|\overline{\Delta}\rangle = [\Upsilon^\dagger W_t\Upsilon]^{-1}, \qquad \delta\kappa = \kappa - \langle\kappa|\overline{\Delta}\rangle. \tag{17}$$

An example [49] of the use of templates was comparing Wiener maps and various statistical quantities for the SK95 data under the hypothesis that it was pure CMB or CMB plus a "dust template" – a cleaned IRAS

$100\mu m$ map with strong input from the DIRBE 100 and 240 μm maps [48]. A correlated component in the 30–40 GHz SK95 data was found [49], but at a low level compared with the CMB signal, i.e., not much dust contamination, in agreement with what other methods showed [50].

3.7.3 Point Source Separation with Templates: A useful method for finding localized sources is to look for known or parameterized non-Gaussian patterns with the position, and possibly orientation and scale of the templates being allowed to float. An example is (optimal) point source removal, in which the map is modelled as a profile times an amplitude κ plus noise plus the other signals present [51]. As the source position varies, eq.(17) defines a $\overline{\kappa}$–map. The dimensionless map $\bar{\nu} \equiv C_{\kappa\kappa}^{-1/2}\overline{\kappa}$ gives the number of σ a signal with the beam pattern shape has. If $|\bar{\nu}|$ is small, e.g., below 2.5σ, it will be consistent with the Gaussian signals in the map; if large, e.g., above 5σ, it will stick out as a non-Gaussian spike ripe for removal. The question then is how many σ one cleans to, and what impact setting such a threshold has on the statistics of the cleaned map, $\Delta' = \Delta - \Upsilon\langle\kappa|\overline{\Delta}\rangle$ with its corrected weight matrix $\widetilde{W}_t' = W_t - W_t\Upsilon C_\kappa \Upsilon^\dagger W_t$. (Actually $\widetilde{W}_t'$ always acts on Δ, so removing $\Upsilon\langle\kappa|\overline{\Delta}\rangle$ is not necessary in the limit of a wide prior dispersion for κ.)

Fig. 8 shows an application of this method to DMR. A comparison is also made between the mean noise field and the $\overline{\kappa}$ field: they are quite similar.

A similar optimal filter was used in [52] for ideal forecasting and in [53] for SZ source finding (based on the beta-profile). Another approach is to not pay such attention to the detailed optimal filter, but use a Gaussian-based filter with variable scale to try to identify point sources. [54] used a Mexican hat filter, i.e., the second derivative of a Gaussian. This continuum wavelet method is optimal only in the case of a scale invariant signal, a Gaussian beam and no noise. In the pure white noise limit with no signal, the straight Gaussian is optimal. The best choice lies in between.

3.8 Comparing: The simplest way to compare data sets A and B is to interpolate on overlap regions dataset A onto the pixels of dataset B. This of course will not always be possible, especially if the pixels are generalized ones, e.g., modal projections such as in the SK95 dataset. We would also like to compare nearby regions even if overlap is only partial. This requires an extrapolation as well as interpolation. In [55], a simple mechanism was described based upon Bayesian methods and a Gaussian interpolating theory for comparing CMB data sets. The statistics of the two experiments is defined by a joint probability for the two datasets, $\mathcal{P}(A,B|C_T)$, where the theory matrix C_T has AB as well as AA and BB parts connecting it. A probability enhancement factor $\beta = \ln \mathcal{P}(A|B,C_T)/\mathcal{P}(A|C_T)$ was intro-

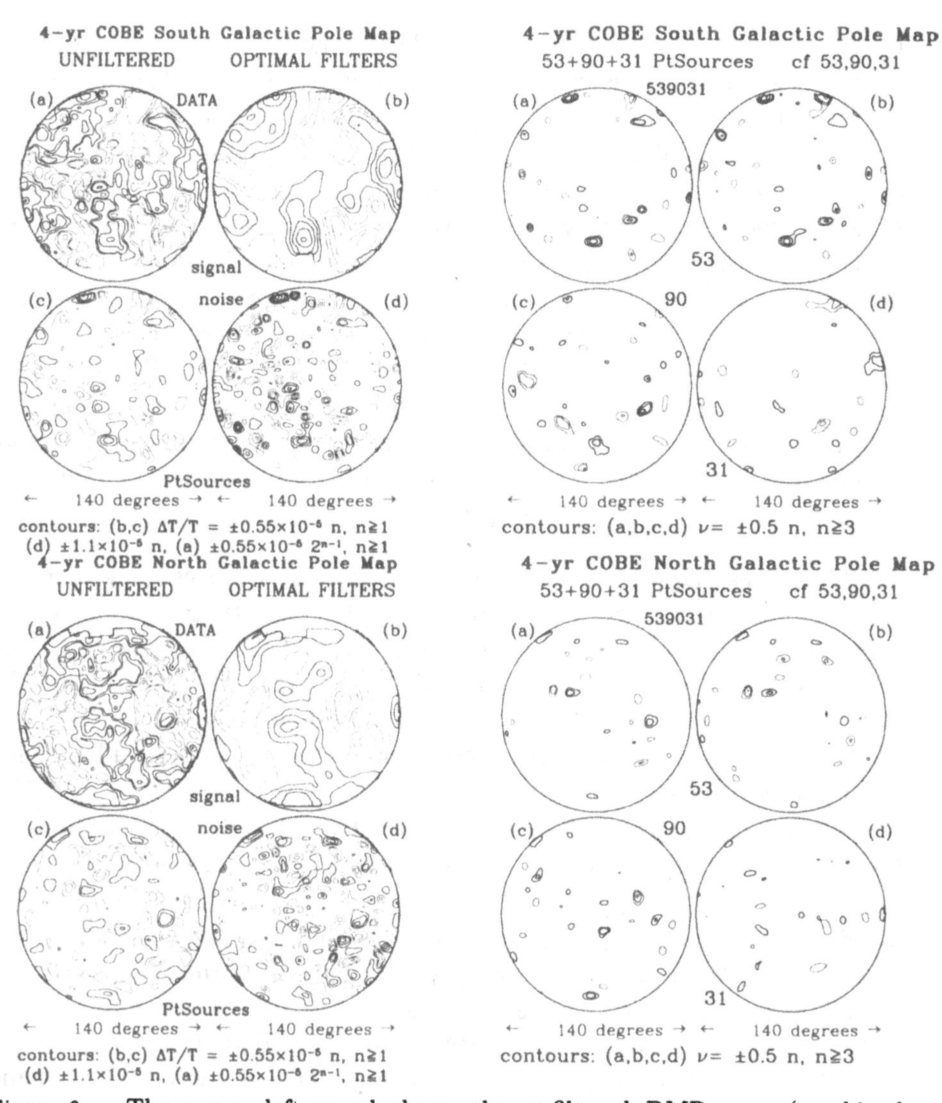

Figure 8. The upper left panel shows the unfiltered DMR map (combined 53+90+31 GHz data) of a 140° diameter region centered on the South Galactic Pole, with the contours indicated, (b) and (d) show the Wiener-filtered signal and noise maps, while (c) shows the fluctuations of height $\nu\sigma$, with $|\nu| > 1.5$. The lower left panel shows the same for the North Galactic Pole. Note the similarity between (c) and (d), that is that the Point Sources found at these low contours are consistent with random concentrations of noise, the logical conclusion within the statistics: there are no rare events that are good source candidates. We chose such a low cut so that one could see the effects of making 2 or 3σ cuts just by counting in, each positive contour being 0.5 higher than the previous. The right panels show the contours of ν rather than $\Delta T/T$ for the individual (A+B) frequency maps. A true source would be expected to persist in the three maps, as well as being in the combined one, the strength variation across channels depending upon the emission mechanism. Again there are no obvious hits. Apart from being a useful way to detect point sources, this is a nice mechanism for looking for anomalous local non-Gaussian patterns, to be investigated to determine if we are dealing with funny noise or true signals by appropriate follow-up investigations.

duced, which is invariant under A, B interchange.

What was visually instructive was to compare the A-data Wiener-filtered onto B-pixels, $\langle s_B|\Delta_A\rangle = C_{T,BA}\widetilde{W}_{t,AA}\overline{\Delta}_A$, with the Wiener-filter of B-data, $\langle s_B|\Delta_B\rangle = C_{T,BB}\widetilde{W}_{t,BB}\overline{\Delta}_B$. They should bear a striking resemblance except for the errors. The example used in [55] was SK95 data for A and MSAM92 for B, which showed remarkable similarities: indeed that the two experiments were seeing the same sky signal, albeit with some deviations near the endpoint of the scan. Although these extrapolations and interpolations are somewhat sensitive to the interpolating theory, if it is chosen to be nearly the best fit the results are robust.

The same techniques were used for an entirely different purpose in [56], testing whether two power spectra with appropriate errors, in this case from Boomerang and Maxima, could have been drawn from the same underlying distribution. The conclusion was yes.

3.9 Compressing: We would like to represent in a lossless way as much information as we can in much smaller datasets. Timestreams to maps (and map-orthogonal noise) are a form of compression, from N_{tbits} to N_{pix}. Map manipulations, even for Gaussian theories, are made awkward because the total weight matrix involves C_N, often simplest in structure in the pixel basis, and C_T, which is often naturally represented in a spherical harmonic basis. Signal-to-noise eigenmodes ξ_k are a basis in which $\widetilde{C}_N^{-1/2} C_T \widetilde{C}_N^{-1/2}$ is diagonal, hence are statistically independent of each other if they are Gaussian. They are a complete representation of the map. Another S/N basis of interest is one in which $C_T^{1/2}\widetilde{C}_N^{-1}C_T^{1/2}$ is diagonal, since $\widetilde{C}_N^{-1}$ comes out directly from the signal/noise extraction step. Finding the S/N modes is another $O(N_{pix}^3)$ problem. Typically, $\langle \xi_k^2\rangle$ falls off dramatically at some k, and higher modes of smaller eigenvalues can be cut out without loss of information. These truncated bases can be used to test the space of theoretical models by Bayesian methods e.g., [64], and determine cosmic parameters. Further compression to bandpowers is even better for more rapid determination of cosmic parameters.

The related compression concept of parameter eigenmodes mentioned in § 1.4 finds linear combinations of the cosmological or other variables y_α that we are trying to determine which are locally statistically-independent on the multidimensional likelihood surface. It provides a nice framework for dealing with near-degeneracies.

3.10 Power Spectra and Parameter Estimation: Determining the statistical distributions of any target parameters characterizing the theories, whether they are cosmological in origin or power amplitudes in bands or correlation function in angular bins, is a major goal of CMB analysis.

For power spectra and correlation functions, it is natural that these would involve pairs of pixels. Operators linear in pixel pairs are called *quadratic estimators*. Maximum likelihood expressions of parameters in Gaussian theories also often reduce to calculating such quadratic forms, albeit as part of iteratively-convergent sequences. Hence the study of quadratic estimators has wide applicability.

3.10.1 Quadratic Estimators: Estimating power spectra and correlation functions of the data has traditionally been done by minimizing a χ^2 expression,

$$\chi^2 = Tr[W(\overline{\Delta\Delta^\dagger} - C(y))W(\overline{\Delta\Delta^\dagger} - C(y))]\,/\,4, \tag{18}$$

where W is some as yet unspecified weight matrix. The 1/4 is from the pair sum. The critical element is to make a model $C(y)$ for the pixel pair correlation $\overline{\Delta\Delta^\dagger}$ which is as complete a representation as possible. For example, if we adopt a linear dependence about some fiducial $C_* = C(y_*)$, i.e., $C(y) = C_* + \sum_\beta C_\beta \delta y^\beta$ then

$$\sum_\beta \mathcal{F}_{\alpha\beta}\delta y^\beta = \tfrac{1}{2}\mathrm{Tr}[WC_\alpha W(\overline{\Delta\Delta^\dagger} - C_*)], \quad C_\alpha \equiv \partial C/\partial y^\alpha\,, \tag{19}$$

$$\mathcal{F}_{\alpha\beta} \equiv \tfrac{1}{2}\mathrm{Tr}[WC_\alpha WC_\beta] - \tfrac{1}{2}\mathrm{Tr}[WC_{\alpha\beta}W(\overline{\Delta\Delta^\dagger} - C_*)], \quad C_{\alpha\beta} \equiv \frac{\partial^2 C}{\partial y^\alpha \partial y^\beta},$$

$$\text{Fisher information matrix}: \quad F_{\alpha\beta} \equiv \langle \mathcal{F}_{\alpha\beta}\rangle = \tfrac{1}{2}\mathrm{Tr}[WC_\alpha WC_\beta]\,. \tag{20}$$

Since $\mathcal{F}_{\alpha\beta}$ has Δ dependence, this is not strictly a quadratic estimator. However, it is if the Fisher matrix $F_{\alpha\beta}$, the ensemble average of $\mathcal{F}_{\alpha\beta}$, is used. Iterations with F rather than $\mathcal{F}$ converge to the same y_* if y is continually updated by the δy. $C_* = C_N + C_T$ has been assumed.

The correlation in the parameter errors can be estimated from the average of a quartic combination of pixel values,

$$\begin{aligned}\langle \delta y \delta y^\dagger\rangle = &\ \mathcal{F}^{-1}\tfrac{1}{2}\mathrm{Tr}[WC_\alpha WCWC_\beta WC]\mathcal{F}^{-1} \\ &+\mathcal{F}^{-1}\tfrac{1}{4}\mathrm{Tr}[WC_\alpha W(\delta C)]\mathrm{Tr}[WC_\beta W(\delta C)]\mathcal{F}^{-1}\,,\end{aligned}$$

where $(\delta C) \equiv C - C_*$. Note, $\langle \delta y \delta y^\dagger\rangle = \mathcal{F}^{-1}$ if $C = C_*$ and $W = C_*^{-1}$.

Another approach is to estimate the errors by taking the second derivative of χ^2, $\partial\chi^2/\partial y^\alpha \partial y^\beta = \mathcal{F}_{\alpha\beta}$:

$$\begin{aligned}\langle \partial^2\chi^2/\partial y_\alpha \partial y_\beta\rangle &= \tfrac{1}{2}\mathrm{Tr}[WC_\alpha WC_\beta] - \tfrac{1}{2}\mathrm{Tr}[WC_{\alpha\beta}W(\delta C)] \\ &\to F_{\alpha\beta} \text{ as } (\delta C) \to 0.\end{aligned}$$

These two limiting cases hold for the maximum likelihood solution as we now show.

3.10.2 Maximum Likelihood Estimators & Iterative Quadratics: The maximum likelihood solution for a Gaussian signal plus noise is of the form eq.(19) with $W = \widetilde{W}_t = (\widetilde{C}_N + C_T(y))^{-1}$, the "optimal" weight which gives the minimum error bars. This is found by differentiating $-2\ln\mathcal{L} = \mathrm{Tr}[\widetilde{W}_t\overline{\Delta\Delta^\dagger})] - \mathrm{Tr}[\ln\widetilde{W}_t] + N_{pix}\ln(2\pi)$. Of course the y_β dependence of $\widetilde{W}_t$ means the expression is not really a quadratic estimator. A solution procedure is the Newton-Raphson method, with the weight matrix updated in each iterative improvement δy_β to $y_{*\beta}$, at a cost of a matrix inversion. At each step we are doing a quadratic estimation, and when the final state has been converged upon, the weight matrix can be considered as fixed. Solving for the roots of $\partial\ln\mathcal{L}/\partial y_\alpha$ using the Newton-Raphson method requires that we calculate $\partial^2\ln\mathcal{L}/\partial y_\alpha\partial y_\beta$, the curvature of the likelihood function. Other matrices have been used for expediency, since one still converges on the maximum likelihood solution; in particular the curvature matrix expectation value, i.e., the Fisher matrix $F_{\alpha\beta}$, which is easier to calculate than $\mathcal{F}_{\alpha\beta}$. Further, to ensure stability of the iterative Newton-Raphson method, some care must be taken to control the step size each time the y_β are updated. Since the $\langle\delta y_\alpha\delta y_\beta\rangle$ estimate is just $[\mathcal{F}^{-1}]_{\alpha\beta}$, this matrix must at least be determined after convergence to characterize the correlated errors.

Note that only in the Gaussian and uniform prior cases is the integration over $s_{(j)}$ analytically calculable, although maximum likelihood equations can be written for simple forms that can be solved iteratively, e.g., maximum entropy priors.

3.10.3 The Bandpower Case: The bandpower associated with a given window function $\varphi_{b\ell}$ of a theory with spectrum $C_{T\ell}$ is defined as the average power,

$$\langle C_\ell\rangle_b \equiv \mathcal{I}[\varphi_{b\ell}C_{T\ell}]/\mathcal{I}[\varphi_{b\ell}], \tag{21}$$

with $\mathcal{I}$ defined by eq.(7). Many choices are possible for $\varphi_{b\ell}$, but the simplest is probably the best: e.g., $\chi_b(\ell)$ which is unity in the ℓ-band, zero outside. Another example uses the average window function of the experiment, $\overline{\mathcal{W}}_\ell\chi_b(\ell)$; others use window functions related to relative amounts of signal and noise. There is some ambiguity in the choice for the filter $\varphi_{b\ell}$ [57, 58], but often not very much sensitivity to its specific form.

We want to estimate bandpowers q^b taken relative to a general shape $\mathcal{C}_\ell^{(s)}$ rather than the flat shape: $q^b \equiv \mathcal{I}(\varphi_{b\ell}C_{T\ell})/\mathcal{I}(\varphi_{b\ell}\mathcal{C}_\ell^{(s)})$, so $\langle\mathcal{C}\rangle_b = \bar{q}_b\langle\mathcal{C}^{(s)}\rangle_b$. We therefore model $\overline{\Delta\Delta^\dagger}$ by

$$C_{pp'}(q) = \widetilde{C}_{N,pp'} + \sum_b C^{(s)}_{b,pp'}q^b \equiv C_{*,pp'} + \sum_b C^{(s)}_{b,pp'}\delta q^b,$$
$$C^{(s)}_{b,pp'} = \mathcal{I}[\mathcal{W}_{pp',\ell}\chi_b(\ell)\mathcal{C}^{(s)}_\ell], \qquad \delta q^b = q^b - q^b_*. \tag{22}$$

Here the q_*^b are the bandpowers on a last iteration, ultimately converging to maximum likelihood bandpowers if the relaxation is allowed to go to completion, $\delta q^b \to 0$. The pixel-pixel correlation matrices $C_{Tb,pp'} = C_{b,pp'}^{(s)} q^b$ for the bandpowers b follow from eq.(6).

The usual technique [14] is to use this linear expansion in the q^b, the downside being that the bandpowers could be negative – when there is little signal and the pixel pair estimated signal plus noise is actually less than the noise as estimated from C_N. By choosing $\exp(q^b)$ rather than q^b as the variables, positivity of the signal bandpowers can be assured. (The summation convention for repeated indices is often used here.)

The full maximum likelihood calculation has been used for bandpower estimates for COBE, SK95, Boomerang and Maxima, among others. The cost is $\mathcal{O}(N_{pix}^3)$ for each parameter, so this limits the number of pixels that can be treated. For $N_{pix} \gtrsim 2 \times 10^5$ or so, the required computing power becomes prohibitive, requiring 640 Gb of memory and of order 3×10^{17} floating-point operations, which translates to a week or more even spread over a Cray T3E with $\sim$ 1024 900-MHz processors. This will clearly be impossible for the ten million pixel Planck data set. Numerical roundoff would be prohibitive anyway.

The compressed bandpowers estimated from all current data and the satellite forecasts shown in Fig. 3 were also computed using the maximum likelihood Newton-Raphson iteration method (for rather fewer "pixels").

3.10.4 A Wavenumber-basis Example: There is an instructive case in which to unravel the terms of eq.(19). If we assume the noise and signal-shape correlation matrices in the wavenumber basis $\mathbf{Q}$ of § 3.4, $\widetilde{C}_{N,\mathbf{QQ}'}$ and $C_{Tb,\mathbf{QQ}'}^{(s)}$, are functions only of the magnitude of the wavenumber, $Q \approx \ell + \frac{1}{2}$, then

$$\sum_{b'} F_{bb'} q^{b'} = \tfrac{1}{2} \sum_{Q} g_Q W^2(Q) C_{Tb}^{(s)}(Q) (C_{est}(Q) - \widetilde{C}_N(Q)) \, , \tag{23}$$

$$F_{bb'} = \tfrac{1}{2} \sum_{Q} g_Q W^2(Q) C_{Tb}^{(s)}(Q) C_{Tb'}^{(s)}(Q) \, ,$$

$$W(Q) = (\widetilde{C}_N(Q) + \sum_{b} C_{Tb}^{(s)}(Q) q_*^b)^{-1} \quad C_{est}(Q) \equiv \int \frac{d\phi_{\mathbf{Q}}}{2\pi} \, |\widetilde{\Delta}_{\mathbf{Q}}|^2 \, .$$

Here g_Q is the effective number of modes contributing to the ΔQ=1 wavenumber interval.

In the all-sky homogeneous noise limit, this of course gives the correct result, with $g_Q = f_{sky}(2\ell+1)$ and $C_{Tb}^{(s)}(Q) = C_\ell^{(s)} \overline{\mathcal{W}}_\ell \chi_b(\ell)$, where $\overline{\mathcal{W}}_\ell$ is the window (the angle-average of $|u(\mathbf{Q})|^2 \mathcal{B}_\ell^2$ in the language of § 3.4.1). Since $\chi_b(\ell)\chi_{b'}(\ell) = \delta_{bb'}(\ell)\chi_b(\ell)$, $F_{bb'}$ is diagonal, with a value expressed in terms of a band-window $\varphi_{b\ell}$:

$$F_{bb} = \tfrac{1}{2}\mathcal{I}[\varphi_{b\ell}C_\ell^{(s)}], \quad \varphi_{b\ell} = 4\pi(g_Q/(2Q))W^2(Q)C_\ell^{(s)}\overline{\mathcal{W}}_\ell^2\chi_b(\ell). \quad (24)$$

It may now seem that we have identified up to a normalization the correct band-window to use in this idealized case, with a combination of signal and noise weighting. However the ambiguity referred to earlier remains, for what we are actually doing in estimating the observed q^b is building a piecewise discontinuous model relative to the shape $C_\ell^{(s)}$. The analogue for a given $C_{T\ell}$ is the model $\sum_\ell \chi_b(\ell)q_{Tb}C_\ell^{(s)}$. Since for the theory, $q^b = \mathcal{I}[\varphi_{b\ell}C_{T\ell}]/\mathcal{I}[\varphi_{b\ell}C_\ell^{(s)}]$, any $\varphi_{b\ell}$ satisfying $\varphi_{b\ell}\chi_{b'}(\ell) = \varphi_{b\ell}$ will do to have q_{Tb} the ratio of target-bandpower to shape-bandpower.

The downweighting by signal-plus-noise power is just what is required to give the correct variance in the signal power estimate in the limit of small bin size:

$$\langle(\Delta C_{Tb})^2\rangle = F_{bb}^{-1}\langle C_\ell^{(s)}\rangle_b^2 \to \frac{2}{g_\ell}\frac{(C_{T\ell}\overline{\mathcal{W}}_\ell + \tilde{C}_{N\ell})^2}{\overline{\mathcal{W}}_\ell^2} \text{ as } \Delta\ell \to 1. \quad (25)$$

In the sample-variance dominated regime, the error is $C_{T\ell}\sqrt{2/g_\ell}$, while in the noise-dominated regime it is $\tilde{C}_{N\ell}\sqrt{2/g_\ell}/\overline{\mathcal{W}}_\ell$, picking up enormously above the Gaussian beam scale. Recent experiments are designed to have signal-to-noise near unity on the beam scale, hence are sample-variance dominated above. The interpretation of the $(\tilde{C}_{N\ell}+C_{T\ell})^{-1}$ weighting factor is that correlated spatial patterns associated with a wavenumber ℓ that are statistically high just because of sample variance are to be downweighted by $C_{T\ell}$.

Although these results are only rigourous for the all-sky case with homogeneous and isotropic noise, they have been used to great effect for forecasting errors (§3.11) for $f_{sky} < 1$ regions.

3.10.5 Faster Bandpowers: Since the need for speed to make the larger datasets tractable is essential, much attention is now being paid to much faster methods for bandpower estimation.

A case has been made for highly simplified weights (e.g., diagonal in pixel space) that would still deliver the basic results, with only slightly increased error bars [60]. The use of non-optimal weighting schemes has had a long history. For example, a quadratic estimator of the correlation function was used in the very first COBE DMR detection publication, with $diag(W_N)$ for W, a choice we have also used for COBE and the balloon-borne FIRS. Here one is estimating correlation function amplitudes C^α, where α refers to an angular bin: $C_{pp'} = C_{Npp'} + \sum_\alpha C^\alpha\chi_\alpha(\theta_{pp'})$, where $\cos(\theta_{pp'}) = \hat{q}_p \cdot \hat{q}_{p'}$ and $\chi_\alpha(\theta)$ is one inside and zero outside of the bin. As for bandpowers, the distribution is non-Gaussian, so errors were estimated using Monte Carlo simulations of the maps.

Any weight matrix W which is diagonal in pixel space is easy and fast to calculate. In [60] it was shown that making W the identity was adequate for Boomerang-like data. The actual approach used a fast highly binned estimate of $C(\theta)$, which can be computed quickly, which was then lightly smoothed, and finally a Gauss-Legendre integration of $C(\theta)$ was done to estimate $\langle C \rangle_b$.

One can still work with the full $W = \widetilde{W}_t$ if the noise weight matrix C_N^{-1} has a simple form in the pixel basis. For the MAP satellite, it is nearly diagonal, and a method exploiting this has been applied to MAP simulations to good effect [61].

Since many of the signals are most simply described in multipole space, it is natural to try to exploit this basis, especially in the sample-variance dominated limit. In the Boomerang analysis of [4, 38], filtering of form $\mu(\varpi_p)u(\mathbf{Q})\mathcal{B}(Q)$ was imposed. Many of the features of the diagonal ansatz of § 3.10.4 in wavenumber space were exploited, complicated of course by the spatial mask μ, which leads to mode-mode coupling. The main approximation was to use the masked transforms for the basic isotropic Q functions, expressing where needed the masked power spectra as linear transformations on the true underlying power spectrum in ℓ-space that we are trying to find. Monte Carlo simulations were used to evaluate a number of the terms. Details are given in [4, 38] and elsewhere.

More could be said about making power spectrum estimation nearly optimal and also tractable, and much remains to be explored.

3.10.6 Relating Bandpowers to Cosmic Parameters: To make use of the q^b for cosmological parameter estimation, we need to know the entire likelihood function for q^b, and not just assume a Gaussian form using the curvature matrix. This can only be done by full calculation, though there are two analytic approximations that have have been shown to fit the one-point distributions quite well in all the cases tried [57]. The simplest and most often used is an "offset lognormal" distribution, a Gaussian in $z^b = \ln(q^b + q_N^b)$, where the offset q_N^b is related to the noise in the experiment:

$$\begin{aligned}
&\mathcal{P}(q) \propto \exp[-\tfrac{1}{2}(z-\bar{z})^b \mathcal{F}^{(z)}_{bb'} (z-\bar{z})^{b'}], \\
&\mathcal{F}^{(z)}_{bb'} = (\bar{q}^b + q_N^b)\mathcal{F}^{(q)}_{bb'}(\bar{q}^{b'} + q_N^{b'}), \\
&\bar{q}^b / q_N^b = [\mathcal{F}^{(q)0}_{bb'} / \mathcal{F}^{(q)}_{bb'}]^{1/2} - 1\,,
\end{aligned} \tag{26}$$

where $\bar{q}^b$ is the maximum likelihood value. To evaluate q_N^b in eq.(26), the curvature matrix evaluated in the absence of signal, $\mathcal{F}^{(q)0}_{bb'}$, is needed. For Boomerang and Maxima analysis, the two Fisher matrices are standard output of the "MADCAP" maximum likelihood power spectrum estimation code [59]. Another (better) approach if the likelihood functions $\mathcal{L}(\{q^b\})$ are available for each q^b is to fit $\ln \mathcal{L}$ by $-\frac{1}{2}[(z-\bar{z})_b]^2 g_b^2$, and use $g_b G_{bb'} g_{b'}$ in place of $\mathcal{F}^{(z)}_{bb'}$, where $G_{bb'} = diag(\mathcal{F}^{-1})^{1/2} \mathcal{F} diag(\mathcal{F}^{-1})^{1/2}$. Here $diag()$ denotes the diagonal part of the matrix in question. Values of q^b_N and other data for a variety of experiments are given in [57] (where it is called x^b). It is also possible to estimate q^b_N using a quadratic estimator [4]. To compare a given theory with spectrum $C_{T\ell}$ with the data using eq.(26), the q^b's need to be evaluated with a specific choice for $\varphi_{b\ell}$. Although this should be band-limited, $\propto \chi_{b\ell}$, as described in § 3.10.4, the case can be made for a number of alternate forms.

3.11 Forecasting: A typical forecast involves making large simplifications to the full statistical problem, such as those used in § 3.10.4. For example, the noise is homogeneous, possibly white, and the fluctuations are Gaussian. This was applied to estimates of how well cosmic parameters could be determined for the LDBs and MAP and Planck in the 9 parameter case described earlier [13], and for cases where the power spectrum parameters are allowed to open up beyond just tilts and amplitudes to parameterized shapes [62]. The methods were also applied when CMB polarization was included, and when LSS and supernovae were included [63].

The procedure is exactly like that for maximum likelihood parameter estimation, except one often does not bother with a realization of the power spectrum, rather uses the average (input) value and estimates errors from the ensemble average curvature, i.e., the Fisher matrix: F^{-1} is therefore a measure of $\langle \delta y_\alpha \delta y_\beta \rangle$, where $\delta y_\alpha = y_\alpha - \bar{y}_\alpha$ is the fluctuation of y_α about its most probable value, $\bar{y}_\alpha$. As mentioned above, when the y_α are the bandpowers instead of the cosmic parameters, we get the forecasts of power spectra and their errors, as for MAP and Planck in Fig. 3. Those results ignored foregrounds. However other authors have treated foregrounds with the Gaussian and max-ent prior assumptions in this simplified noise case [45, 47].

Clearly forecasting can become more and more sophisticated as realistic noise and other signals are added, ultimately to grow into a full pipeline-testing simulation of the experiment, which all of the CMB teams strive to do to validate their operations.

Challenge: The lesson of the CMB analysis done to date is that almost all of the time and debate is spent on the path to a believable primary power spectrum, with the cosmological parameters quickly dropping out once the spectrum has been agreed upon. The analysis challenge, requiring clever new algorithms, gets more difficult as we move to more LDBs and interferometers, to MAP and to Planck. This is especially so with the ramping-up efforts on primary polarization signals, expected at <10% of the total anisotropies, and the growing necessity to fully confront secondary anisotropies and foregrounds simultaneously with the primary signals.

Acknowledgments: Many people have worked with us at CITA on aspects of the CMB analysis touched on here, Carlo Contaldi, Andrew Jaffe, Lloyd Knox, Steve Myers, Barth Netterfield, Ue-Li Pen, Dmitry Pogosyan, Simon Prunet, Marcelo Ruetalo, Kris Sigurdson, Tarun Souradeep, Istvan Szapudi, and James Wadsley, along with Julian Borrill, Eric Hivon and other members of the Boomerang and CBI teams, and George Efstathiou and Neil Turok at Cambridge.

References

1. Seljak, U. and Zaldarriaga, M. 1996, ApJ 469, 437. (CMBfast)
2. Miller, A.D. *et al.* 1999, ApJ 524, L1. (TOCO)
3. de Bernardis, P. *et al.* 2000, Nature 404, 995.
4. Netterfield, C.B. *et al.* 2001, astro-ph/0104460; de Bernardis, P. *et al.* 2001, astro-ph/0105296. (BOOM)
5. Mauskopf, P. *et al.* 2000, ApJ Lett 536, L59. (BOOM-NA)
6. Hanany, S. *et al.* 2000, ApJ Lett. 545, 5.
7. Lee, A.T. *et al.* 2001, astro-ph/0104459; Stompor, R. *et al.* 2001, astro-ph/0104462. (MAXIMA)
8. Padin, S. *et al.* 2000, ApJ Lett., submitted, astro-ph/0012211; Myers, S. *et al.* 2001, preprint. (CBI)
9. Leitch, E.M. *et al.* 2001, astro-ph/0104488; Halverson, N.W. *et al.* 2001, astro-ph/0104489; Pryke, C. *et al.* 2001, , astro-ph/0104490. (DASI)
10. Bond, J.R. 1996, in *Cosmology and Large Scale Structure*, Les Houches Session LX, eds. R. Schaeffer J. Silk, M. Spiro and J. Zinn-Justin (Elsevier), pp. 469-674
11. Bond, J.R. 1994, Phys. Rev. Lett. 74, 4369.
12. e.g., Efstathiou, G. & Bond, J.R. 1999, M.N.R.A.S. 304, 75, where many other near-degeneracies between cosmological parameters are also discussed.
13. Bond, J.R., Efstathiou, G. & Tegmark, M. 1997, M.N.R.A.S. 291, L33.
14. Bond, J.R., Jaffe, A.H. & Knox, L. 1998, Phys. Rev. D 57, 2117.
15. Pen Ue-Li, Seljak U. & Turok N. 1997, Phys. Rev. Lett. 79, 1611.
16. Allen, B., Caldwell, R.R., Dodelson, S., Knox, L., Shellard, E.P.S. & Stebbins, A. 1997, astro-ph/9704160.
17. Frenk, C.S. *et al.* 1999, ApJ 525, 554.
18. Birkinshaw, M. 1999, Phys. Rep. 310, 97.
19. Wadsley, J.W. & Bond, J.R. 2001, preprint.
20. Bond, J.R., Pogosyan, D., Kofman, L. & Wadsley, J.W. 1998, in "Wide Field Surveys in Cosmology", Proc. XIV IAP Colloquium Conf., ed. S. Colombi & Y. Mellier, (Paris: Editions Frontieres), p. 17, astro-ph/9810093.
21. Pogosyan, D., Bond, J.R., Kofman, L. & Wadsley, J.W. 1998, in "Wide Field Surveys in Cosmology", Proc. XIV IAP Colloquium Conf., ed. S. Colombi & Y. Mellier, (Paris: Editions Frontieres), p. 61, astro-ph/9810072.
22. Persi, F.M., Spergel, D.N., Cen, R. & Ostriker, J.P. 1995, ApJ 442, 1.
23. Refregier, A., Komatsu, E., Spergel, D.N. & Pen, Ue-Li 2000, astro-ph/9912180.
24. Scaramella, R., Cen, R. & Ostriker, J.P. 1993, ApJ 416, 399.
25. Seljak, U., Burwell, J. & Pen, Ue-Li 2000, astro-ph/0001120.

26. da Silva, A.C., Barbosa, D., Liddle, A.R. & Thomas, P.A. 2000, MNRAS 317, 37.
27. Springel, V., White, M. & Hernquist, L. 2000, astro-ph/0008133.
28. Bond, J.R. 1998, in *The Early Universe*, Proc. NATO Summer School, Victoria, B.C., Canada, August 1986, p. 283-334, ed. W. Unruh and G. Semenoff, Dordrecht: Reidel.
29. Bond, J.R., Carr, B.J, & Hogan, C.J. ApJ 1991, 367, 420.
30. Bond, J.R. & Myers, S. 1996, ApJ Supp 103, 1, 41; 63.
31. Bond, J.R., Ruetalo, M. & Wadsley, J.W. 2001, in Proc. TAW8 conference on High Redshift CLusters, Missing Baryons and CMB Polarization.
32. Lange, A. *et al.* 2001, Phys. Rev D, in press.
33. Bond, J.R., Crittenden, R., Jaffe, A.H. & Knox, L. 1999, Computing in Science and Engineering 1, 21, astro-ph/9903166, and references therein.
34. Ferreira, P.G. & Jaffe, A.H. 1999, M.N.R.A.S. astro-ph/9909250.
35. Prunet, S. *et al.* 2001,, Proc. Garching Conference on "Analysis of Large Astronomical Data Sets", astro-ph/0101073.
36. Stompor, R. *et al.* 2001, Proc. Garching Conference on "Analysis of Large Astronomical Data Sets", astro-ph/0012418.
37. Dore, O., Teyssier, R., Bouchet, F.R., Vibert, D. & Prunet, S. 2001, AstronAp, astro-ph/0101112.
38. Hivon, E., Gorksi, K.M., Netterfield, C.B., Crill, B.P., Prunet, S., Hansen, F. 2001, ApJ, astro-ph/0105302.
39. Stompor, R. *et al.* 2001, Phys. Rev. D, astro-ph/0106451.
40. Hinshaw, G. *et al.* 2001, Proc. Garching Conference on "Analysis of Large Astronomical Data Sets", astro-ph/0011555; see also E.L. Wright 1996, astro-ph/9612006.
41. P.F. Muciaccia, P. Natoli & N. Vittorio 1998, ApJ 488, L63.
42. K.M. Gorski, E. Hivon & B.D. Wandelt 1998, astro-ph/9812350.
43. Crittenden, R. & Turok, N. 1999, astro-ph/9806374.
44. Hobson, M.P. & Lasenby, A.N. 1998, M.N.R.A.S. 298, 905.
45. Bouchet, F.R. & Gispert, R. 1999, New Astron. 4, 443.
46. Prunet, S. *et al.* 2000b, AstronAp Lett, astro-ph/0012497.
47. Hobson, M.P., Jones, A.W., Lasenby, A.N. & Bouchet, F.R. 1999, M.N.R.A.S. astro-ph/9806387.
48. Schlegel, D.J., Finkbeiner, D.P. & Davis, M. 1998, ApJ 500, 525.
49. Jaffe, A.H., Finkbeiner, D. & Bond, J.R. 1999, preprint.
50. de Oliveira-Costa, A., et al. 1997, ApJ Lett 482, 17.
51. Crittenden, R. & Bond, J.R. 2001, preprint.
52. Tegmark, M. and de Oliveira-Costa, A. 1998, ApJ 500, 83.
53. Haehnelt, M.G. & Tegmark. M. 1996, M.N.R.A.S. 279, 545.
54. Cayon, L. *et al.* 2000, M.N.R.A.S. astro-ph/9912471
55. Knox, L., Bond, J.R., Jaffe, A.H., Segal, M. & Charbonneau, D. 1998, Phys. Rev. D, 58, 1443.
56. Jaffe, A.H. *et al.* 2001, Phys. Rev. Lett., in press.
57. Bond, J.R., Jaffe, A.H. & Knox, L. 2000, ApJ 533, 19.
58. Knox, L. 2000, Phys. Rev. D, astro-ph/9902046.
59. Borrill, J. 1999, Phys. Rev. D 59, 7302.
60. Szapudi, I., Prunet, S., Pogosyan, D., Szalay, A. & Bond, J.R. 2001, ApJ Lett, in press.
61. Oh, S.P., Spergel, D.N. & Hinshaw, G. 1999, ApJ 510, 551.
62. Souradeep, T., Bond, J.R., Knox, L., Efstathiou, G. & Turner, M.S. 1998, in "Proceedings of COSMO-97", ed. L. Roszkowski (World Scientific), astro-ph/9802262.
63. Eisenstein, D., Hu, W. & Tegmark, M. 1998, ApJ, astro-ph/9807130.
64. Bond, J.R. & Jaffe, A. 1998, Phil. Trans. R. Soc. London 357, 57.
65. Bond, J.R. *et al.* 2000, Proc. IAU Symposium 201, astro-ph/0011378; Proc. CAPP 2000, astro-ph/0011379; Proc. Neutrino 2000, astro-ph/0011377.
66. Bond, J.R., Pogosyan, D. and Souradeep, T. 2000, Phys. Rev. D **62**.

Part IV
ENERGY FLOW AND FEEDBACK

ENERGY FLOW IN THE UNIVERSE

CRAIG J. HOGAN
Astronomy and Physics Departments
University of Washington
Seattle, Washington 98195, USA

Abstract. A brief but broad survey is presented of the flows, forms and large-scale transformations of mass-energy in the universe, spanning a range of about twenty orders of magnitude ($\approx m_{Planck}/m_{proton}$) in space, time and mass. Forms of energy considered include electromagnetic radiation, magnetic fields, cosmic rays, gravitational energy and gravitational radiation, baryonic matter, dark matter and vacuum energy, and neutrinos; sources considered include vacuum energy and cosmic expansion, fluctuations and gravitational collapse, AGN and quasars, stars, supernovae and gamma ray bursts.

1. Global Energy

Everything that happens is a transformation of mass-energy. Starting with inflation and the Big Bang, mass-energy flows through and organizes structures spanning an enormous range of scales— lengths and times from billions of years down to milliseconds. These notes survey the main features of cosmic energy cycles on a global scale and trace the causal links between them.

An absolute luminosity limit for anything is imposed by General Relativity. Suppose a sphere of radius R is filled with light of total mass-energy Mc^2 and released an an instant; the energy has left the sphere after a time R/c, with an average luminosity of Mc^3/R. But the gravity of the energy imposes a limit on how small R can be: if it is smaller than the Schwarzschild radius $R_S = 2GM/c^2$, no light can escape at all since it is within the event horizon of a black hole. The maximum luminosity of any source is therefore

$$L_{GR} = c^5/2G = m_{Planck}^2/2 = 1.81 \times 10^{59} \text{erg sec}^{-1}, \tag{1}$$

R.G. Crittenden and N.G. Turok (eds.), Structure Formation in the Universe, 283–293.

independent of mass. (We have expressed G in terms of the Planck mass $m_{Planck} = \sqrt{\hbar c/G} = 1.2 \times 10^{19}$ GeV $\approx 10^{-5}$g, according to the convention $\hbar = c = 1$. It is tempting call L_{GR} the "Planck Luminosity" since it corresponds to a Planck mass per Planck time, but in fact Planck's constant $\hbar$ cancels out when using units of luminosity— L_{GR} does not depend on quantum mechanics.) This luminosity represents an upper bound on the rate of energy transformation of any kind, on any scale.

For objects within the universe, nothing approaching this luminosity has ever been observed— not because of size, but because radiation interacts with matter (indeed, it must interact to be generated) and therefore takes time to "leak out" of a system. The maximal limiting luminosity requires both an efficiency close to unity and an interaction time close to a light-travel time. Even neutrinos interact more strongly than this in dense collapsing cores of stars where they are copiously produced. The only situation where the limit seems likely to be approached is in production of gravitational radiation from merging black holes of comparable mass, events which may eventually be observable in gravitational waves.[1, 2]

Roughly speaking, the Big Bang saturated the absolute bound L_{GR}. At any time during the early radiation-dominated phase of the early universe, this is about equal to the energy of cosmic radiation within a Hubble volume divided by a Hubble time. If the universe today is dominated by vacuum energy (a cosmological constant or some other form with $\Omega_\Lambda \approx 0.7$)[3, 4], then the "$PdV$ work" being done right now in each Hubble volume is also comparable to L_{GR}. (A comparable rate of transformation occurred during inflation and during reheating.) The cosmic background radiation is less than this today by a factor of about 10^4 because of redshifting.

The comparable amount of energy locked up as rest-energy of Dark Matter has not substantially interacted with anything microscopically for a long time— for most candidates, at least since the weak interactions decoupled. However, a significant flow of energy occurs in the dark matter via gravitational collapse. This energy predates even the cosmic radiation, originating in primordial fluctuations in binding energy which produce cosmic structure, which probably date back to inflation. These perturbations are injecting observable energy flows into the universe today, into the non-vacuum components— the Dark Matter, with $\Omega_m \approx 0.3$, and the baryons, with $\Omega_b \approx 0.03$. Their dimensionless amplitude is about 10^{-5}— this is the fraction of total mass-energy available as free energy in this form. There are of course small flows caused by radiation temperature anisotropies but the dominant effect is gravitational collapse, which in turn creates kinetic motion in the dark matter and causes heating of baryonic gas by compression and shocking. This process heats the bulk of cosmic baryonic matter to temperatures of about $10^{-6} m_{proton} \approx 10^7$ K, creating a pressure sufficient to keep most of the baryons from falling into galaxies and stars.[5, 6] The gas achieves a steady state at this characteristic temperature, where the "hierarchical heating" is balanced by adiabatic losses to the expansion (and thereby slightly modify the global expansion rate.)

Roughly speaking, the cosmic fluctuation amplitude of $Q \approx 10^{-5}$ injects $10^{-5}\rho c^2$ of energy per Hubble time in the cosmic web, with typical velocity scale $Q^{1/2} \approx 10^{-2.5}c$ and typical size $Q^{1/2}/H$ corresponding today to galaxy superclusters. The power in this form of heating is substantial, about 10^{52} erg/sec in the baryons and $10^{53.5}$ erg/sec in the dark matter. Because it radiates inefficiently this dominant reservoir of intergalactic matter is practically invisible except as a diffuse soft X-ray background.

2. Stars

The other forms of energy involve the small fraction of baryons which make it into galaxies. The total density of baryons in galaxies, including all stars and their remnants as well as star-forming gas, adds up to less a quarter of the baryons, or less than one percent of the total density. A small fraction of this material makes it into AGN and supernovae. Somehow the activity is coupled so that these things all contribute roughly comparable total energy budgets.

Most light since the Big Bang has been made by ordinary main-sequence stars. They are still forming from gas, although most of the stars in our past light cone formed about 10^{10} years ago— about 10^{10} of them in each of 10^{10} galaxies in our reference volume, formed over roughly 10^{10} years. Each one lasts for a long time; most of the mass-energy budget is in low mass stars which last for billions of years (10^{10}y for 1 $M_{\odot}$). The total power in stars in all galaxies over all of cosmic history is now close to being accounted for in cosmic backgrounds from the optical to the far infrared.[7, 8, 9, 13] The distribution in time is estimated from redshifts of directly imaged galaxies[10] and the backgrounds are now close to being resolved, so the spacetime distribution of the source populations are close to accounted for. The total flux is about 1/30 of the cosmic background radiation, approximately equally distributed between direct light from stars and reradiated light from obscuring dust, and mostly radiated since a redshift of about two, when the universe was three times smaller than today.

This energy ultimately derives mostly from the conversion of hydrogen to helium, with some contribution from synthesis to heavier elements. The enrichment history is recorded as fossil abundances in stars and can also be traced directly in high-redshift absorption systems[11]. The total amount of light agrees with the total production of elements by stars[13] if the metals are mostly ejected from the stellar parts of galaxies and join the dominant intergalactic gas. This large-scale sharing of metals is confirmed from X-ray line emission in galaxy clusters[12].

Stellar formation as well as stellar instabilities (important especially at the end of the stable nuclear burning stage) all occur at roughly the Chandrasekhar mass, corresponding to a number of protons

$$N_C = 3.1(Z/A)^2(m_{Planck}/m_{proton})^3 \approx 10^{57} \tag{2}$$

(shown here with its classical definition as the limiting mass of an electron-degeneracy-supported dwarf; Z and A are the average charge and mass of the ions, typically $Z/A \approx 0.5$ and $M_C = 1.4M_\odot$, where $M_\odot = 1.988 \times 10^{33}g \approx 0.5M_*$ is the mass of the Sun.) One way or another the large numbers in Table 1 for baryonic flows all ultimately derive from the large number $(m_{Planck}/m_{proton}) \approx 10^{19}$; this is true even for the global cosmological quantities, since the age of the universe now is about the lifetime of a star.

3. Quasars

In the centers of galaxies, a small fraction of material (up to about 10^{-2} of the stellar mass, so on the order of 10^{-4} of the total mass-energy) accumulates and organizes itself into a different kind of engine, called quasars, active galactic nuclei or AGN, which generate their enormous power from gas interacting with massive black holes. Although events can occur in quasars quickly (on timescales of days or less, the Schwarzchild time for the massive holes) the bright phase lasts for tens of millions of years, determined by the behavior of the surrounding gas. The bright activity of quasars peaked at redshifts of about two and is much less today[14], but the black hole remnants of 10^6 to $10^9 M_\odot$ reside still in the centers of most galaxies including our own.[15] The total number N of quasars is thus about the same as the number of galaxies, with a tendency for large bright ones to lie in the biggest galaxies. The overall energy from quasars is not much less than that from stars, due to their large efficiency in converting rest-mass into energy. They dominate the energy budget of the universe for most hard radiation such as X-rays and gamma rays (with some competition from supernovae), and are mainly responsible for ionizing the intergalactic gas.[16]

Active nuclei derive all of their electromagnetic energy from gravity—either from the binding energy of the infalling material, or from the rotational energy of the black hole.[17, 18, 19] Gas falls in and forms a disk near the hole, fattened by heating into a torus or corona. Magnetic fields help to extract the orbital and spin energy and also to channel some of it into "Poynting jets" of relativistic matter moving so close to the speed of light that a particle's kinetic energy is many times its rest mass, with Lorentz factors of $\Gamma = E/mc^2 \approx 10$. Light emerges at all wavelengths, from radio to gamma rays, reflecting activity on many scales and nonthermal radiative processes involving relativistic particles, magnetic fields and bulk kinetic energy of matter. There should be a comparable luminosity in cosmic rays, a small portion of which is channelled into high energy neutrinos.

The accretion rate of matter and hence the luminosity is approximately regulated by feedback on the gas accretion. For both AGN sources and massive stars the characteristic luminosity is the Eddington limit, $L_E = 3GMm_pc/2r_e^2 = 1.25 \times 10^{38}(M/M_\odot)$ erg/sec (where the classical electron radius $r_e = e^2/m_ec^2$), above which radiation pressure outwards on ionized gas exceeds gravitational attraction; brighter sources tend to disassemble themselves. An Eddington-limited source lasts a Salpeter time $Mc^2\epsilon/L_E = 4 \times 10^8\epsilon$ y, which is independent of mass but does depend on the overall efficiency ϵ of extracting rest mass Mc^2, which may be as large as tens of percent for material near a black hole. Significant variability occurs on all timescales down to the Schwarzschild time of the black hole, $R_S/c = 10^{-5}\text{sec}(M/M_\odot) = 100\text{sec}(M/10^7M_\odot)$.

Because mergers of galaxies are common, it is likely that mergers of their central holes are common. If 10^{10} galaxies within the Hubble volume each merge about once per Hubble time, there is about one such event per year in our past light cone, releasing about 10^{62} ergs for a $10^8M_\odot$ hole. Such an event, with a luminosity of $\approx L_{GR}$, far outshines all other sources put together for a Schwarzschild time (on the order of minutes for a $10^8M_\odot$ hole). Even at an average rate of one per year, the gravitational wave luminosity of the universe radiated from these mergers is on average comparable to all the other forms of energy combined, stars and everything. These waves have not yet been detected, but they are in principle easily detectable with space-borne gravitational wave detectors such as LISA[21].

4. Superstars and Supernovae

Smaller but equally violent energy releases occur as byproducts of instabilities in dead or dying stars. When a star exhausts the nuclear fuel in its core, it is no longer stable; the core collapses seeking a new equilibrium, and the release of energy from this collapse blows off the enveloping material. The outcome depends on the mass and composition of the star. A small star like our Sun will blow off about half of its mass, the rest of it left behind in a white dwarf, a glowing ember of still-unburned nuclear fuel (e.g. He, C, N, O, Ne,...), about 10,000 km diameter (about the size of the Earth) and a million times the density of ordinary matter, stabilized by electron degeneracy pressure against gravity. More massive stars create iron cores above the Chandrasekhar limit of $1.4M_\odot$ at which electron degeneracy support fails, and collapse to a neutron star with a diameter of only about 10 km and the same density as an atomic nucleus. Massive cores above a few solar masses cannot be supported by neutron degeneracy or gluon pressure, and collapse all the way to black holes.

Collapse of these remnants releases gravitational binding energy. Smaller and denser objects create more ($\propto 1/r$) and faster ($\propto \rho^{-1/2}$) energy release. White dwarf formation ejects a planetary nebula at high velocity; neutron star formation leads to a Type II supernova explosion.[20] Other spectacular effects occur when remnants live in binary systems and perform a whirling dance with normal stars or with each other. Accretion onto compact remnants from companion stars leads to cataclysmic X-ray sources, and accretion onto a white dwarf can trigger a nearly complete nuclear deflagration and disruption, leading to a Type Ia supernova.[22]

The scale of energy budgets of cataclysms derives from the mass-energy of the stellar remnants, with a basic scale set by the rest mass of the sun $M_\odot c^2 = 1.8 \times 10^{54}$ ergs. The nuclear energy available from a white dwarf is about 10^{51} ergs, most of which is liberated when a Type Ia supernova explodes, mostly as blast energy. The binding energy of a neutron star is about $10^{53.5}$ ergs, almost all of which is radiated as neutrinos during a Type II supernova. (These were directly detected from supernova 1987a; the sum of such events over the Hubble volume leads to a soon-to-be-detectable neutrino background[23, 24].) A small fraction of the neutrinos as well as a bounce shock from the neutron star couple to the enveloping material, dumping heat which ejects it at high velocity. About 10^{51} erg emerges as blast energy, less than ten percent of this as light. Heat, light, heavy elements, magnetic fields, kinetic motion and cosmic rays all carry a substantial amount of energy far away and spread over a volume vastly larger than their source, providing a regulatory cycle and coupling of energy and material flow.

The overall energy budgets of the supernovae are again surprisingly close to that of the stars and AGNs, although most of the SN energy budget is emitted in neutrinos, with a small fraction as blast energy and an even smaller fraction as light. Even though only about a percent of stellar mass participates in supernovae, the ν production efficiency is much higher than nuclear energy (≈ 0.1 as opposed to 0.007). Although the energy sources for the different types of supernovae are completely different (and the Type Ia even leaves no compact remnant), their blast energies are similar. Both eject a substantial mass of chemically enriched and freshly-made radioactive material which powers a glow lasting for months. Also by coincidence, the cosmic rates of the different types of supernovae are comparable in spite of the very different progenitors; SNeIa are just a few times brighter, and a few times rarer, than SNeII.

5. Fireballs and Hypernovae

Long shrouded in mystery, gamma ray bursts now seem to make use of the same compact remnants and combine features of both supernovae and of quasars. Apparently, even these most exotic of sources do not require new cast— only new roles, new settings and combinations for the familiar players, neutron stars and black holes. They are a kind of naked spinning supernova and miniature quasar wrapped into one.

The main event in a gamma ray burst is a "relativistic fireball."[25, 26] A burst of much energy in a small space results in an expanding plasma of photons, electrons and positrons, neutrinos and antineutrinos. There is enough scattering for matter to behave like a fluid, though few enough baryons (less than $10^{-4}M_\odot$) not to inhibit acceleration with particle rest energy, so the relativistic fluid expands very close to the speed of light, with a Lorentz factor $\Gamma \geq 100$. The kinetic energy is dissipated in shocks and radiated as gamma rays at radii of $10^{13}-10^{15}$ cm, the scale of the solar system. Because of Lorentz beaming any observer sees at most a small relativistically-blueshifted patch of the fireball, allowing rapid variability. (The fireball may also itself be beamed, visible only from some directions.) The interaction with the environment as well as the beaming leads to a wide variety of events, with variability down to milliseconds but a duration up to hundreds of seconds, and optical afterglows which remain observably bright for a few days— as large a temporal dynamic range as in a quasar, but scaled small. The energy of the brightest gamma ray bursts is typically estimated in both gamma rays and optical afterglow energy to be $10^{53.5}$ erg[28] (or in extreme cases $10^{54.5}$ erg, assuming isotropic emission; allowing for anisotropic beaming, the total energy budget could well be less than this.)

The fireball, like a supernova, is created by a cataclysmic combination of stellar remnants. A favorite current model invokes a stellar-mass black hole or neutron star surrounded by a torus of neutron-density material— essentially, a donut-shaped neutron star surrounding a more massive black hole.[29, 30] This donut+hole system resembles a very dense, scaled-down version of the quasars, with magnetic fields, now in combination with the highly dissipative neutrinos, extracting energy from the spin of a black hole and/or the orbital energy of the torus. The Schwarzchild time for a $10M_\odot$ hole is 10^{-4} sec so the smallest timescales can be explained; as in quasars the disk and the event last for much longer than this dynamical time.

A spinning black hole can liberate up to $0.29Mc^2$ or 5×10^{54} ergs for a $10M_\odot$ hole; material in a disk can liberate up to $0.42Mc^2$ or 10^{53} ergs for a $1M_\odot$ disk. There is thus ample energy in a stellar-mass "microquasar" to power the high-Γ burst of gamma rays and a lower-Γ optical afterglow.

It seems likely that beaming should often produce the latter without the former— a new population of objects which would lack gamma rays, perhaps appearing like short-lived, very bright SNeII. These may have already been noticed in distant supernova surveys [3, 4] but in any case are not common (at most about 10^3 per day) so they are likely not important in the overall energy budget.

It is not clear exactly how this configuration is produced, but several ideas fit well into stellar evolution scenarios. One model is a "hypernova"[31], which is like a particularly massive Type II supernova core collapse but with a collapse of some of the material inhibited or delayed by rotation. The middle forms a black hole, some of the rest forms the neutron torus. Alternatively, two neutron stars in a close binary (themselves formed from earlier supernova explosions) might eventually coalesce by gravitational radiation of their orbital energy; the mass can exceed the maximum mass of a stable neutron star, leading to a black hole surrounded by a dense neutron torus.[32] Either of these scenarios plausibly leads to the enormous magnetic fields required to form a microquasar. One way to distinguish them observationally is by observing where the bursts occur: the first picture produces a burst after only tens of millions of years, while the second may be after billions of years, and should produce bursts far from star-formation regions.[31]

TABLE 1. Scales of Time, Energy, and Power in the Universe

Energy source	Luminosity L/ object (erg/sec)	Duration D/ flash	Rate $R = NH$	$N_{active} = NHD$	Energy $E = DL$ (erg)	Power $I_{tot} = RE$ (erg/sec)	total N in $V_0/H_0 = R/H$
Big Bang	$\approx 10^{59}$	10^{12}sec	$1/10^{12}$sec	1	10^{71}	10^{59}	1
Λ	10^{59}	10^{10} y	$1/10^{10}$ y	1	10^{76}	10^{59}	1
$Q = 10^{-5}$	10^{46}	10^{10} y	$1/10^{2.5}$y	$10^{7.5}$	10^{65}	10^{54}	$10^{7.5}$
Stars ($1M_\odot$)	$10^{33.5}$	10^{10} y	10^{10}/y	10^{20}	10^{51}	$10^{53.5}$	10^{20}
AGN($10^7 M_\odot$)	10^{45}	10^{15} sec	1/y	$10^{7.5}$	10^{60}	$10^{52.5}$	10^{10}
AGN($10^9 M_\odot$)	10^{47}	10^{15} sec	0.01/y	$10^{5.5}$	10^{62}	$10^{52.5}$	10^{8}
GW($10^8 M_\odot$)	10^{59}	1000 sec	1/y	$10^{-4.5}$	10^{62}	$10^{54.5}$	10^{10}
SNeII(ν)	10^{53}	seconds	1/sec	1	10^{53}	10^{53}	$10^{17.5}$
SNe(O/IR)	10^{43}	$10^{6.5}$ sec	1/sec	$10^{6.5}$	$10^{49.5}$	$10^{49.5}$	$10^{17.5}$
GRB(γ)	10^{53}	seconds	1/day	10^{-5}	10^{53}	10^{48}	$10^{12.5}$
GRB(O/IR)	10^{48}	10^{5}sec	1/day	1	10^{53}	10^{48}	$10^{12.5}$

6. Energy Budgets

Table 1 shows a broad summary of energy flows of various kinds: the cosmic microwave background; the accelerating universe; the free energy injected by cosmological fluctuations and gravitational instability; normal stars in galaxies; small and large active galactic nuclei; gravitational waves from mergers of AGN engines in galaxy mergers; neutrinos from core-collapse supernovae; optical radiation and blast energy from Type I and Type II supernovae; gamma ray and optical emission from gamma ray bursts. Except for the Big Bang (which refers to the radiation-dominated epoch of the universe), entries in the table refer to events at moderate redshift (less than a few) out to distances of the order of the Hubble distance cH_0^{-1} or about 14 billion light-years for a Hubble constant $H_0 = 70\text{km sec}^{-1}\ \text{Mpc}^{-1}$, in a reference "Hubble Volume" $V_0 \equiv 4\pi c^3 H_0^{-3}/3 = 3 \times 10^{11}\text{Mpc}^3$, containing about $10^{9.5}$ giant galaxies (with a luminosity $\approx 2 \times 10^{10} L_\odot$ each), and a spacetime volume V_0/H_0. Within this spacetime volume there have been about N events of each kind, from one Big Bang to 10^{20} stars. The numbers represent order-of-magnitude averages over moderate redshifts $0 \leq z \leq 3$, which includes the bright epochs of star formation and quasars. The entries show the luminosity L of each single event; the typical duration D of each event; the rate R at which new events appear; the number N_{active} of events active at any given time; the energy E released in the designated form by each object; the power I_{tot} produced by each population in the entire Hubble volume; and the total number N in the Hubble spacetime volume. The ubiquitous appearance of numbers like $\approx 10^{20}$ can be traced in all cases to $m_{Planck}/m_{proton} \approx 10^{19}$. Not shown are the timescales of most rapid variation; for each source this has a dynamic range of order $(m_{Planck}/m_{proton})^{1/2}$, extending for quasars down to less than a day and for compact sources down to small fractions of a second. Recall that one day$=10^5$ sec, one month$=10^{6.5}$ sec, one year$=10^{7.5}$ sec. For AGN and GRB, the energy budgets in kinetic energy, Poynting flux, and cosmic rays are comparable to the nonthermal electromagnetic budgets shown; for supernovae, these forms are somewhat less. The largest energy by far is the work being done by the cosmological constant negative pressure in creating new vacuum energy, which replaces the bulk of the entire mass-energy content of the universe in a Hubble time.

7. Cosmic Ecology

A glance at I_{tot} in table 1 shows a remarkable coincidence: in spite of 20 orders of magnitude variation in mass and timescale (and N), the integrated power is comparable for all of these populations if we count the GRB's as a subclass of supernovae. This coincidence can be understood if there are feedback loops controlling the release of energy— a globally regulated choreography coupling the formation rate of stars, the events leading to their death and the transformation and ejection of the elements, the formation of galaxies and quasars and the feeding of their central engines.

This may be just a coincidence, or it may be a hint that stars, galaxies and indeed the universe behave as "whole systems" controlled by nonlocal interactions between interdependent parts spanning a large range of scales. Many mechanisms are available to provide the coupling: radiation, magnetic fields[33], cosmic rays, and fast fluid flows. Although the scales of the individual sources all derive directly from fundamental physics, their frequency in the cosmos depends on this poorly understood "cosmo-ecology" of interacting systems.

Another point is the sheer dynamism of the sky on all timescales. The rate R of many new events include a range, from seconds to years, accessible to direct surveys. Somewhere in the sky a new observable supernova appears every second, with over a million brightly shining at any time; on average a new quasar appears every year, with tens of millions shining at any time. The dynamic range of variability includes a range, from milliseconds (for GRB's) to years (for quasars), accessible to direct monitoring. Astronomical and data exploration techniques have hardly started to sample what is happening.

Have we seen it all, thought of everything that could happen, already explored the entire range of things that could be happening out there? These questions hang in the air whenever new experiments are contemplated, and for some large projects, such as gravitational wave detectors, are major strategic concerns.[1, 2, 21, 34, 35] As the experience with gamma ray bursts shows, for even the most exotic sources the ancient optical band still holds vital information for uncovering the physics of the sources. There are almost certainly new combinations of familiar players (such as flares of stars being eaten by dead quasar black holes) which exist but are not yet found. Microlensing[36] and supernova surveys[3, 4] have shown what CCD arrays and data-mining can do; these technologies promise to expand the scope, depth and precision of the digital exploration of the time domain by orders of magnitude in the next few years and reveal a still richer phenomenology.[37]

Acknowledgements

This work was supported at the University of Washington by NSF and NASA, and at the Max-Planck-Institute für Astrophysik by a Humboldt Research Award. I am grateful for the hospitality of the Isaac Newton Institute for Mathematical Sciences, Cambridge.

References

1. Haehnelt, M. G., Mon. Not. Roy. Astr. Soc. 269, 199-208 (1994)
2. Nakamura, T., Sasaki, M., Tanaka, T.,& Thorne, K. S., Astrophys. J. 487, L139-L142 (1997)
3. Riess, A. G. et al., Astron.J. 116, 1009-1038 (1998)
4. Perlmutter, S., Astrophys. J. 517, 565 (1999)
5. Cen, R. & Ostriker, J. P., Astrophys. J., in press (astro-ph/9806281) (1999)
6. Fukugita, M., Hogan, C. J. & Peebles, P. J. E., Astrophys. J. 503, 518-530 (1998)
7. Bernstein, R., Ph. D. Thesis, Caltech (1998)
8. Schlegel, D. J., Finkbeiner, D. P., & Davis, M., Astrophys. J. 500, 525-553 (1998)(astro-ph/9710327)
9. Hauser, M. G., et al., Astrophys. J., 508, 25-43 (1998) (astro-ph/9806167)
10. Madau, P., in Xth Rencontres de Blois meeting "The Birth of Galaxies", eds. B. Guiderdoni, F. R. Bouchet, Trinh X. Thuan, & Tran Thanh Van (Gif-sur-Yvette: Edition Frontieres, 1999))(astro-ph/9812087)
11. Pettini, M., astro-ph/9902173, in "Chemical Evolution from Zero to High Redshift", ed. J. Walsh and M. Rosa (Berlin: Springer)
12. Renzini, A., astro-ph/9902361, ibid.
13. Pei, Y. C., Fall, M., & Hauser, M. G., Astrophys. J., in press (1999)(astro-ph/9812182)
14. Schmidt, M., Schneider, D. P. & Gunn, J. E., Astron. J. 110, 68-77 (1995)
15. Magorrian, J., et al., Astron. J. 115, 2285-2305 (1998)
16. Haardt, F. & Madau, P., Astrophys. J. 461, 20-37 (1996)
17. Blandford, R. D. & Znajek, R. L., Mon. Not. R. Astron. Soc. 179, 433 (1977)
18. Rees, M. J., Ann. Rev. Astron. Astrophys. 22, 471-506 (1984)
19. Begelman, M. C. & Rees, M. J., *Gravity's Fatal Attraction: Black Holes in the Universe*, W. H. Freeman (1996)
20. Arnett, W. D., *Supernovae and Nucleosynthesis: An Investigation of the History of Matter, from the Big Bang to the Present*, Princeton (1996)
21. See http://lisa.jpl.nasa.gov/, http://ligo.caltech.edu/
22. Nomoto, K., Iwamoto, K. & Kishimoto, N., Science 276, 1378-1382 (1997)
23. Totani, T. & Sato, K., Astropart. Phys. 3, 367-376 (1995)
24. Hartmann, D. H. & Woosley, S. E., Astropart. Phys. 7, 137-146 (1997)
25. Meszaros, P., & Rees, M., Astrophys. J. 405, 278-284 (1993)
26. Meszaros, P., Rees, M. and Wijers, R. A. M. J., New Astronomy, in press (1998)(astro-ph/9808106)
27. Mezger, M. R. et al., Nature 387, 878-880 (1997)
28. Kulkarni, S. R. et al., Nature, 393, 35-39 (1998); see also http://gcn.gsfc.nasa.gov/gcn/gcn3_archive.html
29. Paczynski, B., Acta Astron. 41, 257 (1991)
30. Narayan, R., Paczynski, B.,& Piran, T., Astrophys. J. 395, L83-L86 (1992)
31. Paczynski, B. Astrophys. J. 494, L45-L48 (1998)
32. Ruffert, M. & Janka, H.- Th., Astron. Astrophys. in press (1999)(astro-ph/9809280)
33. Kronberg, P. P., Rep. Prog. Phys. 57, 325-382 (1994)
34. Abramovici, A. et al., Science 256, 325 (1989)
35. Thorne, K., in *Three Hundred Years of Gravitation* ed. S. W. Hawking & W. Israel, 330-446 (Cambridge, 1987)
36. Alcock, C. et al., Astrophys. J. 479, 119-146 (1997)
37. Stubbs, C. W., preprint (1998)(astro-ph/9810488)

FEEDBACK IN GALAXY FORMATION

JOSEPH SILK
Astrophysics, University of Oxford
NAPL, Keble Road, Oxford OX1 3RH, UK

Abstract. The role of feedback in galaxy formation is reviewed. Feedback occurs via outflows and photoionization produced by massive stars, supernova explosions and tidal interactions between protodisks and dark matter, and the interaction of supermassive black holes with their environment in galactic nuclei. Feedback may account for the star formation rate in, and sizes of, galactic disks, and the conspiracy between the dark and baryonic matter in galaxies.

1. Introduction

Large-scale structure formation is rapidly becoming a mature subject. Successes including the modelling of the clustering of galaxies and the properties of galaxy clusters, such as their frequency, mass function, structure and gas content. The abundance of massive galaxies and their rotation curves are understood, as is the distribution of both clusters and galaxies with redshift. The mass fraction and density distribution of intergalactic gas can be satisfactorily explained.

However theory has had less success in accounting for the properties of the luminous, baryon-rich regions of galaxies. Inclusion of baryons into large-scale simulations, along with star formation, generates galaxies that superficially resemble the observed systems. Cold dark matter simulations generically predict a galaxy mass function with a low mass tail $dN/dM \propto M^{-2}$, whereas the observed luminosity function of galaxies, $dN/dL \propto L^{-\alpha}$ at low L, has $\alpha \approx 1.3$. This difference is generally cured by introducing feedback from star formation. However once feedback is tuned to account for the galaxy luminosity function, one no longer has much freedom to adjust feedback physics, as most simply formulated, to address other problems, such as the normalisation of the Tully-Fisher relation and the issue of disk

R.G. Crittenden and N.G. Turok (eds.), Structure Formation in the Universe, 295–302.

sizes. High resolution simulations find that disk sizes are a factor ~ 10 too small as compared to observed L_* galaxies.

Feedback is currently at work in accounting for the low star formation efficiency in galactic disks. Feedback may have been responsible in galaxy spheroids for the correlation between spheroid luminosity and central black hole mass. Understanding ongoing feedback may provide insights into the role of feedback in galaxy formation. In this assessment of feedback in galaxy formation, I will describe how realistic models may be developed that could account for some of the presently unresolved aspects of the origin of such key galaxy structural features as disk sizes and the conspiracy between dark and baryonic matter.

2. Feedback and the Efficiency of Star Formation

The star formation efficiency in galactic disks, defined to be the fraction of gas converted into stars per dynamical time-scale, is a few percent in such systems as the Milky Way. Detailed simulations, hitherto only performed in 2-D, follow a two-phase interstellar medium, in which supernova explosions drive bubbles of hot, $\sim 10^6$ K gas that sweep up shells of cold gas. The shells eventually interact and fragment, generating random motions of cold gas clouds and forming stars. Most of the ejected energy is radiated away. The resulting inefficiency of star formation may be simply interpreted in terms of momentum input into the interstellar medium via OB stellar winds and especially supernova remnants, with the efficiency for star formation being of order σ_{gas}/v_{SN}. Here σ_{gas} is the rms interstellar cloud velocity dispersion and v_{SN} is the specific momentum from supernovae. The simulations demonstrate that one can account for the velocity and mass distributions of the gas clumps and for the power-law spatial correlation function of the T-Tauri stars in star-forming clouds (Scalo and Chappell 1999).

The low efficiency suggests that self-regulation of star formation will allow star formation to continue until a substantial fraction of the gas reservoir is expended. Gas sinks may however consist not just of low mass stars and remnants of massive stars, but may also arise due to gas driven out as a consequence of the porosity of the cold gas. This is likely to be especially important in a direction perpendicular to the galactic plane, where the gas-scale height is only ~ 100 pc.

A simple model of a multiphase interstellar medium appeals to an explicit formulation of the porosity parameter:

$$P = -\ln(1 - f_{hot}) = G^{-\frac{1}{2}} \dot{\rho}_* \rho_{gas}^{-\frac{3}{2}} (\sigma_f / \sigma_{gas})^{2.7}.$$

Here f_{hot} is the hot gas volume fraction, $\dot{\rho}_*$ is the star formation rate per unit volume, ρ_{gas} is the cold gas density, σ_{gas} is the gas velocity dispersion, predominantly due to the interstellar cloud motions, and σ_f is a fiducial velocity dispersion that depends on the specific momentum input from supernovae. One can imagine two limiting cases: either $f_{hot} \ll 1$ or $1 - f_{hot} \ll 1$, the disk volume being dominated either by cold or by hot gas, respectively. Only in the latter case is supernova feedback important. One finds for $f_{hot} \sim 1$, that $P \sim 1$ and the star formation rate is

$$\dot{\rho}_* \approx \epsilon \rho_{gas}^{\frac{3}{2}}.$$

Here

$$\epsilon \equiv G^{\frac{1}{2}} (\sigma_{gas}/\sigma_f)^{2.7}$$

may be defined to be the efficiency of star formation, namely the mass fraction of gas converted into stars per disk dynamical time.

This situation is stable if the feedback is negative: that is if the hot phase gas suppresses rather than simulates star formation. One might imagine this to be the case if the hot gas, as it is over-pressurized, shreds cold cloud complexes by Rayleigh-Taylor or Kelvin-Helmholtz instabilities, and breaks out of the disk via chimneys and eruptions that entrain cold gas within the outflows. Only in the most symmetric of situations would the hot phase envelop an entire cold cloud, whose dimension is typically of order the disk scale height, and induce its collapse. The star formation efficiency is predicted to be low, but in the vicinity of a bar or in a merger, where there will be tidally enhanced gas-streaming motions, the efficiency can be greatly increased. The star formation rate will be self-regulated and steady.

One manifestation of high porosity will be gas outflows from disks. Hot gas is ejected via superbubbles and chimneys at a rate of order *(bubble volume) (hot gas density) (flow time)*$^{-1}$. The ratio of the hot gas ejection rate relative to the star formation rate can be expressed as

$$f_{hot} p_{gas} (\epsilon \mu_{gas} \Omega v_{flow})^{-1} \sim \epsilon^{-1} f_{hot}.$$

Here μ_{gas} is the disk gas surface density, p_{gas} is the interstellar gas pressure, ϵ is the star formation efficiency, and Ω is the disk rotation rate. This ratio will be of order 5-10 for galaxies with $f_{hot} \sim 0.5$: these are typical, star-forming disks according to the present model. Similar outflow rates are actually observed (Martin 1999).

In the opposite case, when the disk volume is dominated by cold gas, feedback from star formation is unimportant. In this case, runaway star formation occurs. A situation where cold gas dominates would occur during galaxy formation, especially in the dense core where gas cooling predominates. This is reminiscent of a star-burst at the site of protospheroid formation.

3. Feedback and the Sizes of Galactic Disks

Angular momentum generated by tidal torques in the early universe provides a natural explanation of disk sizes, provided that the contracting and cooling gas approximately conserves angular momentum. This appeared to be a good approximation until the advent of high resolution numerical simulations of galaxy formation. It has now been confirmed by several groups that both the cold dark matter and the baryons are extremely clumpy. One consequence is that most of the angular momentum of the baryons is lost during the collapse. The resulting disks are a factor of at least 5 or 10 too small.

Suggested solutions generally incorporate feedback to retard the gas contraction and angular momentum loss. Delay of gas cooling (and associated star formation) to, say, $z = 1$, can solve the angular momentum problem, but at the probably unacceptable price of excessive late disk evolution. A more sophisticated approach is needed.

One example of a possible resolution is to develop a viscous disk model that retards gas contraction, allowing stars to form before excessive angular momentum loss occurs. Suppose the galactic disk consists of a multiphase medium, with clouds that form by gravitational instability and are accelerated both tidally and by supernova momentum input, and a hot intercloud medium that is heated by the supernova explosions. The viscosity is due to cloud-cloud collisions. In such a self-gravitating disk, an exponential surface density profile is generated for the stars if the viscous drag and star formation timescales are assumed to be similar (Lin and Pringle 1987). If star deaths provide an important source of cloud acceleration, such a coincidence would arise naturally.

The characteristic star formation time t_{sf} in the Milky Way disk is presently about 10^9 years. This is of order the viscous time-scale t_ν, by assumption, and yields the duration of the bulk of disk star formation. The disk scale is expected to be of order (μ is the disk surface density and v_{rot} is the maximum rotation velocity)

$$r_d \sim \frac{\sigma}{\Omega}\left(\frac{v_{SN}}{v_{rot}}\right)^{\frac{1}{3}}\left(\frac{\Omega v_{rot}}{G\mu}\frac{t_\nu}{t_{sf}}\right)^{\frac{1}{3}}\left(\frac{\mu}{\mu_{gas}}\right)^{\frac{1}{3}}.$$

For typical values, $r_d \sim 3$ kpc, which is comparable to the observed value for Milky Way-type disks.

4. Feedback, Quasars and Spheroids

There is no accepted theory for spheroid formation. The stars may form in small units and assemble via mergers of substructures, or else spheroids may form more monolithically. Nearby examples of ultraluminous infrared galaxies ($L \gtrsim 10^{12} L_\odot$) almost invariably are merger-triggered starbursts that seem to rapidly develop, as imaged in nearby examples, a de Vaucouleurs light profile in the near infrared.

Modelling of 175μm ISO counts and of the far infrared background, as well as SCUBA identifications of submillimeter sources, suggest that there is a substantial high redshift ($z > 3$) ULIG population, which accounts for at least half of the integrated diffuse background light density. Whether this translates into star formation, and whether the starburst interpretation of ULIGS holds at high z, is unclear.

Quasar studies suggest that active galactic nuclei may be intimately related to spheroid formation for at least two reasons. A high inferred local star formation rate is required in order to account for the very high metallicities observed in the circumquasar environment. The quasar luminosity density tracks that of the star formation rate history to redshifts up to 5. The bulk of high redshift star formation (beyond $z \sim 3$) is presumably associated with spheroid formation. The morphologies, star formation rates and luminosity functions of Lyman-break galaxies support such an interpretation.

A third, indirect but even more compelling, connection between quasars and spheroids has come from the empirical correlation between central blackhole mass and spheroid luminosity or stellar mass: $M_b \approx 0.006 M_{sph}$. The inferred mass density in black holes is

$$\rho_{bh} = 3.3 \times 10^6 h \left(\frac{M_{bh}/M_{sph}}{0.006}\right) \left(\frac{\Omega_{sph}}{0.002h^{-1}}\right) M_\odot \,\mathrm{Mpc}^{-3}.$$

The diffuse background light generated via the process of quasar formation can be estimated, if we adopt an accretion efficiency f, generally considered to be around ten percent, as

$$\varepsilon = f \rho_{bh} c^2 (1+z_q)^{-1} = 2.6 \times 10^{-3} \left(\frac{f}{0.1}\right) \left(\frac{5}{1+z_q}\right) \mathrm{eV\ cm}^{-3}.$$

Here the mean redshift of quasar formation is denoted by z_q. For comparison, the diffuse background light intensity, predominantly at far infrared wavelengths, amounts to $10 - 20$ nw/m^2 or $(2.6 - 5.2) \times 10^{-3}$ eV cm^{-3}.

Hence quasar luminosity can plausibly be a dominant contribution to the diffuse background light. This suggests that many high redshift SCUBA sources may be quasar-dominated, and that the processes of quasar and spheroid formation may be intimately connected. Feedback may also help account for the relation between black hole mass and spheroid mass.

Quasars plausibly form as an early phase in protogalaxies. Quasar-driven outflows may drive galactic winds and suppress protogalactic star formation, thereby producing almost coeval spheroid formation (Granato et al 1999) because of the relatively short duration of the luminous quasar phase. The feedback may operate as follows (Silk and Rees 1998). Suppose a significant fraction of the luminosity of a quasar drives an outflow. This occurs during the major merger that forms the galaxy. Gas is being fed during the merger into the inner galaxy to feed the central engine at a rate that cannot exceed $\sim \sigma^3/G$, where σ is the halo circular velocity. If the quasar luminosity is a fraction f_E of the Eddington luminosity $L_E = 4\pi c G m_p M_{bh} \sigma_T^{-1}$, one might expect the black hole to grow until it disrupts the feeding process, thereby limiting the protogalactic gas reservoir and ultimately the stellar mass. This leads to the expectation that

$$M_{bh} = \frac{\sigma_T}{4\pi c G f_E G^2 m_p} \sigma^5 = 1.2 \times 10^8 f_E^{-1} (\sigma/250 \mathrm{km\ s}^{-1})^5 M_\odot .$$

This is close to the observed correlation, both in dependence on spheroid luminosity (which scales approximately as σ^4) and in normalization.

A more precise comparison between black hole and spheroid luminosity is obtained by using the fundamental plane correlation, $L \propto \sigma^{2.54} I_e^{-0.62}$ (Jorgensen et al. 1996) to yield $M_{bh} \propto L_{sph}^2 I_e^{-5/4}$. It would be of interest to see whether the scatter in the (M_{bh}, M_{sph}) diagram is reduced by including a dependence on surface brightness, as is suggested by the feedback model.

5. Feedback and Conspiracy

The dark matter and baryonic components appear to have some underlying interconnections. Oone can always argue that the global near-coincidence between $\Omega_m \approx 0.3$ and $\Omega_b \approx 0.03$ is an accident, if not due to some early universe fine-tuning effect. It is not impossible that one could increase Ω_b and reduce Ω_m to allow a universe containing predominantly baryons.

However, on large scales, the preferred global ratio $\Omega_m/\Omega_b \sim 5 - 10$ means that one can treat the two components independently in the linear regime of fluctuation growth. A baryonic imprint on $\delta\rho/\rho$, via so-called Sakharov oscillations, only arises if $\Omega_b/\Omega_m \gtrsim 0.3$. On galactic scales, the situation is reversed. The baryonic component cools and contracts relative to the dark matter, and the inner protogalaxy is baryon-dominated. The dark matter acquires a universal density profile, as demonstrated by N-body simulations, with a central density cusp approximately proportional to r^{-1}. The mean density is a factor $\sim$ 200 larger than the cosmological density at turn-around in a simple top-hat model of a spherically symmetric overdensity. The shape of the predicted rotation curve is in good agreement with observations for normal disk galaxies.

The normalisation cannot however be matched to the observed rotation velocities. Excessive dark matter is inferred within the optical radius amounting to a factor of about 3 (Navarro and Steinmetz 1999). This is inferred from high resolution simulations of the Milky Way galaxy, where the baryon mass and dark mass are accurately measured within the solar circle. Independent, and more general, confirmation comes from studies of the normalisation of the Tully-Fisher correlation, which shows little scatter in the relation between luminosity and maximum rotational velocity, and hence dynamical mass within the half-light radius. The baryonic mass is actually observed to be comparable to the halo dark matter mass within the luminous regions of disk galaxies. This has been dubbed the disk-halo conspiracy.

There are other, possibly related, problems that stem from high resolution CDM modelling of dark matter halos. The number of satellites predicted is some two orders of magnitude larger than is observed (Moore et al. 1999). Another difficulty arises in low surface brightness dwarf spirals, which are everywhere dark matter-dominated. One infers a soft core, and not the predicted cuspy core, from rotation curve observations. As much as 10 percent of the dark halos are in clumps. The dark halo substructure contains the remnants of the earlier stages of the hierarchy of merging substructures.

Can stellar feedback solve these problems? Indeed, feedback is invoked to suppress star formation in dwarf galaxies, via gas ejection in early winds. However while this can reconcile the number of observed satellites, the number of dark satellites is not diminished. There may be serious difficulties in overheating the disks of spirals via tidal interactions of halo clumps.

Dynamical feedback is more promising. This might operate as follows. The protodisk is generically asymmetric, and simulations indeed suggest a transient phase of a massive gas-rich bar following the last gas-rich massive merger. While this is normally thought to be the precursor phase of an elliptical, it is by no means excluded that spirals also form this way. The tidal interactions between such a bar and the dark halo have not hitherto been studied. It is tempting, however, to make the following speculations. The tidal torques exerted by the bar will help homogenise the inner dark halo. Gas streaming along the bar will concentrate the gas and provoke bulge formation, the disk forming later as the residual high angular momentum gas settles into the disk. The angular momentum transferred by the bar must not of course be so great as to aggravate the disk angular momentum problem. Provided, say, two-thirds of the baryonic angular momentum is transferred to the dark halo, the dark matter to baryon ratio in the bar-dominated region would be reduced by a third. The reduction in baryon angular momentum by only a factor of 3 is far less than the reduction factor found in the high resolution N-body simulations where such bar physics is normally not incorporated. Some of the high angular momentum gas ejected from the bar will later cool and settle into the disk, even losing some angular momentum via viscous cloud-cloud interactions, while its dark matter counterpart will remain in the outer halo. This scenario associates early disk formation with a massive merger, required in order to set the initial conditions for formation of a massive gaseous bar which later forms a bulge. It may be that the more frequent minor mergers are responsible for lesser bulge formation.

6. Conclusions

Galaxy formation is clearly an unsolved problem, and it is likely to remain so for the foreseeable future. Nevertheless we can hope that by suitably parametrising the star formation processes we observe at low redshift, we may be able to predict some properties of the high redshift universe. The danger is that under extreme conditions achievable during the violent processes of galaxy origin, our local prototypes may be inadequate. The relevant physics or astrophysics may be lacking. One such possibility is the possible role of quasars in inducing galaxy formation. As observations of the distant universe probe ever greater depths, we can hope that phenomenological modelling of star formation in remote galaxies will eventually supplement our local templates. The theory of galaxy formation is likely to advance phenomenologically, and this may ultimately guide us towards an improved understanding of the fundamental processes whereby baryons are converted into stars.

References

1. Granato, G. et al. (1999), *MNRAS*, in press
2. Jorgensen, I., Franx, M. and Kjaergaard, P. (1996), *MNRAS*, **280**, 167.
3. Lin, D. and Pringle, J. (1987), *Astrophysical Journal*, **320**, 87.
4. Martin, C. (1999), *Astrophysical Journal*, **513**, 156.
5. Moore, B. et al. (1999), *Astrophysical Journal*, **524**, 19.
6. Navarro, J. and Steinmetz, M. (1999), *Astrophysical Journal*, in press.
7. Silk, J. and Rees, M.J. (1998), *Astronomy and Astrophysics*, **331**, L1.

Part V

THE LARGE SCALE DISTRIBUTION OF MATTER AND GALAXIES

CLUSTERING OF MASS AND GALAXIES

J.A. PEACOCK
Institute for Astronomy, University of Edinburgh
Royal Observatory, Edinburgh EH9 3HJ, UK

Abstract. These lectures cover various aspects of the statistical description of cosmological density fields. Observationally, this consists of the point process defined by galaxies, and the challenge is to relate this to the continuous density field generated by gravitational instability in dark matter. The main topics discussed are (1) nonlinear structure in CDM models; (2) statistical measures of clustering; (3) redshift-space distortions; (4) small-scale clustering and bias. The overall message is optimistic, in that simple assumptions for where galaxies should form in the mass density field allow one to understand the systematic differences between galaxy data and the predictions of CDM models.

1. Preamble

The subject of large-scale structure is in a period of very rapid development. For many years, this term would have meant only one thing: the distribution of galaxies. However, we are increasingly able to probe the primordial fluctuations through the CMB, so that the problem of galaxy formation and clustering is now only one aspect of the general picture of structure formation. The rationale for studying the large-scale distribution of galaxies is therefore altering. Ten years ago, we were happy to produce samples based on a rather sparse random sampling of the galaxy distribution, with the main aim of tying down statistics such as the large-scale power spectrum of number-density fluctuations. A major goal of the subject remains the measurement of the fluctuation spectrum for wavelengths $\gtrsim 100$ Mpc, and the demonstration that this agrees in shape with what can be inferred from the CMB. Nevertheless, we are now increasingly interested in studying the pattern of galaxies with the highest possible fidelity – demanding deep, fully-sampled surveys of the local universe. Such studies will tell us

R.G. Crittenden and N.G. Turok (eds.), Structure Formation in the Universe, 305–340.

much about the processes by which galaxies formed and evolved within the distribution of dark matter. The aim of these lectures is therefore to look both backwards and forwards: reviewing the foundations of the subject and looking forward to the future issues.

2. The CDM family album

2.1. THE LINEAR SPECTRUM

The basic picture of inflationary models (but also of cosmology before inflation) is of a primordial power-law spectrum, written dimensionlessly as the logarithmic contribution to the fractional density variance, σ^2:

$$\Delta^2(k) = \frac{d\sigma^2}{d\ln k} \propto k^{3+n}, \tag{1}$$

where n stands for n_S hereafter. This undergoes linear growth

$$\delta_k(a) = \delta_k(a_0) \left[\frac{D(a)}{D(a_0)}\right] T_k, \tag{2}$$

where the linear growth law is

$$D(a) = a\, g[\Omega(a)] \tag{3}$$

in the matter era, and the growth suppression for low Ω is

$$\begin{aligned} g(\Omega) &\simeq \Omega^{0.65} \text{ (open)} \\ &\simeq \Omega^{0.23} \text{ (flat)} \end{aligned} \tag{4}$$

The transfer function T_k depends on the dark-matter content as shown in figure 1.

Note the baryonic oscillations in figure 1; these can be significant even in CDM-dominated models when working with high-precision data. Eisenstein & Hu (1998) are to be congratulated for their impressive persistence in finding an accurate fitting formula that describes these wiggles. This is invaluable for carrying out a search of a large parameter space.

The state of the linear-theory spectrum after these modifications is illustrated in figure 2. The primordial power-law spectrum is reduced at large k, by an amount that depends on both the quantity of dark matter and its nature. Generally the bend in the spectrum occurs near $1/k$ of order the horizon size at matter-radiation equality, $\propto (\Omega h^2)^{-1}$. For a pure CDM universe, with scale-invariant initial fluctuations ($n = 1$), the observed spectrum depends only on two parameters. One is the shape $\Gamma = \Omega h$, and the other is a normalization. On the shape front, a government health warning

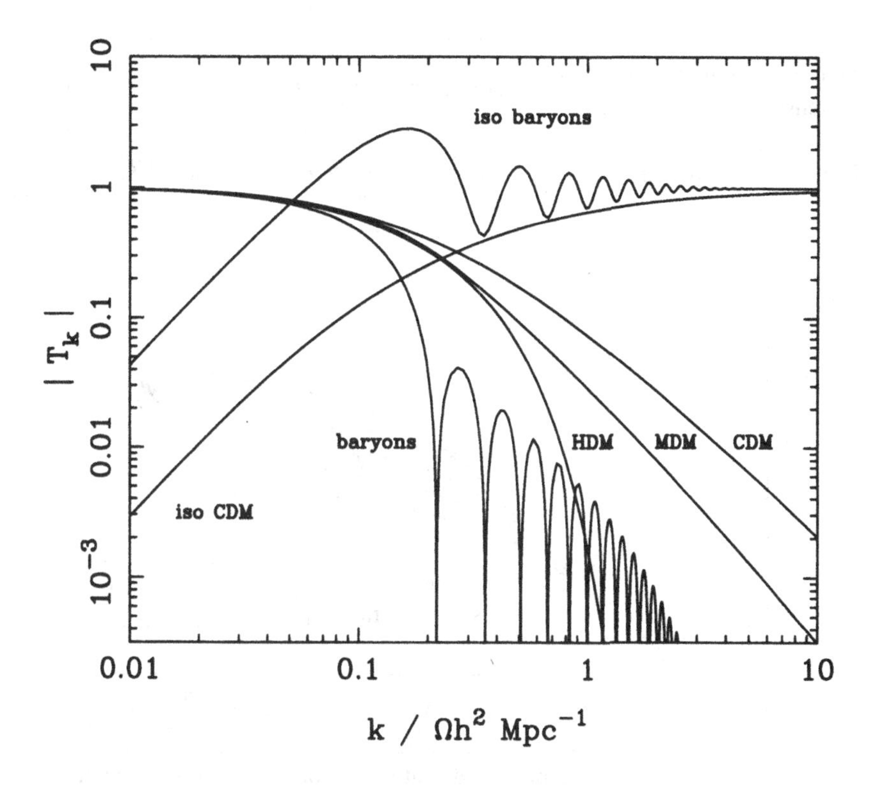

Figure 1. Transfer functions for various dark-matter models. The scaling with Ωh^2 is exact only for the zero-baryon models; the baryon results are scaled from the particular case $\Omega_B = 1$, $h = 1/2$.

is needed, as follows. It has been quite common to take Γ-based fits to observations as indicating a *measurement* of Ωh, but there are three reasons why this may give incorrect answers:

(1) The dark matter may not be CDM. An admixture of HDM will damp the spectrum more, mimicking a lower CDM density.

(2) Even in a CDM-dominated universe, baryons can have a significant effect, making Γ lower than Ωh. An approximate formula for this is given in figure 2 (Peacock & Dodds 1994; Sugiyama 1995).

(3) The strongest (and most-ignored) effect is tilt: if $n \neq 1$, then even in a pure CDM universe a Γ-model fit to the spectrum will give a badly incorrect estimate of the density (the change in Ωh is roughly $0.3(n-1)$; Peacock & Dodds 1994).

2.2. NORMALIZATION

The other parameter is the normalization. This can be set at a number of points. The COBE normalization comes from large angle CMB anisotropies,

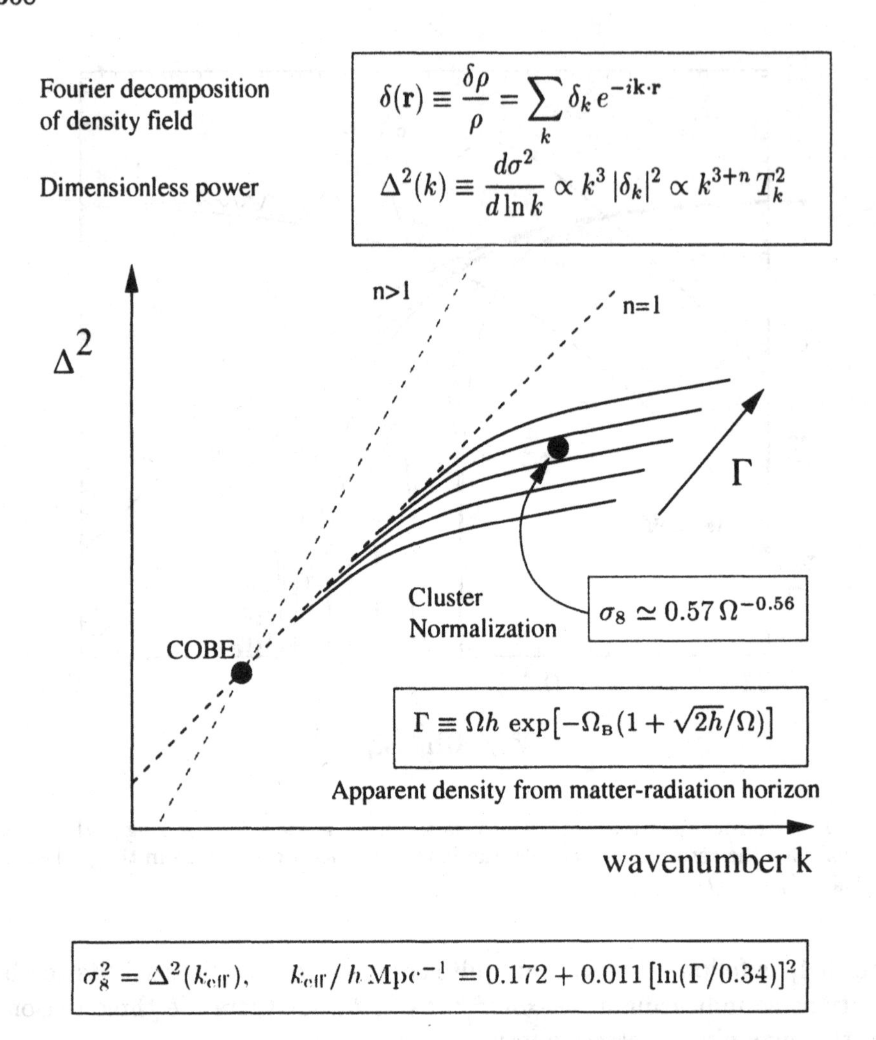

Figure 2. This figure illustrates how the primordial power spectrum is modified as a function of density in a CDM model. For a given tilt, it is always possible to choose a density that satisfies both the COBE and cluster normalizations.

and is sensitive to the power spectrum at $k \simeq 10^{-3}\,h\,\mathrm{Mpc}^{-1}$. The alternative is to set the normalization near the quasi-linear scale, using the abundance of rich clusters. Many authors have tried this calculation, and there is good agreement on the answer:

$$\sigma_8 \simeq (0.5 - 0.6)\,\Omega_m^{-0.6}. \tag{5}$$

(White, Efstathiou & Frenk 1993; Eke et al. 1996; Viana & Liddle 1996). In many ways, this is the most sensible normalization to use for LSS studies, since it does not rely on an extrapolation from larger scales.

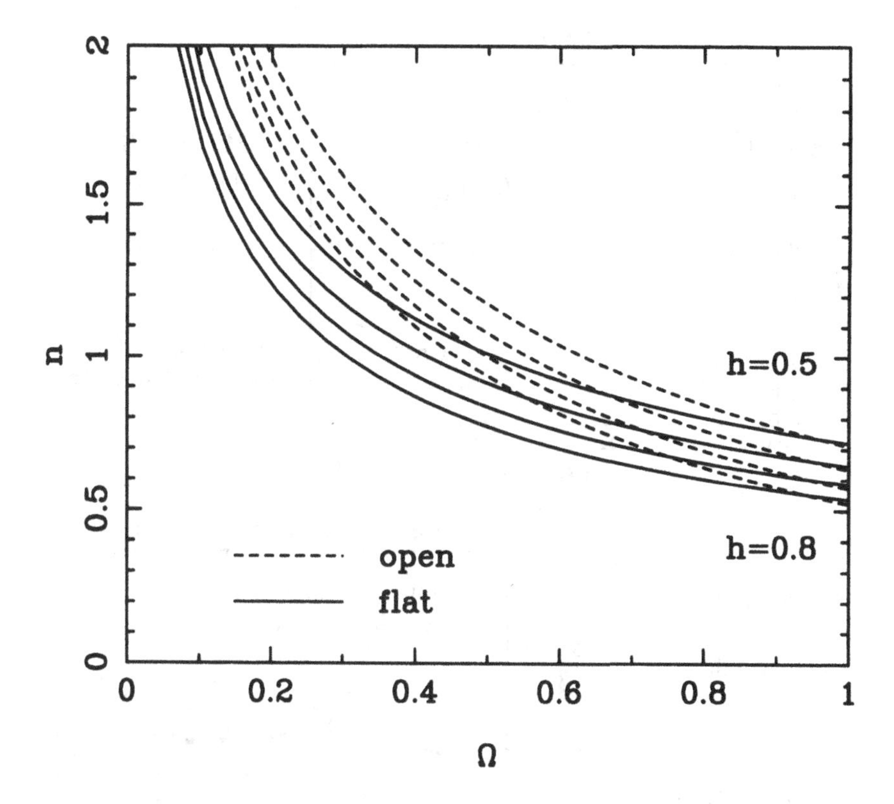

Figure 3. For 10% baryons, the value of n needed to reconcile COBE and the cluster normalization in CDM models.

Within the CDM model, it is always possible to satisfy both these normalization constraints, by appropriate choice of Γ and n. This is illustrated in figure 3. Note that vacuum energy affects the answer; for reasonable values of h and reasonable baryon content, flat models require $\Omega_m \simeq 0.3$, whereas open models require $\Omega_m \simeq 0.5$.

2.3. THE NONLINEAR SPECTRUM

On smaller scales ($k \gtrsim 0.1$), nonlinear effects become important. These are relatively well understood so far as they affect the power spectrum of the mass (e.g. Hamilton et al. 1991; Jain, Mo & White 1995; Peacock & Dodds 1996). Based on a fitting formula for the similarity solution governing the evolution of scale-free initial conditions, it is possible to predict the evolved spectrum in CDM universes to a few per cent precision (e.g. Jenkins et al. 1998).

These methods can cope with most smoothly-varying power spectra, but they break down for models with a large baryon content. Figure 1 shows

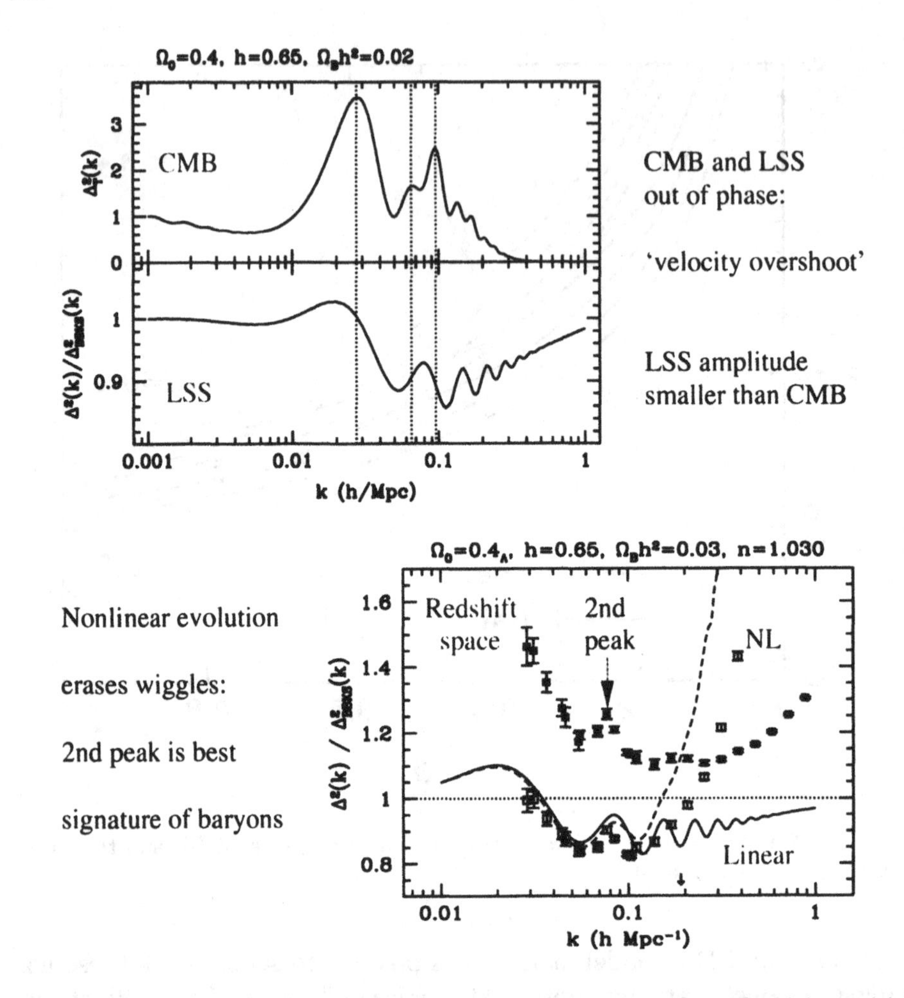

Figure 4. Baryonic fluctuations in the spectrum can become significant for high-precision measurements. Although such features are much less important in the density spectrum than in the CMB (first panel), the order 10% modulation of the power is potentially detectable. However, nonlinear evolution has the effect of damping all beyond the second peak. This second feature is relatively narrow, and can serve as a clear proof of the past existence of oscillations in the baryon-photon fluid (Meiksin, White & Peacock 1999).

that rather large oscillatory features would be expected if the universe was baryon dominated. The lack of observational evidence for such features is one reason for believing that the universe might be dominated by collisionless non-baryonic matter (consistent with primordial nucleosynthesis if $\Omega_m \gtrsim 0.1$).

Nevertheless, baryonic fluctuations in the spectrum can become significant for high-precision measurements. Figure 4 shows that order 10% modulation of the power may be expected in realistic baryonic models (Eisenstein & Hu 1998; Goldberg & Strauss 1998). Most of these features are however removed by nonlinear evolution. The highest-k feature to survive is usually the second peak, which almost always lies near $k = 0.05\,\mathrm{Mpc}^{-1}$ (no h, for a change). This feature is relatively narrow, and can serve as a clear proof of the past existence of baryonic oscillations in forming the mass distribution (Meiksin, White & Peacock 1999). However, figure 4 emphasizes that the easiest way of detecting the presence of baryons is likely to be through the CMB spectrum. The oscillations have a much larger 'visibility' there, because the small-scale CMB anisotropies come directly from the coupled radiation-baryon fluid, rather than the small-scale dark matter perturbations.

3. Statistics

Statistical measures of the cosmological density field relate to properties of the dimensionless density perturbation field

$$\delta(\mathbf{x}) \equiv \frac{\rho(\mathbf{x}) - \langle\rho\rangle}{\langle\rho\rangle}, \qquad (6)$$

although δ need not be assumed to be small.

3.1. CORRELATION FUNCTIONS

The simplest measure is the autocorrelation function of the density perturbation

$$\xi_{\mathrm{A}}(\mathbf{r}) \equiv \langle\delta(\mathbf{x})\delta(\mathbf{x}+\mathbf{r})\rangle\,, \qquad (7)$$

This is a straightforward statistical measure that can also be computed for the dark-matter distribution in N-body simulations. Formally, the averaging operator here is an ensemble average, but one generally appeals to the ergodic nature of the density field to replace this with a volume average.

However, galaxies are a point process, so what astronomers can measure in practice is the two-point correlation function, which gives the excess probability for finding a neighbour a distance r from a given galaxy. By regarding this as the probability of finding a pair with one object in each of the volume elements dV_1 and dV_2,

$$dP = \rho_0^2\,[1 + \xi_2(r)]\,dV_1\,dV_2. \qquad (8)$$

Is it true that $\xi_{\mathrm{A}}(r) = \xi_2(r)$? Life would certainly be simple if so, and much work on large-scale structure has implicitly assumed the Poisson clustering hypothesis, in which galaxies are assumed to be sampled at random from some continuous underlying density field. Many of the puzzles in the field can however be traced to the fact that this hypothesis is probably false, as discussed below.

A related quantity is the cross-correlation function. Here, one considers two different classes of object (a and b, say), and the cross-correlation function ξ_{ab} is defined as the (symmetric) probability of finding a pair in which dV_1 is occupied by an object from the first catalogue and dV_2 by one from the second. Both cross- and auto-correlation functions are readily extended to higher orders and considerations of n-tuples of points in a given geometry.

3.2. FOURIER SPACE

For the Fourier counterpart of this analysis, we assume that the field is periodic within some box of side L, and expand as a Fourier series:

$$\delta(\mathbf{x}) = \sum \delta_{\mathbf{k}} e^{-i\mathbf{k}\cdot\mathbf{x}}. \tag{9}$$

For a real field, $\delta_{\mathbf{k}}(-\mathbf{k}) = \delta^*_{\mathbf{k}}(\mathbf{k})$. Using this definition in the correlation function, most cross terms integrate to zero through the periodic boundary conditions, giving

$$\xi(\mathbf{r}) = \frac{V}{(2\pi)^3} \int |\delta_{\mathbf{k}}|^2 e^{-i\mathbf{k}\cdot\mathbf{r}} d^3k. \tag{10}$$

In short, the correlation function is the Fourier transform of the power spectrum.

We shall usually express the power spectrum in dimensionless form, as the variance per $\ln k$ ($\Delta^2(k) = d\langle\delta^2\rangle/d\ln k \propto k^3 P[k]$):

$$\Delta^2(k) \equiv \frac{V}{(2\pi)^3} 4\pi k^3 P(k) = \frac{2}{\pi} k^3 \int_0^\infty \xi(r) \frac{\sin kr}{kr} r^2 \, dr. \tag{11}$$

This gives a more easily visualizable meaning to the power spectrum than does the quantity $VP(k)$, which has dimensions of volume: $\Delta^2(k) = 1$ means that there are order-unity density fluctuations from modes in the logarithmic bin around wavenumber k. $\Delta^2(k)$ is therefore the natural choice for a Fourier-space counterpart to the dimensionless quantity $\xi(r)$.

In the days before inflation, the primordial power spectrum was chosen by hand, and the minimal assumption was a featureless power law:

$$\left\langle |\delta_k|^2 \right\rangle \equiv P(k) \propto k^n \tag{12}$$

The index n governs the balance between large- and small-scale power. Similarly, a power-law spectrum implies a power-law correlation function.

If $\xi(r) = (r/r_0)^{-\gamma}$, with $\gamma = n + 3$, the corresponding 3D power spectrum is

$$\Delta^2(k) = \frac{2}{\pi}\,(kr_0)^\gamma\,\Gamma(2-\gamma)\,\sin\frac{(2-\gamma)\pi}{2} \equiv \beta(kr_0)^\gamma \tag{13}$$

($= 0.903(kr_0)^{1.8}$ if $\gamma = 1.8$). This expression is only valid for $n < 0$ ($\gamma < 3$); for larger values of n, ξ must become negative at large r (because $P(0)$ must vanish, implying $\int_0^\infty \xi(r)\, r^2\, dr = 0$). A cutoff in the spectrum at large k is needed to obtain physically sensible results.

The most interesting value of n is the scale-invariant spectrum, $n = 1$, *i.e.* $\Delta^2 \propto k^4$. To see how the name arises, consider a perturbation $\delta\Phi$ in the gravitational potential:

$$\nabla^2\delta\Phi = 4\pi G\rho_0\delta \quad \Rightarrow \quad \delta\Phi_k = -4\pi G\rho_0\delta_k/k^2. \tag{14}$$

The two powers of k pulled down by ∇^2 mean that, if $\Delta^2 \propto k^4$ for the power spectrum of density fluctuations, then Δ^2_Φ is a constant. Since potential perturbations govern the flatness of spacetime, this says that the scale-invariant spectrum corresponds to a metric that is a fractal: spacetime has the same degree of 'wrinkliness' on each resolution scale. The total curvature fluctuations diverge, but only logarithmically at either extreme of wavelength.

3.3. ERROR ESTIMATES

A key question for these statistical measures is how accurate they are – i.e. how much does the result for a given finite sample depart from the ideal statistic averaged over an infinite universe? Terminology here can be confusing, in that a distinction is sometimes made between sampling variance and cosmic variance. The former is to be understood as arising from probing a given volume only with a finite number of galaxies (e.g. just the bright ones), so that $\sqrt{N}$ statistics limit our knowledge of the mass distribution within that region. The second term concerns whether we have reached a fair sample of the universe, and depends on whether there is significant power in density perturbation modes with wavelengths larger than the sample depth. Clearly, these two aspects are closely related.

The quantitative analysis of these errors is most simply performed in Fourier space, and was given by Feldman, Kaiser & Peacock (1994). The results can be understood most simply by comparison with an idealized complete and uniform survey of a volume L^3, with periodicity scale L. For an infinite survey, the arbitrariness of the spatial origin means that different modes are uncorrelated:

$$\langle \delta_k(\mathbf{k}_i)\delta_k^*(\mathbf{k}_j)\rangle = P(k)\delta_{ij}. \tag{15}$$

Each mode has an exponential distribution in power (because the complex coefficients δ_k are 2D Gaussian-distributed variables on the Argand plane), for which the mean and rms are identical. The fractional uncertainty in the mean power measured over some k-space volume is then just determined by the number of uncorrelated modes averaged over:

$$\frac{\delta \bar{P}}{\bar{P}} = \frac{1}{N_{\rm modes}^{1/2}}; \qquad N_{\rm modes} = \left(\frac{L}{2\pi}\right)^3 \int d^3k. \tag{16}$$

The only subtlety is that, because the density field is real, modes at k and $-k$ are perfectly correlated. Thus, if the k-space volume is a shell, the effective number of uncorrelated modes is only half the above expression.

Analogous results apply for an arbitrary survey selection function. In the continuum limit, the Kroneker delta in the expression for mode correlation would be replaced a term proportional to a delta-function, $\delta[\mathbf{k}_i - \mathbf{k}_j])$. Now, multiplying the infinite ideal survey by a survey window, $\rho(\mathbf{r})$, is equivalent to convolution in the Fourier domain, with the result that the power per mode is correlated over k-space separations of order $1/D$, where D is the survey depth.

Given this expression for the fractional power, it is clear that the precision of the estimate can be manipulated by appropriate weighting of the data: giving increased weight to the most distant galaxies increases the effective survey volume, boosting the number of modes. This sounds too good to be true, and of course it is: the above expression for the fractional power error applies to the sum of true clustering power and shot noise. The latter arises because we transform a point process. Given a set of N galaxies, we would estimate Fourier coefficients via $\delta_k = (1/N)\sum_i \exp(-i\mathbf{k} \cdot x_i)$. From this, the expectation power is

$$\langle |\delta_k|^2 \rangle = P(k) + 1/N. \tag{17}$$

The existence of an additive discreteness correction is no problem, but the *fluctuations* on the shot noise hide the signal of interest. Introducing weights boosts the shot noise, so there is an optimum choice of weight that minimizes the uncertainty in the power after shot-noise subtraction. Feldman, Kaiser & Peacock (1994) showed that this weight is

$$w = (1 + \bar{n}P)^{-1}, \tag{18}$$

where $\bar{n}$ is the expected galaxy number density as a function of position in the survey.

Since the correlation of modes arises from the survey selection function, it is clear that weighting the data changes the degree of correlation in k

space. Increasing the weight in low-density regions increases the effective survey volume, and so shrinks the k-space coherence scale. However, the coherence scale continues to shrink as distant regions of the survey are given greater weight, whereas the noise goes through a minimum. There is thus a trade-off between the competing desirable criteria of high k-space resolution and low noise. Tegmark (1996) shows how weights may be chosen to implement any given prejudice concerning the relative importance of these two criteria. See also Hamilton (1997b,c) for similar arguments.

3.4. KARHUNEN-LOÈVE AND ALL THAT

Given these difficulties with correlated results, it is attractive to seek a method where the data can be decomposed into a set of statistics that are completely uncorrelated with each other. Such a method is provided by the Karhunen-Loève formalism. Vogeley & Szalay (1996) argued as follows. Define a column vector of data $\underline{d}$; this can be quite abstract in nature, and could be e.g. the numbers of galaxies in a set of cells, or a set of Fourier components of the transformed galaxy number counts. Similarly, for CMB studies, $\underline{d}$ could be $\delta T/T$ in a set of pixels, or spherical-harmonic coefficients $a_{\ell m}$. We assume that the mean can be identified and subtracted off, so that $\langle \underline{d} \rangle = 0$ in ensemble average. The statistical properties of the data are then described by the covariance matrix

$$C_{ij} \equiv \langle d_i d_j^* \rangle \tag{19}$$

(normally the data will be real, but it is convenient to keep things general and include the complex conjugate).

Suppose we seek to expand the data vector in terms of a set of new orthonormal vectors:

$$\underline{d} = \sum_i a_i \underline{\psi}_i; \qquad \underline{\psi}_i^* \cdot \underline{\psi}_j = \delta_{ij}. \tag{20}$$

The expansion coefficients are extracted in the usual way: $a_j = \underline{d} \cdot \underline{\psi}_j^*$. Now require that these coefficients be statistically uncorrelated, $\langle a_i a_j^* \rangle = \lambda_i \delta_{ij}$ (no sum on i). This gives

$$\underline{\psi}_i^* \cdot \langle \underline{d}\, \underline{d}^* \rangle \cdot \underline{\psi}_j = \lambda_i \delta_{ij}, \tag{21}$$

where the dyadic $\langle \underline{d}\, \underline{d}^* \rangle$ is $\underline{\underline{C}}$, the correlation matrix of the data vector: $(\underline{d}\, \underline{d}^*)_{ij} \equiv d_i d_j^*$. Now, the effect of operating this matrix on one of the $\underline{\psi}_i$ must be expandable in terms of the complete set, which shows that the $\underline{\psi}_j$ must be the eigenvectors of the correlation matrix:

$$\langle \underline{d}\, \underline{d}^* \rangle \cdot \underline{\psi}_j = \lambda_j \underline{\psi}_j. \tag{22}$$

Vogeley & Szalay further show that these uncorrelated modes are optimal for representing the data: if the modes are arranged in order of decreasing λ, and the series expansion truncated after n terms, the rms truncation error is minimized for this choice of eigenmodes. To prove this, consider the truncation error

$$\underline{\epsilon} = \underline{d} - \sum_{i=1}^{n} a_i \underline{\psi}_i = \sum_{i=n+1}^{\infty} a_i \underline{\psi}_i. \tag{23}$$

The square of this is

$$\langle \epsilon^2 \rangle = \sum_{i=n+1}^{\infty} \langle |a_i|^2 \rangle, \tag{24}$$

where $\langle |a_i|^2 \rangle = \underline{\psi}_i^* \cdot \underline{\underline{C}} \cdot \underline{\psi}_i$, as before. We want to minimize $\langle \epsilon^2 \rangle$ by varying the $\underline{\psi}_i$, but we need to do this in a way that preserves normalization. This is achieved by introducing a Lagrange multiplier, and minimizing

$$\sum \underline{\psi}_i^* \cdot \underline{\underline{C}} \cdot \underline{\psi}_i + \lambda (1 - \underline{\psi}_i^* \cdot \underline{\psi}_i). \tag{25}$$

This is easily solved if we consider the more general problem where $\underline{\psi}_i^*$ and $\underline{\psi}_i$ are independent vectors:

$$\underline{\underline{C}} \cdot \underline{\psi}_i = \lambda \underline{\psi}_i. \tag{26}$$

In short, the eigenvectors of $\underline{\underline{C}}$ are optimal in a least-squares sense for expanding the data. The process of truncating the expansion is a form of lossy data compression, since the size of the data vector can be greatly reduced without significantly affecting the fidelity of the resulting representation of the universe.

The process of diagonalizing the covariance matrix of a set of data also goes by the more familiar name of principal components analysis, so what is the difference between the KL approach and PCA? In the above discussion, they are identical, but the idea of choosing an optimal eigenbasis is more general than PCA. Consider the case where the covariance matrix can be decomposed into a 'signal' and a 'noise' term:

$$\underline{\underline{C}} = \underline{\underline{S}} + \underline{\underline{N}}, \tag{27}$$

where $\underline{\underline{S}}$ depends on cosmological parameters that we might wish to estimate, whereas $\underline{\underline{N}}$ is some fixed property of the experiment under consideration. In the simplest imaginable case, $\underline{\underline{N}}$ might be a diagonal matrix, so PCA diagonalizes both $\underline{\underline{S}}$ and $\underline{\underline{N}}$. In this case, ranking the PCA modes by eigenvalue would correspond to ordering the modes according to signal-to-noise ratio. Data compression by truncating the mode expansion then does the sensible thing: it rejects all modes of low signal-to-noise ratio.

However, in general these matrices will not commute, and there will not be a single set of eigenfunctions that are common to the $\underline{\underline{S}}$ and $\underline{\underline{N}}$ matrices. Normally, this would be taken to mean that it is impossible to find a set of coordinates in which both are diagonal. This conclusion can however be evaded, as follows. When considering the effect of coordinate transformations on vectors and matrices, we are normally forced to consider only rotation-like transformations that preserve the norm of a vector (e.g. in quantum mechanics, so that states stay normalized). Thus, we write $\underline{d}' = \underline{\underline{R}} \cdot \underline{d}$, where $\underline{\underline{R}}$ is unitary, so that $\underline{\underline{R}} \cdot \underline{\underline{R}}^\dagger = \underline{\underline{I}}$. If $\underline{\underline{R}}$ is chosen so that its columns are the eigenvalues of $\underline{\underline{N}}$, then the transformed noise matrix, $\underline{\underline{R}} \cdot \underline{\underline{N}} \cdot \underline{\underline{R}}^\dagger$, is diagonal. Nevertheless, if the transformed $\underline{\underline{S}}$ is not diagonal, the two will not commute. This apparently insuperable problem can be solved by using the fact that the data vectors are entirely abstract at this stage. There is therefore no reason not to consider the further transformation of scaling the data, so that $\underline{\underline{N}}$ becomes proportional to the identity matrix. This means that the transformation is no longer unitary – but there is no physical reason to object to a change in the normalization of the data vectors.

Suppose we therefore make a further transformation

$$\underline{d}'' = \underline{\underline{W}} \cdot \underline{d}' \tag{28}$$

The matrix $\underline{\underline{W}}$ is related to the rotated noise matrix:

$$\underline{\underline{N}}' = \mathrm{diag}\,(n_1, n_2, \ldots) \quad \Rightarrow \quad \underline{\underline{W}} = \mathrm{diag}\,(1/\sqrt{n_1}, 1/\sqrt{n_2}, \ldots). \tag{29}$$

This transformation is termed prewhitening by Vogeley & Szalay (1996), since it converts the noise matrix to white noise, in which each pixel has a unit noise that is uncorrelated with other pixels. The effect of this transformation on the full covariance matrix is

$$C''_{ij} \equiv \langle d''_i d''^*_j \rangle \quad \Rightarrow \quad \underline{\underline{C}}'' = (\underline{\underline{W}} \cdot \underline{\underline{R}}) \cdot \underline{\underline{C}} \cdot (\underline{\underline{W}} \cdot \underline{\underline{R}})^\dagger \tag{30}$$

After this transformation, the noise and signal matrices certainly do commute, and the optimal modes for expanding the new data are once again the PCA eigenmodes in the new coordinates:

$$\underline{\underline{C}}'' \cdot \underline{\psi}''_i = \lambda \underline{\psi}''_i. \tag{31}$$

These eigenmodes must be expressible in terms of some modes in the original coordinates, $\underline{e}_i$:

$$\underline{\psi}''_i = (\underline{\underline{W}} \cdot \underline{\underline{R}}) \cdot \underline{e}_i. \tag{32}$$

In these terms, the eigenproblem is

$$(\underline{\underline{W}} \cdot \underline{\underline{R}}) \cdot \underline{\underline{C}} \cdot (\underline{\underline{W}} \cdot \underline{\underline{R}})^\dagger \cdot (\underline{\underline{W}} \cdot \underline{\underline{R}}) \cdot \underline{e}_i = \lambda (\underline{\underline{W}} \cdot \underline{\underline{R}}) \cdot \underline{e}_i. \tag{33}$$

This can be simplified using $\underline{\underline{W}}^\dagger \cdot \underline{\underline{W}} = \underline{\underline{N}}'^{-1}$ and $\underline{\underline{N}}'^{-1} = \underline{\underline{R}} \cdot \underline{\underline{N}}^{-1} \underline{\underline{R}}^\dagger$, to give

$$\underline{\underline{C}} \cdot \underline{\underline{N}}^{-1} \cdot \underline{e}_i = \lambda \underline{e}_i, \tag{34}$$

so the required modes are eigenmodes of $\underline{\underline{C}} \cdot \underline{\underline{N}}^{-1}$. However, care is required when considering the orthonormality of the $\underline{e}_i$: $\underline{\psi}_i^\dagger \cdot \underline{\psi}_j = \underline{e}_i^\dagger \cdot \underline{\underline{N}}^{-1} \cdot \underline{e}_j$, so the $\underline{e}_i$ are not orthonormal. If we write $\underline{d} = \sum_i a_i \underline{e}_i$, then

$$a_i = (\underline{\underline{N}}^{-1} \cdot \underline{e}_i)^\dagger \cdot \underline{d} \equiv \underline{\psi}_i^\dagger \cdot \underline{d}. \tag{35}$$

Thus, the modes used to extract the compressed data by dot product satisfy $\underline{\underline{C}} \cdot \underline{\psi} = \lambda \underline{\underline{N}} \cdot \underline{\psi}$, or finally

$$\underline{\underline{S}} \cdot \underline{\psi} = \lambda \underline{\underline{N}} \cdot \underline{\psi}, \tag{36}$$

given a redefinition of λ. The optimal modes are thus eigenmodes of $\underline{\underline{N}}^{-1} \cdot \underline{\underline{S}}$, hence the name signal-to-noise eigenmodes (Bond 1995; Bunn 1996).

It is interesting to appreciate that the set of KL modes just discussed is also the 'best' set of modes to choose from a completely different point of view: they are the modes that are optimal for estimation of a parameter via maximum likelihood. Suppose we write the compressed data vector, $\underline{x}$, in terms of a non-square matrix $\underline{\underline{A}}$ (whose rows are the basis vectors $\underline{\psi}_i^*$):

$$\underline{x} = \underline{\underline{A}} \cdot \underline{d}. \tag{37}$$

The transformed covariance matrix is

$$\underline{\underline{D}} \equiv \langle \underline{x} \underline{x}^\dagger \rangle = \underline{\underline{A}} \cdot \underline{\underline{C}} \cdot \underline{\underline{A}}^\dagger. \tag{38}$$

For the case where the original data obeyed Gaussian statistics, this is true for the compressed data also, so the likelihood is

$$-2 \ln \mathcal{L} = \ln \det \underline{\underline{D}} + \underline{x}^* \cdot \underline{\underline{D}}^{-1} \cdot \underline{x} + \text{constant} \tag{39}$$

The normal variance on some parameter p (on which the covariance matrix depends) is

$$\frac{1}{\sigma_p^2} = \frac{d^2[-2 \ln \mathcal{L}]}{dq^2}. \tag{40}$$

Without data, we don't know this, so it is common to use the expectation value of the rhs as an estimate (recently, there has been a tendency to dub this the 'Fisher matrix').

We desire to optimize σ_p by an appropriate choice of data-compression vectors, $\underline{\psi}_i$. By writing σ_p in terms of $\underline{\underline{A}}$, $\underline{\underline{C}}$ and $\underline{d}$, it may eventually be shown that the desired optimal modes satisfy

$$\left(\frac{d}{dp}\underline{\underline{C}}\right) \cdot \underline{\psi} = \lambda \underline{\underline{C}} \cdot \underline{\psi}. \tag{41}$$

For the case where the parameter of interest is the cosmological power, the matrix on the lhs is just proportional to $\underline{\underline{S}}$, so we have to solve the eigenproblem

$$\underline{\underline{S}} \cdot \underline{\psi} = \lambda \underline{\underline{C}} \cdot \underline{\psi}. \tag{42}$$

With a redefinition of λ, this becomes

$$\underline{\underline{S}} \cdot \underline{\psi} = \lambda \underline{\underline{N}} \cdot \underline{\psi}. \tag{43}$$

The optimal modes for parameter estimation in the linear case are thus identical to the PCA modes of the prewhitened data discussed above. The more general expression was given by Tegmark, Taylor & Heavens (1997), and it is only in this case, where the covariance matrix is not necessarily linear in the parameter of interest, that the KL method actually differs from PCA.

The reason for going to all this trouble is that the likelihood can now be evaluated much more rapidly, using the compressed data. This allows extensive model searches over large parameter spaces that would be unfeasible with the original data (since inversion of an $N \times N$ covariance matrix takes a time proportional to N^3). Note however that the price paid for this efficiency is that a different set of modes need to be chosen depending on the model of interest, and that these modes will not in general be optimal for expanding the dataset itself. Nevertheless, it may be expected that application of these methods will inevitably grow as datasets increase in size. Present applications mainly prove that the techniques work: see Matsubara, Szalay & Landy (1999) for application to the LCRS, or Padmanabhan, Tegmark & Hamilton (1999) for the UZC survey. The next generation of experiments will probably be forced to resort to data compression of this sort, rather than using it as an elegant alternative method of analysis.

4. Redshift-space effects

Peculiar velocity fields are responsible for the distortion of the clustering pattern in redshift space, as first clearly articulated by Kaiser (1987). For a survey that subtends a small angle (i.e. in the distant-observer approximation), a good approximation to the anisotropic redshift-space Fourier spectrum is given by the Kaiser function together with a damping term from nonlinear effects:

$$\delta_k^s = \delta_k^r (1 + \beta\mu^2) D(k\sigma\mu), \tag{44}$$

where $\beta = \Omega_m^{0.6}/b$, b being the linear bias parameter of the galaxies under study, and $\mu = \hat{\mathbf{k}} \cdot \hat{\mathbf{r}}$. For an exponential distribution of relative small-scale peculiar velocities (as seen empirically), the damping function is $D(y) \simeq (1+y^2/2)^{-1/2}$, and $\sigma \simeq 400\,\mathrm{km\,s^{-1}}$ is a reasonable estimate for the pairwise velocity dispersion of galaxies (e.g. Ballinger, Peacock & Heavens 1996).

In principle, this distortion should be a robust way to determine Ω (or at least β). In practice, the effect has not been easy to see with past datasets. This is mainly a question of depth: a large survey is needed in order to beat down the shot noise, but this tends to favour bright spectroscopic limits. This limits the result both because relatively few modes in the linear regime are sampled, and also because local survey volumes will tend to violate the small-angle approximation. Strauss & Willick (1995) and Hamilton (1997a) review the practical application of redshift-space distortions. In the next section, preliminary results are presented from the 2dF redshift survey, which shows the distortion effect clearly for the first time.

5. The state of the art in LSS

5.1. THE APM SURVEY

In the past few years, much attention has been attracted by the estimate of the galaxy power spectrum from the APM survey (Baugh & Efstathiou 1993, 1994; Maddox et al. 1996). The APM result was generated from a catalogue of $\sim 10^6$ galaxies derived from UK Schmidt Telescope photographic plates scanned with the Cambridge Automatic Plate Measuring machine; because it is based on a deprojection of angular clustering, it is immune to the complicating effects of redshift-space distortions. The difficulty, of course, is in ensuring that any low-level systematics from e.g. spatial variations in magnitude zero point are sufficiently well controlled that they do not mask the cosmological signal, which is of order $w(\theta) \lesssim 0.01$ at separations of a few degrees.

The best evidence that the APM survey has the desired uniformity is the scaling test, where the correlations in fainter magnitude slices are expected to move to smaller scales and be reduced in amplitude. If we increase the depth of the survey by some factor D, the new angular correlation function will be

$$w'(\theta) = \frac{1}{D}\, w(D\theta). \tag{45}$$

The APM survey passes this test well; once the overall redshift distribution is known, it is possible to obtain the spatial power spectrum by inverting a convolution integral:

$$w(\theta) = \int_0^\infty y^4 \phi^2 \, dy \int_0^\infty \pi \, \Delta^2(k) \, J_0(ky\theta) \, dk/k^2 \tag{46}$$

(where zero spatial curvature is assumed). Here, $\phi(y)$ is the comoving density at comoving distance y, normalized so that $\int y^2 \phi(y)\, dy = 1$.

This integral was inverted numerically by Baugh & Efstathiou (1993), and gives an impressively accurate determination of the power spectrum. The error estimates are derived empirically from the scatter between independent regions of the sky, and so should be realistic. If there are no undetected systematics, these error bars say that the power is very accurately determined. The APM result has been investigated in detail by a number of authors (e.g. Gaztañaga & Baugh 1998; Eisenstein & Zaldarriaga 1999) and found to be robust; this has significant implications if true.

5.2. PAST REDSHIFT SURVEYS

Because of the sheer number of galaxies, plus the large volume surveyed, the APM survey outperforms redshift surveys of the past, at least for the purpose of determining the power spectrum. The largest surveys of recent years (CfA: Huchra et al. 1990; LCRS: Shectman et al. 1996; PSCz: Saunders et al. 1999) contain of order 10^4 galaxy redshifts, and their statistical errors are considerably larger than those of the APM. On the other hand, it is of great importance to compare the results of deprojection with clustering measured directly in 3D.

This comparison was carried out by Peacock & Dodds (1994; PD94). The exercise is not straightforward, because the 3D results are affected by redshift-space distortions; also, different galaxy tracers can be biased to different extents. The approach taken was to use each dataset to reconstruct an estimate of the linear spectrum, allowing the relative bias factors to float in order to make these estimates agree as well as possible (figure 5). To within a scatter of perhaps a factor 1.5 in power, the results were consistent with a $\Gamma \simeq 0.25$ CDM model. Even though the subsequent sections will discuss some possible disagreements with the CDM models at a higher level of precision, the general existence of CDM-like curvature in the spectrum is likely to be an important clue to the nature of the dark matter.

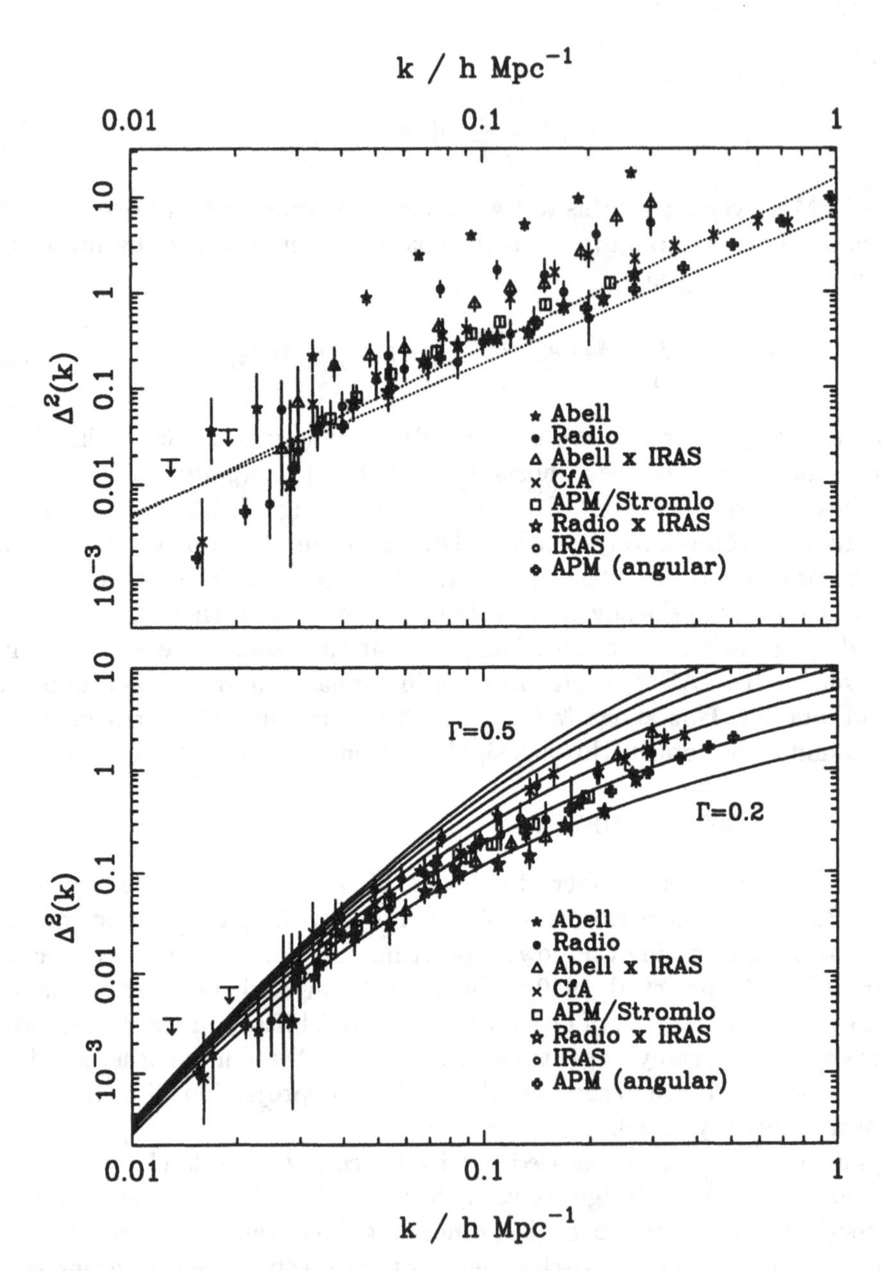

Figure 5. The PD94 compilation of power-spectrum measurements. The upper panel shows raw power measurements; the lower shows these data corrected for relative bias, nonlinear effects, and redshift-space effects.

5.3. THE 2DF SURVEY

The proper resolution of many of the observational questions regarding the large-scale distribution of galaxies requires new generations of redshift

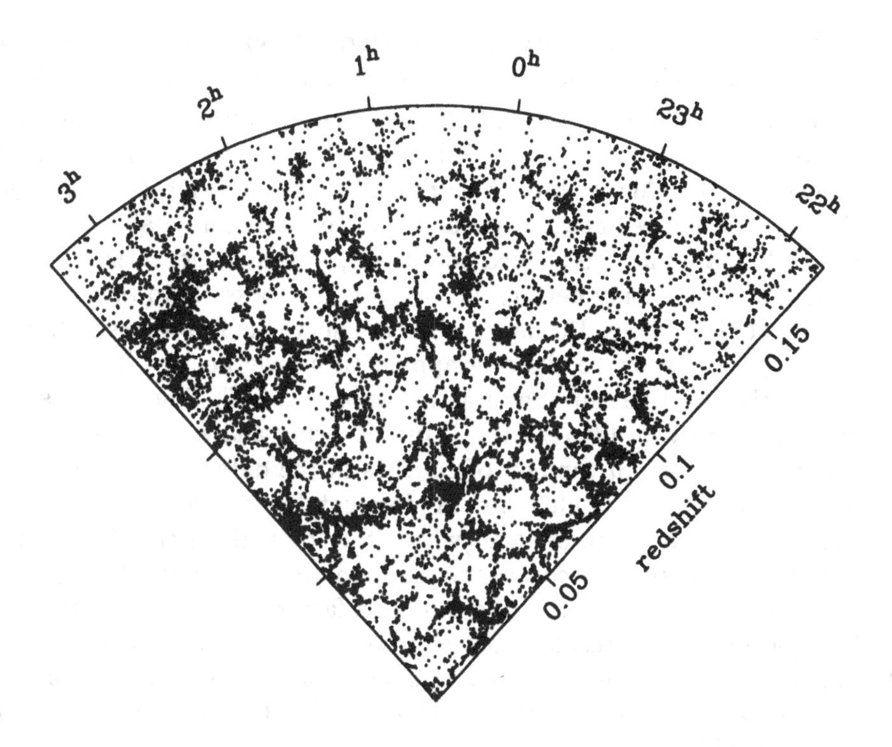

Figure 6. A 4-degree thick slice of the Southern strip of the 2dF redshift survey. This restricted region alone contains 16,419 galaxies.

survey that push beyond the $N = 10^5$ barrier. Two groups are pursuing this goal. The Sloan survey (e.g. Margon 1999) is using a dedicated 2.5-m telescope to measure redshifts for approximately 700,000 galaxies to $r = 18.2$ in the North Galactic Cap. The 2dF survey (e.g. Colless 1999) is using a fraction of the time on the 3.9-m Anglo-Australian Telescope plus Two-Degree Field spectrograph to measure 250,000 galaxies from the APM survey to $B_J = 19.45$ in the South Galactic Cap. At the time of writing, the Sloan spectroscopic survey has yet to commence. However, the 2dF project has measured 77,000 redshifts, and some preliminary clustering results are given below. For more details of the survey, particularly the team members whose hard work has made all this possible, see `http://www.mso.anu.edu.au/2dFGRS/`.

One of the advantages of 2dF is that it is a fully sampled survey, so that the space density out to the depth imposed by the magnitude limit (median $z = 0.12$) is as high as nature allows: apart from a tail of low surface brightness galaxies (inevitably omitted from any spectroscopic survey), the 2dF measure all the galaxies that exist over a cosmologically representative volume. It is the first to achieve this goal. The fidelity of the resulting map of the galaxy distribution can be seen in figure 6, which shows a small subset of the data: a slice of thickness 4 degrees, centred at declination $-27°$.

An issue with using the 2dF data in their current form is that the sky has to be divided into circular 'tiles' each two degrees in diameter ('2dF' = 'two-degree field', within which the AAT is able to measure 400 spectra simultaneously; see `http://www.aao.gov.au/2df/` for details of the instrument). The tiles are positioned adaptively, so that larger overlaps occur in regions of high galaxy density. It this way, it is possible to place a fibre on $> 95\%$ of all galaxies. However, while the survey is in progress, there exist parts of the sky where the overlapping tiles have not yet been observed, and so the effective sampling fraction is only $\simeq 50\%$. These effects can be allowed for in two different ways. In clustering analyses, we compare the counts of pairs (or n-tuplets) of galaxies in the data to the corresponding counts involving an unclustered random catalogue. The effects of variable sampling can therefore be dealt with either by making the density of random points fluctuate according to the sampling, or by weighting observed galaxies by the reciprocal of the sampling factor for the zone in which they lie. The former approach is better from the point of view of shot noise, but the latter may be safer if there is any suspicion that the sampling fluctuations are correlated with real structure on the sky. In practice, both strategies give identical answers for the results below.

At the two-point level, the most direct quantity to compute is the redshift-space correlation function. This is an anisotropic function of the orientation of a galaxy pair, owing to peculiar velocities. We therefore evaluate ξ as a function of 2D separation in terms of coordinates both parallel and perpendicular to the line of sight. If the comoving radii of two galaxies are y_1 and y_2 and their total separation is r, then we define coordinates

$$\pi \equiv |y_1 - y_2|; \qquad \sigma = \sqrt{r^2 - \pi^2}. \tag{47}$$

The correlation function measured in these coordinates is shown in figure 7. In evaluating $\xi(\sigma, \pi)$, the optimal radial weight discussed above has been applied, so that the noise at large r should be representative of true cosmic scatter.

The correlation-function results display very clearly the two signatures of redshift-space distortions discussed above. The fingers of God from small-scale random velocities are very clear, as indeed has been the case from the first redshift surveys (e.g. Davis & Peebles 1983). However, this is arguably the first time that the large-scale flattening from coherent infall has been really obvious in the data.

A good way to quantify the flattening is to analyze the clustering as a

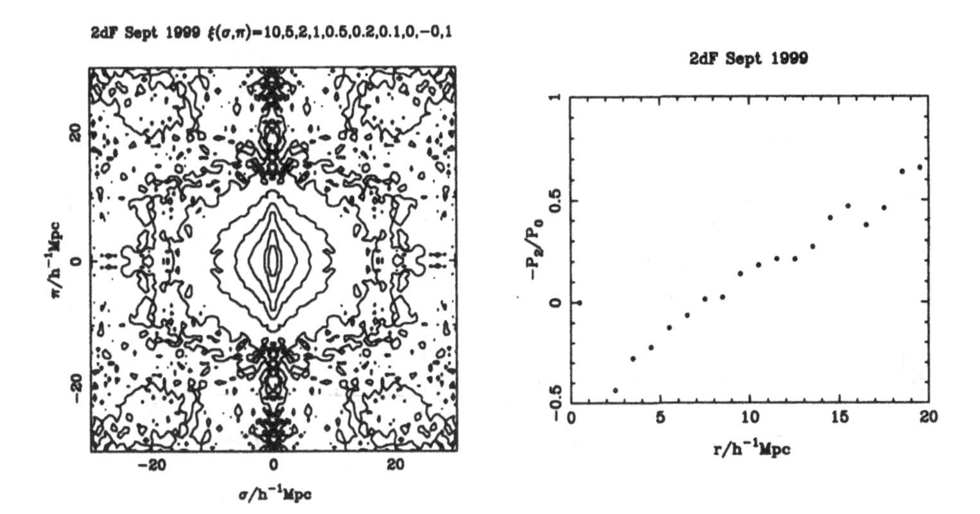

Figure 7. The redshift-space correlation function from the 2dF data, $\xi(\sigma, \pi)$, with a bin size of $0.6\,h^{-1}$ Mpc. σ is the pair separation transverse to the line of sight; π is the radial separation. This plot clearly displays redshift distortions, with 'fingers of God' at small scales and the coherent Kaiser squashing at large σ. The distortions are quantified via the quadrupole-to-monopole ratio of ξ as a function of radius in the second panel. The contours are round at $r = 7\,h^{-1}$ Mpc, but flatten progressively thereafter.

function of angle into Legendre polynomials:

$$\xi_\ell(r) = \frac{2\ell + 1}{2} \int_{-1}^{1} \xi(\sigma = r\sin\theta, \pi = r\cos\theta)\, P_\ell(\cos\theta)\, d\cos\theta. \tag{48}$$

The quadrupole-to-monopole ratio should be a clear indicator of coherent infall. In linear theory, it is given by

$$\frac{\xi_2}{\xi_0} = f(n)\,\frac{4\beta/3 + 4\beta^2/7}{1 + 2\beta/3 + \beta^2/5}, \tag{49}$$

where $f(n) = (3+n)/n$ (Hamilton 1992). On small and intermediate scales, the effective spectral index is negative, so the quadrupole-to-monopole ratio should be negative, as observed.

However, it is clear that the results on the largest scales are still significantly affected by finger-of-God smearing. The best way to interpret the observed effects is to calculate the same quantities for a model. To achieve this, we use the observed APM 3D power spectrum, plus the distortion model discussed above. This gives the plots shown in figure 8. The free parameter is β, and this is set at a value of 0.5, approximately consistent with other arguments for a universe with $\Omega = 0.3$ and little large-scale bias (e.g. Peacock 1997). Although a quantitative comparison has not yet been carried out, it is clear that this plot closely resembles the observed data.

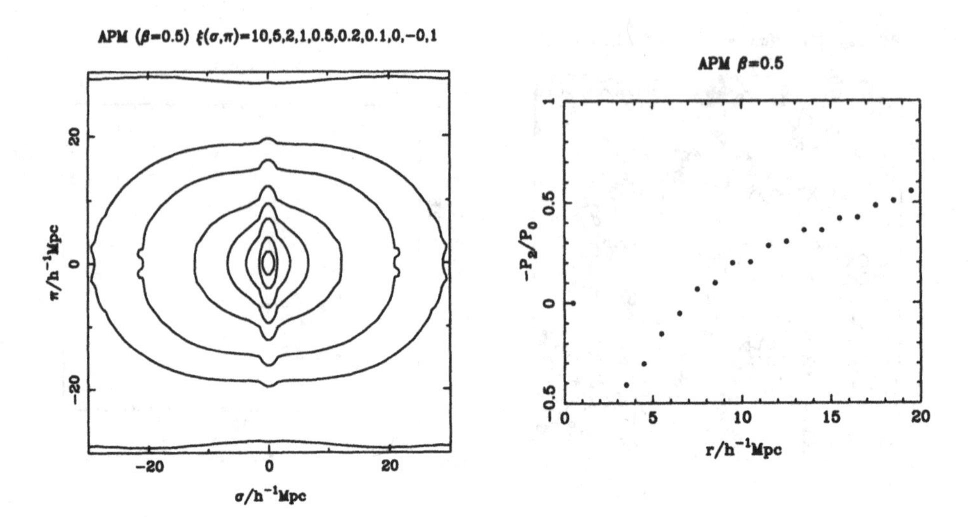

Figure 8. The redshift-space correlation function predicted from the real-space APM power spectrum, assuming the model of Ballinger, Peacock & Heavens (1996), with $\beta = 0.5$.

By the end of 2001, the size of the 2dF survey should have expanded by a factor 3, increasing the pair counts tenfold. It should then be possible to trace the correlations well beyond the present limit, and follow the redshift-space distortion well into the linear regime. However, the biggest advantage of a survey of this size and uniformity is the ability to subdivide it. All analyses to date have lumped together very different kinds of galaxies, whereas we know from morphological segregation that different classes of galaxy have spatial distributions that differ from each other. The homogeneous 2dF data allow classification into different galaxy types (representing, physically, a sequence of star-formation rates), from the spectra alone (Folkes et al. 1999). It will be a critical test to see if the distortion signature can be picked up in each type individually. Although the large-scale behaviour of each galaxy type will probably be quite similar, differences in the clustering properties will inevitably arise on smaller scales, giving important information about the sequence of galaxy formation.

6. Small-scale clustering

6.1. HISTORY

One of the earliest models to be used to interpret the galaxy correlation function was to consider a density field composed of randomly-placed independent clumps with some universal density profile (Neyman, Scott & Shane 1953; Peebles 1974). Since the clumps are placed at random, the only correlations arise from points in the same clump. The correlations are easily deduced by using statistical isotropy: calculate the excess number of pairs separated by a distance r in the z direction (chosen as some arbitrary polar axis in a spherically-symmetric clump). For power-law clumps, with $\rho = nBr^{-\epsilon}$, truncated at $r = R$, this model gives $\xi \propto r^{3-2\epsilon}$ in the limit $r \ll R$, provided $3/2 < \epsilon < 3$. Values $\epsilon > 3$ are unphysical, and require a small-scale cutoff to the profile. There is no such objection to $\epsilon < 3/2$, and the expression for ξ tends to a constant for small r in this case (see Yano & Gouda 1999).

A long-standing problem for this model is that the correlation function in this case is much flatter than is observed for galaxies: $\xi \propto r^{-1.8}$ is the canonical slope, requiring $\epsilon = 2.4$. The first reaction may be to say that the model is incredibly naive by comparison with our sophisticated present understanding of the nonlinear evolution of CDM density fields. However, as will be shown below, it may after all contain more than a grain of truth.

6.2. THE CDM CLUSTERING PROBLEMS

A number of authors have pointed out that the detailed spectral shape inferred from galaxy data appears to be inconsistent with that of nonlinear evolution from CDM initial conditions. (e.g. Efstathiou, Sutherland & Maddox 1990; Klypin, Primack & Holtzman 1996; Peacock 1997). Perhaps the most detailed work was carried out by the VIRGO consortium, who carried out $N = 256^3$ simulations of a number of CDM models (Jenkins et al. 1998). Their results are shown in figure 9, which gives the nonlinear power spectrum at various times (cluster normalization is chosen for $z = 0$) and contrasts this with the APM data. The lower small panels are the scale-dependent bias that would required if the model did in fact describe the real universe, defined as

$$b(k) \equiv \left(\frac{\Delta^2_{\rm gals}(k)}{\Delta^2_{\rm mass}} \right)^{1/2} . \qquad (50)$$

In all cases, the required bias is non-monotonic; it rises at $k \gtrsim 5\,h^{-1}$ Mpc, but also displays a bump around $k \simeq 0.1\,h^{-1}$ Mpc. If real, this feature seems impossible to understand as a genuine feature of the mass power spectrum; certainly, it is not at a scale where the effects of even a large baryon fraction would be expected to act (Eisenstein et al. 1998; Meiksin, White & Peacock 1999).

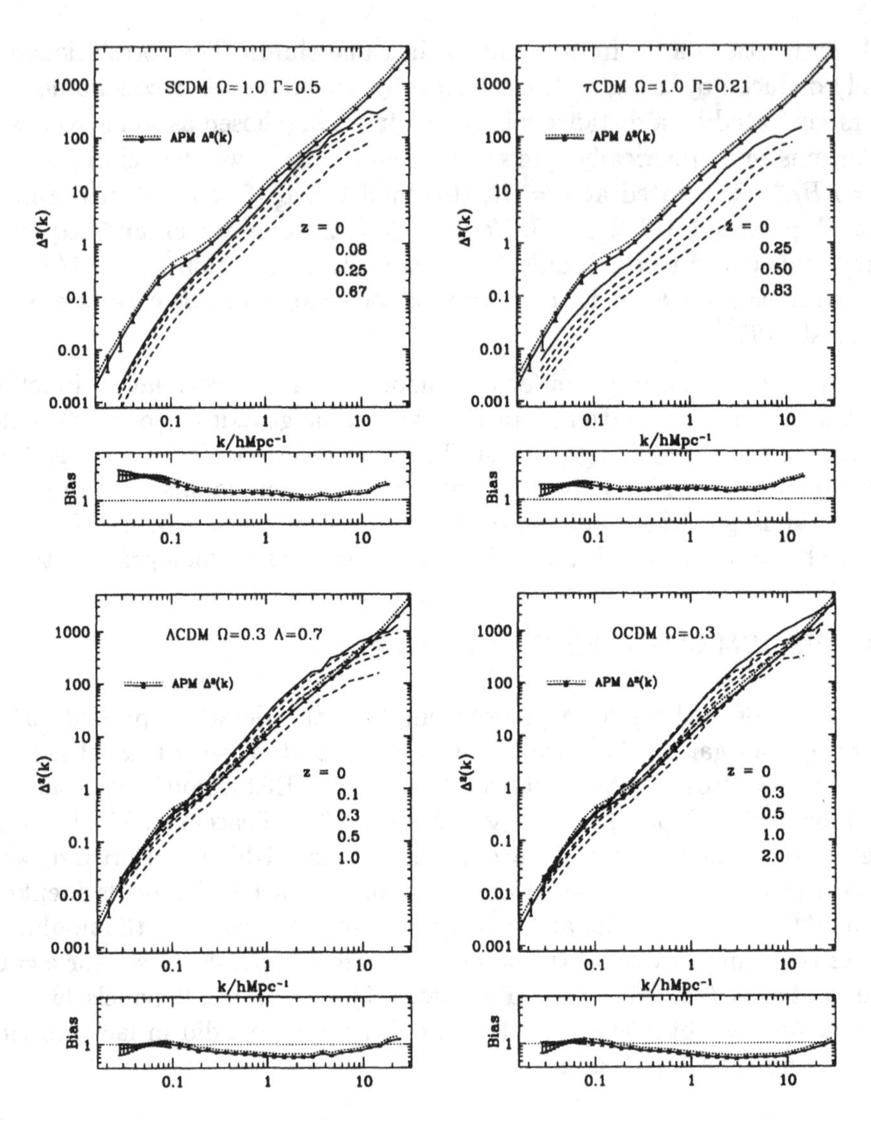

Figure 9. The nonlinear evolution of various CDM power spectra, as determined by the Virgo consortium (Jenkins et al. 1998).

7. Bias

The conclusions from the above discussion are either that the physics of dark matter and structure formation are more complex than in CDM models, or that the relation between galaxies and the overall matter distribution is sufficiently complicated that the effective bias is not a simple slowly-varying monotonic function of position.

7.1. SIMPLE BIAS MODELS

The simplest assumption is that all the complicated physical effects leading to galaxy formation depend in a causal (but nonlinear) way on the local mass density, so that we write

$$\rho_{\rm light} = f(\rho_{\rm mass}). \tag{51}$$

Coles (1993) showed that, under rather general assumptions, this equation would lead to an effective bias that was a monotonic function of scale. This issue was investigated in some detail by Mann, Peacock & Heavens (1998), who verified Coles' conclusion in practice for simple few-parameter forms for f, and found in all cases that the effective bias varied rather weakly with scale. The APM results thus are either inconsistent with a CDM universe, or require non-local bias.

A puzzle with regard to this conclusion is provided by the work of Jing, Mo & Börner (1998). They evaluated the projected real-space correlations for the LCRS survey (see figure 10). This statistic also fails to match the prediction of CDM models, but this can be amended by introducing a simple *antibias* scheme, in which galaxy formation is suppressed in the most massive haloes. This scheme should in practice be very similar to the Mann, Peacock & Heavens recipe of a simple weighting of particles as a function of the local density; indeed, the main effect is a change of amplitude, rather than shape of the correlations. The puzzle is this: if the APM power spectrum is used to predict the projected correlation function, the result agrees almost exactly with the LCRS. Either projected correlations are a rather insensitive statistic, or perhaps the Baugh & Efstathiou deconvolution procedure used to get $P(k)$ has exaggerated the significance of features in the spectrum. The LCRS results are one reason for treating the apparent conflict between APM and CDM with caution.

7.2. HALO CORRELATIONS

In reality, bias is unlikely to be completely causal, and this has led some workers to explore stochastic bias models, in which

$$\rho_{\rm light} = f(\rho_{\rm mass}) + \epsilon, \tag{52}$$

where ϵ is a random field that is uncorrelated with the mass density (Pen 1998; Dekel & Lahav 1999). Although truly stochastic effects are possible in galaxy formation, a relation of the above form is expected when the galaxy and mass densities are filtered on some scale (as they always are, in practice). Just averaging a galaxy density that is a nonlinear function of the mass will lead to some scatter when comparing with the averaged mass

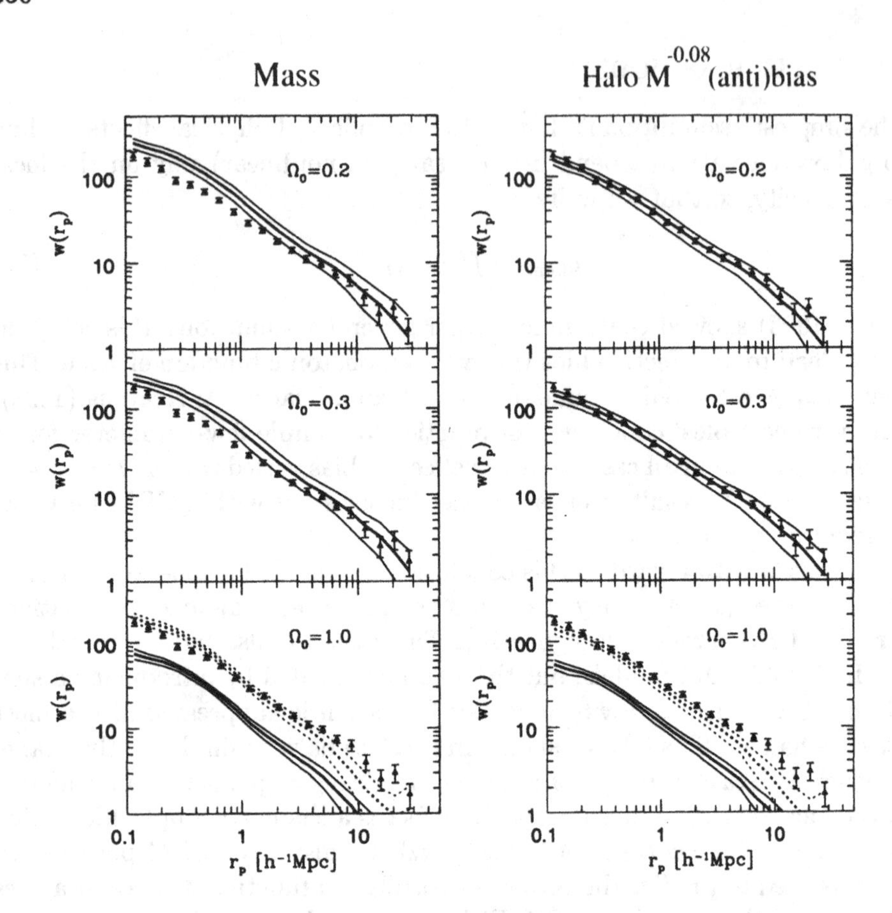

Figure 10. **The projected correlation function from the LCRS fails to match CDM models when comparison is made to just the mass distribution. However, the agreement is excellent when allowance is made for a small degree of scale-dependent antibias; galaxy formation is suppressed in the most massive haloes (Jing, Mo & Börner 1998).**

field; a scatter will also arise when the relation between mass and light is non-local, however, and this may be the dominant effect.

The simplest and most important example of non-locality in the galaxy-formation process is to recognize that galaxies will generally form where there are galaxy-scale haloes of dark matter. In the past, it was generally believed that dissipative processes were critically involved in galaxy formation, since pure collisionless evolution would lead to the destruction of galaxy-scale haloes when they are absorbed into the creation of a larger-scale nonlinear system such as a group or cluster. However, it turns out that this *overmerging problem* was only an artefact of inadequate resolution. When a simulation is carried out with $\sim 10^6$ particles in a rich cluster, the cores of galaxy-scale haloes can still be identified after many crossing times (Ghigna et al. 1997). Furthermore, if catalogues of these 'sub-haloes' are created within a cosmological-sized simulation, their correlation function is quite different from that of the mass, resembling the single power law seen in galaxies (e.g. Klypin et al. 1999; Ma 1999).

These are very important results, and they hold out the hope that many of the issues concerning where galaxies form in the cosmic density field can be settled within the domain of collisionless simulations. Dissipative physics will still be needed to understand in detail the star-formation history within a galaxy-scale halo. Nevertheless, the idea that there may be a one-to-one correspondence between galaxies and galaxy-scale dark-matter haloes is clearly an enormous simplification – and one that increases the chance of making robust predictions of the statistical properties of the galaxy population.

7.3. NUMERICAL GALAXY FORMATION

The formation of galaxies must be a non-local process to some extent. The modern paradigm was introduced by White & Rees (1978): galaxies form through the cooling of baryonic material in virialized haloes of dark matter. The virial radii of these systems are in excess of 0.1 Mpc, so there is the potential for large differences in the correlation properties of galaxies and dark matter on these scales.

A number of studies have indicated that the observed galaxy correlations may indeed be reproduced by CDM models. The most direct approach is a numerical simulation that includes gas, and relevant dissipative processes. This is challenging, but just starting to be feasible with current computing power (Pearce et al. 1999). The alternative is 'semianalytic' modelling, in which the merging history of dark-matter haloes is treated via the extended Press-Schechter theory (Bond et al. 1991), and the location of galaxies within haloes is estimated using dynamical-friction arguments (e.g. Cole et al. 1996; Kauffmann et al. 1996; Somerville & Primack 1997). Both these approaches have yielded similar conclusions, and shown how CDM models can match the galaxy data: specifically, the low-density flat ΛCDM model that is favoured on other grounds can yield a correlation function that is close to a single power law over $1000 \gtrsim \xi \gtrsim 1$, even though the mass correlations show a marked curvature over this range (Pearce et al. 1999; Benson et al. 1999; see figure 11). These results are impressive, yet it is frustrating to have a result of such fundamental importance emerge from a complicated calculational apparatus. There is thus some motivation for constructing a simpler heuristic model that captures the main processes at work in the full semianalytic models. The following section describes an approach of this sort (Peacock & Smith, in preparation).

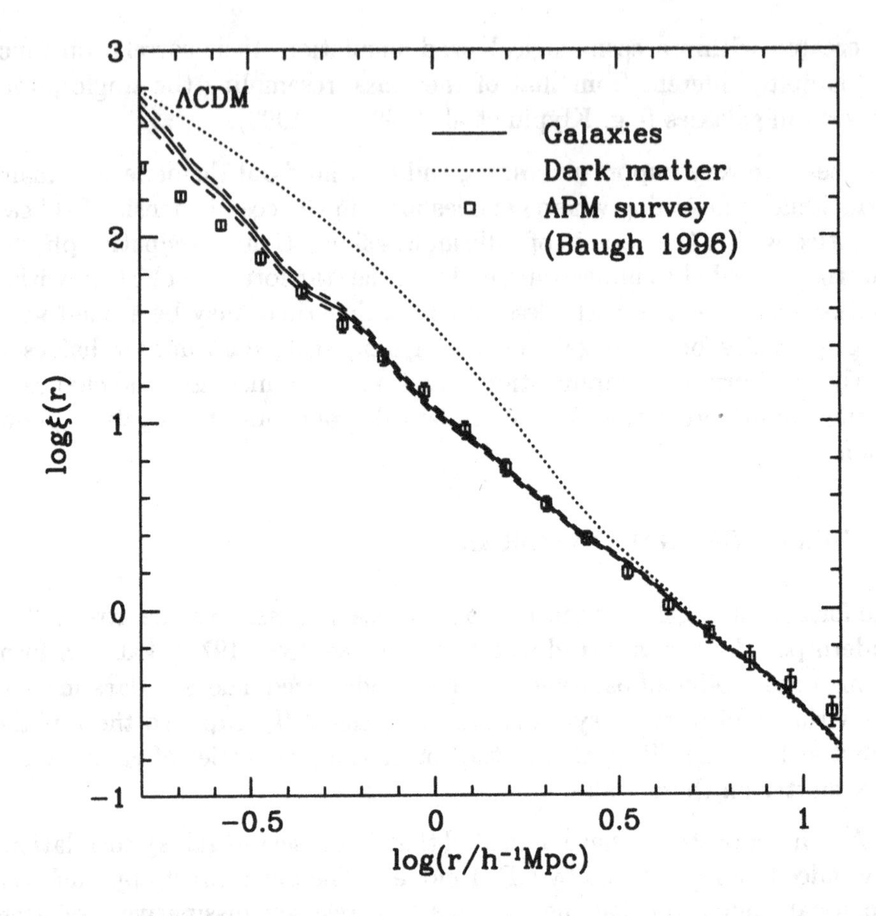

Figure 11. The correlation function of galaxies in the semianalytical simulation of an LCDM universe by Benson et al. (1999).

7.4. HALO-OLOGY AND BIAS

We mentioned above the early model of Neyman, Scott & Shane (1953), in which the nonlinear density field was taken to be a superposition of randomly-placed clumps. With our present knowledge about the evolution of CDM universes, we can make this idealised model considerably more realistic: hierarchical models are expected to contain a distribution of masses of clumps, which have density profiles that are more complicated than isothermal spheres. These issues are well studied in N-body simulations, and highly accurate fitting formulae exist, both for the mass function and for the density profiles. Briefly, we use the mass function of Sheth & Tormen

(1999; ST) and the halo profiles of Moore et al. (1999; M99).

$$\begin{aligned} f(\nu) &= 0.21617[1+(\sqrt{2}/\nu^2)^{0.3}]\exp[-\nu^2/(2\sqrt{2})] \\ \Rightarrow F(>\nu) &= 0.32218[1-\mathrm{erf}(\nu/2^{3/4})] \\ &\quad + 0.14765\Gamma[0.2,\nu^2/(2\sqrt{2})], \end{aligned} \tag{53}$$

where Γ is the incomplete gamma function.

Recently, it has been claimed by Moore et al. (1999; M99) that the commonly-adopted density profile of Navarro, Frenk & White (1996; NFW) is in error at small r. M99 proposed the alternative form

$$\rho/\rho_b = \frac{\Delta_c}{y^{3/2}(1+y^{3/2})}; \quad (r < r_{\rm vir}); \quad y \equiv r/r_c. \tag{54}$$

Using this model, it is then possible to calculate the correlations of the nonlinear density field, neglecting only the large-scale correlations in halo positions. The power spectrum determined in this way is shown in figure 12, and turns out to agree very well with the exact nonlinear result on small and intermediate scales. The lesson here is that a good deal of the nonlinear correlations of the dark matter field can be understood as a distribution of random clumps, provided these are given the correct distribution of masses and mass-dependent density profiles.

How can we extend this model to understand how the clustering of galaxies can differ from that of the mass? There are two distinct ways in which a degree of bias is inevitable:

(1) Halo occupation numbers. For low-mass haloes, the probability of obtaining an L^* galaxy must fall to zero. For haloes with mass above this lower limit, the number of galaxies will in general not scale with halo mass.

(2) Nonlocality. Galaxies can orbit within their host haloes, so the probability of forming a galaxy depends on the overall halo properties, not just the density at a point. Also, the galaxies will end up at special places within the haloes: for a halo containing only one galaxy, the galaxy will clearly mark the halo centre. In general, we expect one central galaxy and a number of satellites.

The numbers of galaxies that form in a halo of a given mass is the prime quantity that numerical models of galaxy formation aim to calculate. However, for a given assumed background cosmology, the answer may be determined empirically. Galaxy redshift surveys have been analyzed via grouping algorithms similar to the 'friends-of-friends' method widely employed to find virialized clumps in N-body simulations. With an appropriate correction for the survey limiting magnitude, the observed number of

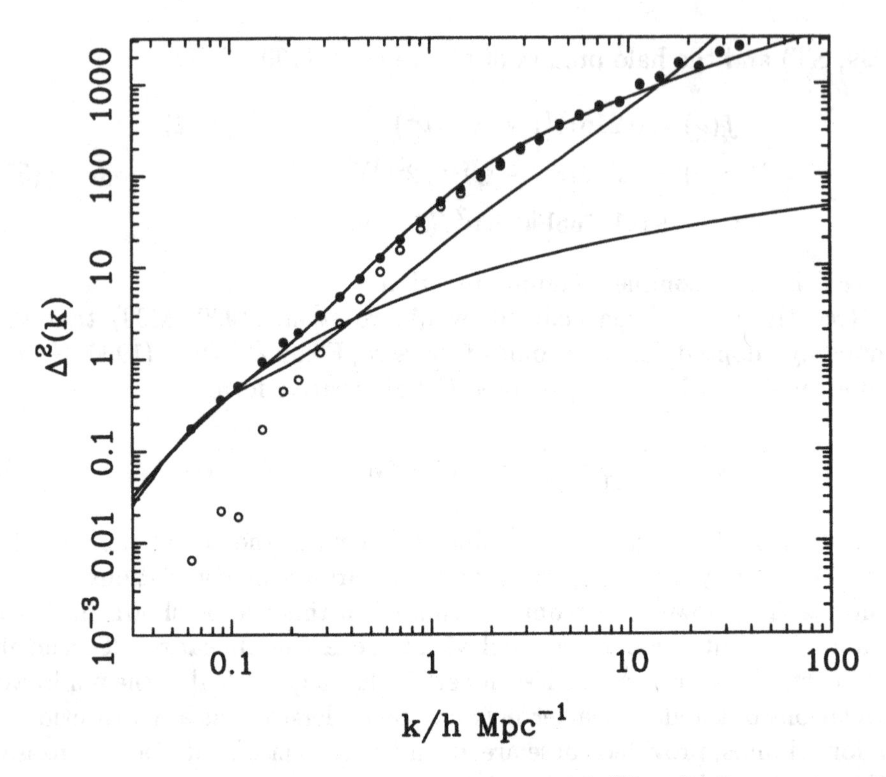

Figure 12. The power spectrum for the ΛCDM model. The solid lines contrast the linear spectrum with the nonlinear spectrum, calculated according to the approximation of PD96. The spectrum according to randomly-placed haloes is denoted by open circles; if the linear power spectrum is added, the main features of the nonlinear spectrum are well reproduced.

galaxies in a group can be converted to an estimate of the total stellar luminosity in a group. This allows a determination of the All Galaxy System (AGS) luminosity function: the distribution of virialized clumps of galaxies as a function of their total luminosity, from small systems like the Local Group to rich Abell clusters.

The AGS function for the CfA survey was investigated by Moore, Frenk & White (1993), who found that the result in blue light was well described by

$$d\phi = \phi^* \left[(L/L^*)^\beta + (L/L^*)^\gamma\right]^{-1} dL/L^*, \tag{55}$$

where $\phi^* = 0.00126h^3\text{Mpc}^{-3}$, $\beta = 1.34$, $\gamma = 2.89$; the characteristic luminosity is $M^* = -21.42 + 5\log_{10} h$ in Zwicky magnitudes, corresponding to $M_B^* = -21.71 + 5\log_{10} h$, or $L^* = 7.6 \times 10^{10} h^{-2} L_\odot$, assuming $M_B^\odot = 5.48$. One notable feature of this function is that it is rather flat at low luminosities, in contrast to the mass function of dark-matter haloes (see Sheth

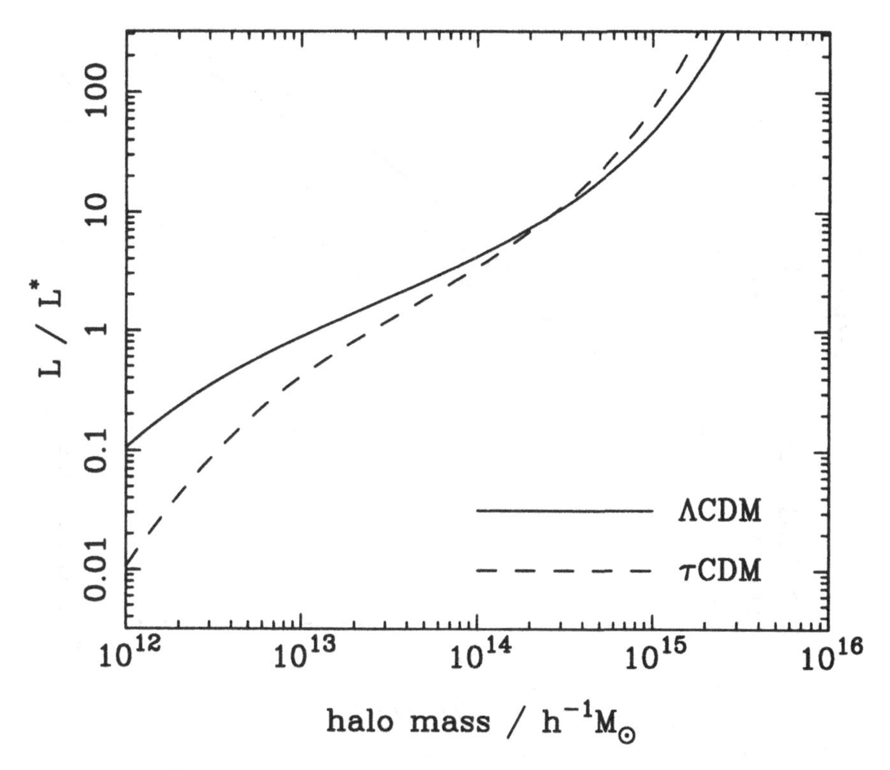

Figure 13. The empirical luminosity–mass relation required to reconcile the observed AGS luminosity function with two variants of CDM. L^* is the characteristic luminosity in the AGS luminosity function ($L^* = 7.6 \times 10^{10} h^{-2} L_\odot$). Note the rather flat slope around $M = 10^{13}$ to $10^{14} h^{-1} M_\odot$, especially for ΛCDM.

& Tormen 1999). It is therefore clear that any fictitious galaxy catalogue generated by randomly sampling the mass is unlikely to be a good match to observation. The simplest cure for this deficiency is to assume that the stellar luminosity per virialized halo is a monotonic, but nonlinear, function of halo mass. The required luminosity–mass relation is then easily deduced by finding the luminosity at which the integrated AGS density $\Phi(> L)$ matches the integrated number density of haloes with mass $> M$. The result is shown in figure 13.

We can now return to the halo-based galaxy power spectrum and use the correct occupation number, N, as a function of mass. This is needs a little care at small numbers, however, since the number of haloes with occupation number unity affects the correlation properties strongly. These haloes contribute no correlated pairs, so they simply dilute the signal from the haloes with $N \geq 2$. The existence of antibias on intermediate scales can probably be traced to the fact that a large fraction of galaxy groups contain only one $> L_*$ galaxy. Finally, we need to put the galaxies in the correct

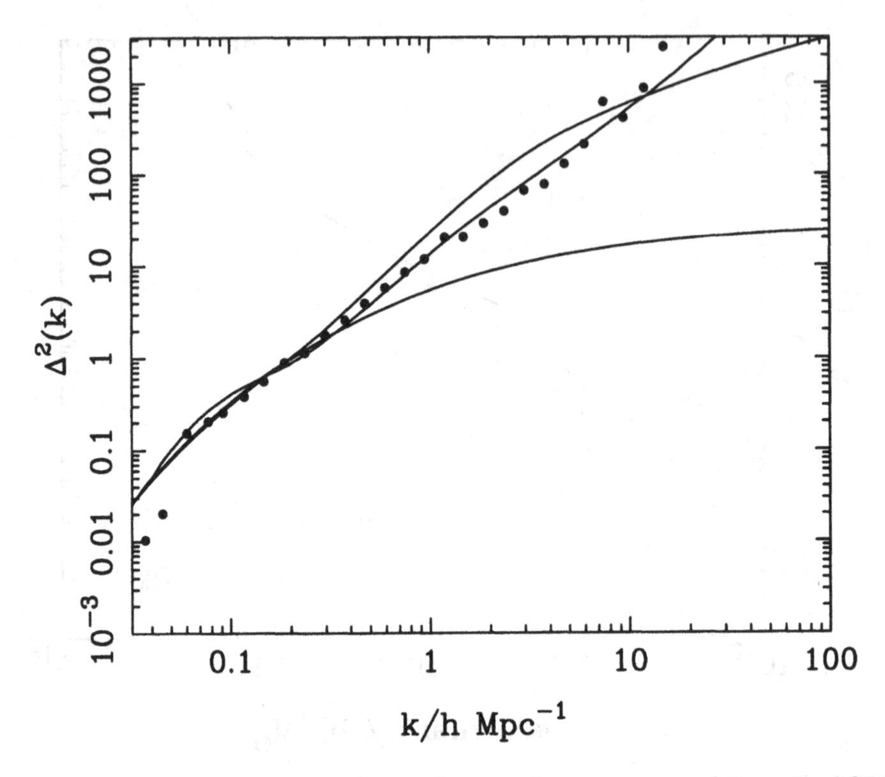

Figure 14. The power spectrum for a galaxy catalogue constructed from the ΛCDM model. A reasonable agreement with the APM data (solid line) is achieved by simple empirical adjustment of the occupation number of galaxies as a function of halo mass, plus a scheme for placing the haloes non-randomly within the haloes.

location, as discussed above. If one galaxy always occupies the halo centre, with others acting as satellites, the small-scale correlations automatically follow the slope of the halo density profile, which keeps them steep. The results of this exercise are shown in figure 14.

Although it is encouraging that it is possible to find simple models in which it is possible to understand the observed correlation properties of galaxies, there are other longstanding puzzles concerning the galaxy distribution. Arguably the chief of these concerns the dynamical properties of galaxies, in particular the pairwise peculiar velocity dispersion. This statistic has been the subject of debate, and preferred values have crept up in recent years, to perhaps 450 or 500 km s^{-1} at projected separations around 1 Mpc (e.g. Jing, Mo & Börner 1998), most simple models predict a higher figure. Clearly, the amplitude of peculiar velocities depends on the normalization of the fluctuation spectrum; however, if this is set from the abundance of rich clusters, then Jenkins et al. (1998) found that reasonable values were predicted for large-scale streaming velocities, independent of Ω.

However, Jenkins et al. also found a robust prediction for the pairwise peculiar velocity dispersion around 1 Mpc of about $800\,\mathrm{km\,s^{-1}}$. The observed galaxy velocity field appears to have a higher 'cosmic Mach number' than the predicted dark-matter distribution.

This difficulty is also solved by the simple bias model discussed here. Two factors contribute: the variation of occupation number with mass downweights the contribution of more massive groups, with larger velocity dispersions. Also, where one galaxy is centred on a halo, it gains a peculiar velocity which is that of the centre of mass of the halo, but does not reflect the internal velocity dispersion of the halo. Given a full N-body simulation, it is easy enough to predict what would be expected for a realistic bias model: one needs to construct a halo catalogue, calculating the peculiar velocities and internal velocity dispersions of each halo. Knowing the occupation number as a function of mass, a montecarlo catalogue of 'galaxies' complete with peculiar velocities can be generated. As shown in figure 15, the effect of the empirical bias recipe advocated here is sufficient to reduce the predicted dispersion into agreement with observation. The simple model outlined here thus gives a consistent picture, and it is tempting to believe that it may capture some of the main features of realistic models for galaxy bias.

8. Conclusions

It should be clear from these lectures that large-scale structure has advanced enormously as a field in the past two decades. Many of our longstanding ambitions have been realised; in some cases, much faster than we might have expected. Of course, solutions for old problems generate new difficulties. We now have good measurements of the clustering spectrum and its evolution, and it is arguable that the discussion of section 7.4 captures the main features of the placement of galaxies with respect to the mass. However, a fairly safe bet is that one of the major results from new large surveys such as 2dF and Sloan will be a heightened appreciation of the subtleties of this problem.

Nevertheless, we should not be depressed if problems remain. Observationally, we are moving from an era of 20% – 50% accuracy in measures of large-scale structure to a future of pinpoint precision. This maturing of the subject will demand more careful analysis and rejection of some of our existing tools and habits of working. The prize for rising to this challenge will be the ability to claim a real understanding of the development of structure in the universe. We are not there yet, but there is a real prospect that the next 5–10 years may see this remarkable goal achieved.

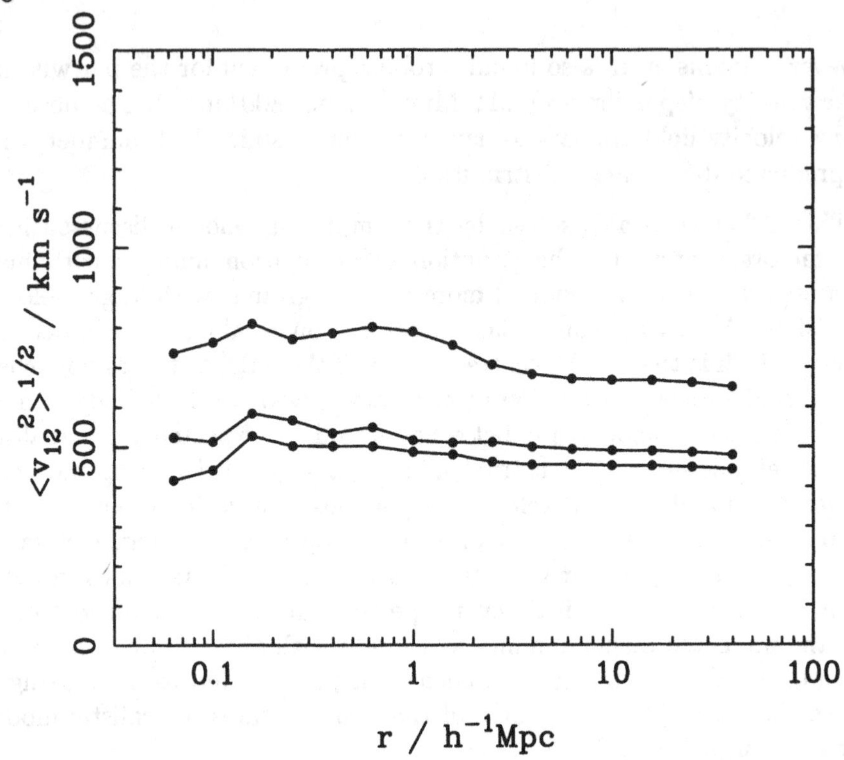

Figure 15. The line-of-sight pairwise velocity dispersion for the ΛCDM model. The top curve shows the results for all the mass; the lower pair of curves shows the predicted galaxy results, with and without assuming that one galaxy occupies the halo centre (the former case gives the lowest curve).

Acknowledgements

I thank my colleagues in the 2dF Galaxy Redshift Survey for permission to reproduce our joint results in section 5.3, and Robert Smith for the joint work reported in section 7.4.

References

Ballinger W.E., Peacock J.A., Heavens A.F., 1996, MNRAS, 282, 877
Baugh C.M., Efstathiou G., 1993, MNRAS, 265, 145
Baugh C.M., Efstathiou G., 1994, MNRAS, 267, 323
Benson A.J., Cole S., Frenk C.S., Baugh C.M., Lacey C.G., 1999, astro-ph/9903343
Bond J.R., Cole S., Efstathiou G., Kaiser N., 1991, Astrophys. J., 379, 440
Bond J.R., 1995, Phys. Rev. Lett., 74, 4369
Bunn E.F., 1995, PhD thesis, Univ. of California, Berkeley
Coles P., 1993, MNRAS, 262, 1065

Colless M., 1999, Phil. Trans. R. Soc. Lond. A, 357, 105
Davis M., Peebles P.J.E., 1983, Astrophys. J., 267, 465
Dekel A., Lahav O., 1999, Astrophys. J., 520, 24
Efstathiou G., Sutherland W., Maddox S.J., 1990, Nature, 348, 705
Eisenstein D.J., Hu W., 1998, ApJ, 496, 605
Eisenstein D.J., Zaldarriaga M., 1999, astro-ph/9912149
Eke V.R., Cole S., Frenk C.S., 1996, Mon. Not. R. Astr. Soc., 282, 263
Feldman H.A., Kaiser N., Peacock J.A., 1994, Astrophys. J., 426, 23
Folkes S., et al., 1999, Mon. Not. R. Astr. Soc., 308, 459
Gaztañaga E., Baugh C.M. 1998, Mon. Not. R. Astr. Soc., 294, 229
Ghigna S., Moore B., Governato F., Lake G., Quinn T., Stadel J., 1998, Mon. Not. R. Astr. Soc., 300, 146
Goldberg D.M., Strauss M.A., 1998, ApJ, 495, 29
Hamilton A.J.S., Kumar P., Lu E., Matthews A., 1991, Astrophys. J., 374, L1
Hamilton A.J.S., 1992, Astrophys. J., 385, L5
Hamilton A.J.S., 1997a, astro-ph/9708102
Hamilton A.J.S., 1997b, Mon. Not. R. Astr. Soc., 289, 285
Hamilton A.J.S., 1997c, Mon. Not. R. Astr. Soc., 289, 295
Huchra J.P., Geller M.J., de Lapparant V., Corwin H.G., 1990, Astrophys. J. Suppl., 72, 433
Jain B., Mo H.J., White S.D.M., 1995, Mon. Not. R. Astr. Soc., 276, L25
Jenkins A., Frenk C.S., Pearce F.R., Thomas P.A., Colberg J.M., White S.D.M., Couchman H.M.P., Peacock J.A., Efstathiou G., Nelson A.H., 1998, ApJ, 499, 20
Jing Y.P., Mo H.J., Börner G., 1998, ApJ, 494, 1
Kaiser N., 1987, MNRAS, 227, 1
Klypin A., Primack J., Holtzman J., 1996, Astrophys. J., 466, 13
Klypin A., Gottloeber S., Kravtsov A.V., Khokhlov A.M., 1999, 516, 530
Ma C.-P., 1999, Astrophys. J., 510, 32
Maddox S. Efstathiou G., Sutherland W.J., 1996, MNRAS, 283, 1227
Mann R.G., Peacock J.A., Heavens A.F., 1998, MNRAS, 293, 209
Margon B., 1999, Phil. Trans. R. Soc. Lond. A, 357, 93
Matsubara T., Szalay A.S., Landy S.D., 1999, astro-ph/9911151
Meiksin A.A., White M., Peacock J.A., MNRAS, 1999, 304, 851
Moore B., Frenk C.S., White S.D.M., 1993, Mon. Not. R. Astr. Soc., 261, 827
Moore B., Quinn T., Governato F., Stadel J., Lake G., 1999, astro-ph/9903164
Navarro J.F., Frenk C.S., White S.D.M., 1996, ApJ, 462, 563
Neyman, Scott & Shane 1953, ApJ, 117, 92
Padmanabhan N., Tegmark M., Hamilton A.J.S., 1999, astro-ph/9911421
Peacock J.A., Dodds S.J., 1994, MNRAS, 267, 1020

Peacock J.A., Dodds S.J., 1996, MNRAS, 280, L19
Peacock J.A., 1997, Mon. Not. R. Astr. Soc., 284, 885
Pearce F.R., et al., 1999, astro-ph/9905160
Peebles P.J.E., 1974, A&A, 32, 197
Pen, W.-L., 1998, ApJ, 504, 601
Saunders W., et al., 1999, astro-ph/9909191
Shectman S.A., Landy S.D., Oemler A., Tucker D.L., Lin H., Kirshner R.P., Schechter P.L., 1996, Astrophys. J., 470, 172
Sheth R.K., Tormen G., 1999, astro-ph/9901122
Strauss M.A., Willick J.A., 1995, Physics Reports, 261, 271
Sugiyama N., 1995, Astrophys. J. Suppl., 100, 281
Tegmark M., 1996, Mon. Not. R. Astr. Soc., 280, 299
Tegmark M., Taylor A.N., Heavens A.F., 1997, Astrophys. J., 480, 22
Tegmark M., Hamilton A.J.S., Strauss M.A., Vogeley M.S., Szalay A.S., 1998, Astrophys. J., 499, 555
Viana P.T., Liddle A.R., 1996, MNRAS, 281, 323
Vogeley M.S., Szalay, A.S., 1996, Astrophys. J., 465, 34
White S.D.M., Rees M., 1978, Mon. Not. R. Astr. Soc., 183, 341
White S.D.M., Efstathiou G., Frenk C.S., 1993, Mon. Not. R. Astr. Soc., 262, 1023
Yano T., Gouda N., 1999, astro-ph/9906375

LARGE SCALE STRUCTURE OF THE UNIVERSE

A.G. DOROSHKEVICH
Theoretical Astrophysics Center,
Juliane Maries Vej 30, DK-2100 Copenhagen Ø, Denmark

1. Introduction

Now it is generally recognized that on large scales the universe is isotropic and homogeneous while on smaller scales it is extremely clumpy. Indeed, detected small variations of CMB anisotropy by COBE - $\Delta T/T \sim 10^{-5}$- restrict the possible inhomogeneities on larger scales, while existence of so prominent galaxy concentrations as, for example, groups and clusters of galaxies, superclusters similar to the Perseus – Pisces or the Great Wall (de Lapparent, Geller & Huchra 1988), and voids similar to the Böotes void (Kirshner et al. 1983) illustrates the degree of inhomogeneities observed on scales $\leq 100h^{-1}$Mpc. These inhomogeneities are usually called the Large Scale Structure of the universe (LSS).

The observed large scale galaxy distribution can be roughly described as a random set of wall-like elements surrounding large under dense regions. These walls are linked to the general LSS by a random broken network of filaments. Numerous high density clouds are embedded within filaments and walls. However, the walls, filaments and high density clouds represent only the rare, most prominent, structure elements and the use of these terms allows us to emphasize the dominant type of discussed LSS elements. More correct description implies however a continuous distribution of structure elements together with their main properties, such as, the richness, shape or morphology, etc.

Now it is generally accepted that formation of the LSS is caused by gravitational instability of the homogeneous matter distribution and majority of the observed properties of the LSS can be derived from the properties of initial perturbations. But the matter concentrated within denser walls and richer clouds is partly relaxed and the internal properties of such structure elements only weakly depend on the initial perturbations. This general statement is consistent with results of numerous numerical simulations.

R.G. Crittenden and N.G. Turok (eds.), Structure Formation in the Universe, 341–350.

Observationally the LSS is usually related to the galaxy distribution but now it is commonly recognized that the dark matter (DM) is the main constituent in the universe while the baryonic component contributes only about 5 – 10 per cent to the total density, and the fraction of luminous matter concentrated within observed galaxies is even smaller. A priori, it is not obvious that the spatial distributions of DM and luminous matter are the same and, moreover, there are good observational reasons to think that these distributions are different (biased) both in small and even in large scales. On small scales the formation of galaxies is sensitive to dissipative, gas dynamics, and thermal processes, which distort the degree of concentration of DM and gaseous fractions of matter and lead to existence of extended halos of DM around galaxies. But even on large scale, when the gravitational forces dominate and it is not expected that the spatial distributions of DM and baryonic fractions of matter should differ, an essential bias between the spatial distribution of luminous matter (galaxies) and the joint distribution of DM and baryonic matter is possible. In this case it is generated by the spatial variation of the process of formation of the LSS and galaxies.

The formation of the LSS is a complicated multistep process. In the most popular models with CDM-like broad band power spectra the structure formation begins at large redshifts first at very small scales, comparable with the globular clusters and dwarf galaxies, and only at $z = 0$ it reaches scales 20 – $30h^{-1}$Mpc, typical for the observed wall-like superclusters of galaxies. During all period of structure evolution the interaction of small and large scale perturbations is important. At high redshifts it is seen as a spatial modulation of the rate of formation of high density clouds. At small redshifts it is seen as a concentration of high density clouds and filaments within richer walls and as appearance of large under dense regions. At the period of reheating this interaction can be responsible for the origin of large scale bias between spatial distributions of DM and luminous components of matter in the universe (Dekel & Silk 1986; Dekel & Rees 1987).

Now the gravitational clustering of DM is well reproduced in numerical simulations which provide us with examples of matter distribution for various redshifts and initial power spectra. They take into account, with a reasonable precision, the mentioned above interaction of small and large scale perturbations, the relaxation of compressed matter, and other factors. Nonetheless, in order to gain more clarity in our understanding of the process of structure formation the approximate theoretical description of expected structure properties should be used.

Of course, in such theoretical models we cannot hope to take into account the influence of all important factors together and therefore different theoretical approaches are required to describe, for example, the main properties of high density clouds and under dense regions. This means that with such theoretical models we can characterize statistically the influence of several factors on the formation and evolution of structure, underline the main tendencies of evolution and obtain the approximate statistical description of expected characteristics of different elements of the LSS. Further on, predictions of such statistical models can be tested and improved by confrontation with simulated and observed matter distribution what allows us to estimate the actual role of the considered and omitted factors on the evolution. In this sense, such theoretical models provide the expected fitting relations for characteristics of the LSS rather then the precise description of the evolutionary process.

To characterize the observed and simulated matter distribution, to discriminate the LSS elements and to determine their properties and interconnections appropriate statistical methods should be used. Now several such methods are proposed and applied for the description of galaxy and DM distribution in observed and simulated catalogues. The most popular methods are the correlation analysis, power spectrum and count–in–cell analysis, which are closely connected to each other and describe successfully the matter distribution on scales $\leq 10h^{-1}$Mpc. In this range of scales the influence of relaxation is especially important and the shape of measured correlation function is not directly connected with properties of the initial perturbations. The refined theoretical description of these methods and their application to observed and simulated matter distribution can be found, for example, in Pebbles (1993), Efstathiou (1996), Schaeffer (1996).

The two–point correlation function falls faster on scales greater than $\sim 10h^{-1}$Mpc and other methods and statistics should be used to characterize the LSS. In this case we are interested in the description of general geometry of the LSS, the selection of individual LSS elements and statistics of their richness, morphology and internal properties. Among the proposed methods there are the percolation analysis (Zel'dovich, Einasto & Shandarin 1982), the Minimal Spanning Tree technique (Barrow, Bhavsar & Sonoda 1985; van de Weygaert 1991), the Inertia Tensor approach (Vishniac 1986; Babul & Starkman 1992) and the Minkowski Functional and genus statistics (Bardeen et al. 1986; Gott, Melott & Dickinson 1986; Mecke, Buchert & Wagner 1994; Coles, Davies & Pearson 1996). More detailed review of some of these methods can be found in Coles (1992). Applications of these methods for the observed Las Campanas Redshift Survey (Shectman et al. 1996, hereafter LCRS) and Durham/UKST Redshift Survey (Ratcliffe et al. 1998, hereafter DURS) as well as for simulated catalogues can be found in Doroshkevich et al. (1996, 1999 a, b, c), Shandarin & Yess (1998), Sahni, Sathyaprakash & Shandarin (1997), Demiański & Doroshkevich (1999), Schmalzing et al. (1999), and Demiański et al. (1999).

2. Statistical characteristics of DM structure

For the Harrison – Zel'dovich primordial power spectrum with the CDM-like transfer function, $T(k/k_0)$, the parameters of DM structure are suitably expressed through the typical length scale, l_0, and typical dimensionless 'time', $\tau(z)$, as follows:

$$l_0^{-2} = \int_0^\infty kT^2(k/k_0)dk = m_{-2}k_0^2, \; \tau = \frac{m_{-2}\sigma_\rho(z)}{\sqrt{3m_0}}, \; k_0 = \Gamma h\mathrm{Mpc}^{-1}, \quad (2.1)$$

$$l_0 \approx \frac{6.6}{\Gamma}\sqrt{\frac{0.023}{m_{-2}}}h^{-1}\mathrm{Mpc}, \; m_{-2} = \int_0^\infty xT^2(x)dx, \; m_0 = \int_0^\infty x^3T^2(x)dx$$

where $\Gamma \approx \Omega_m h$ is the spectral parameter, σ_ρ^2 is the variance of density perturbations as described by the linear theory, and m_{-2} & m_0 are the moments of initial power spectrum. The 'time', τ, can be connected with such characteristics as the amplitude of quadrupole component of CMB anisotropy, ΔT_Q, the autocorrelation function of galaxies, $\xi(r)$, or with σ_8^2, that is the variance of mass within a sphere with radius $8h^{-1}$Mpc.

For the open CDM and flat ΛCDM models the amplitude of perturbations is expressed through the quadrupole component of temperature anisotropy (Bunn & White 1997) as follows:

$$\tau_0 = \tau_T = 0.29\left(\frac{h}{0.6}\right)^2\left(\frac{\Omega_m}{0.5}\right)^{1.8-0.2\ln(\Omega_m/0.5)}\left(\frac{\Delta T_Q}{20\mu K}\right), \quad \Omega_\Lambda = 0, \quad (2.2)$$

$$\tau_0 = \tau_T = 0.21\left(\frac{h}{0.6}\right)^2\left(\frac{\Omega_m}{0.3}\right)^{1.275-0.05\ln(\Omega_m/0.3)}\left(\frac{\Delta T_Q}{20\mu K}\right), \quad \Omega_\Lambda = 1-\Omega_m.$$

The amplitude of perturbations, τ_0, can be directly expressed through the two point autocorrelation function, $\xi(r)$, and for $\xi(r) = (r_\xi/r)^\gamma$, $r \le r_0$, we have

$$\sigma_s^2 = \int_0^r dx(1-x/r)x\xi(x)|_{r\to\infty} \approx \frac{r_0^{2-\gamma}r_\xi^\gamma}{(2-\gamma)(3-\gamma)}, \; \tau_0 = \tau_\xi = \frac{\sigma_s}{\sqrt{3}l_0} \quad (2.3)$$

where r_0 is the first zero-point of autocorrelation function. The parameter r_0 is usually found with a small precision but for $\gamma \approx 1.5 - 1.7$, $1-\gamma/2 \approx 0.25 - 0.15$ even essential variations of r_0 do not change significantly the final estimate of τ_ξ. This method is sensitive to the matter distribution at scales $r \approx 5 - 20h^{-1}$Mpc.

For the CDM-like power spectrum σ_8 links with τ_0 as

$$\tau_0 = \tau_8 \approx \frac{\sigma_8}{3}\left(\frac{\Gamma}{0.3}\right)^{0.44}\left[1+\left(\frac{\Gamma}{0.3}\right)^{1.4}\right]^{0.56}, \quad (2.4)$$

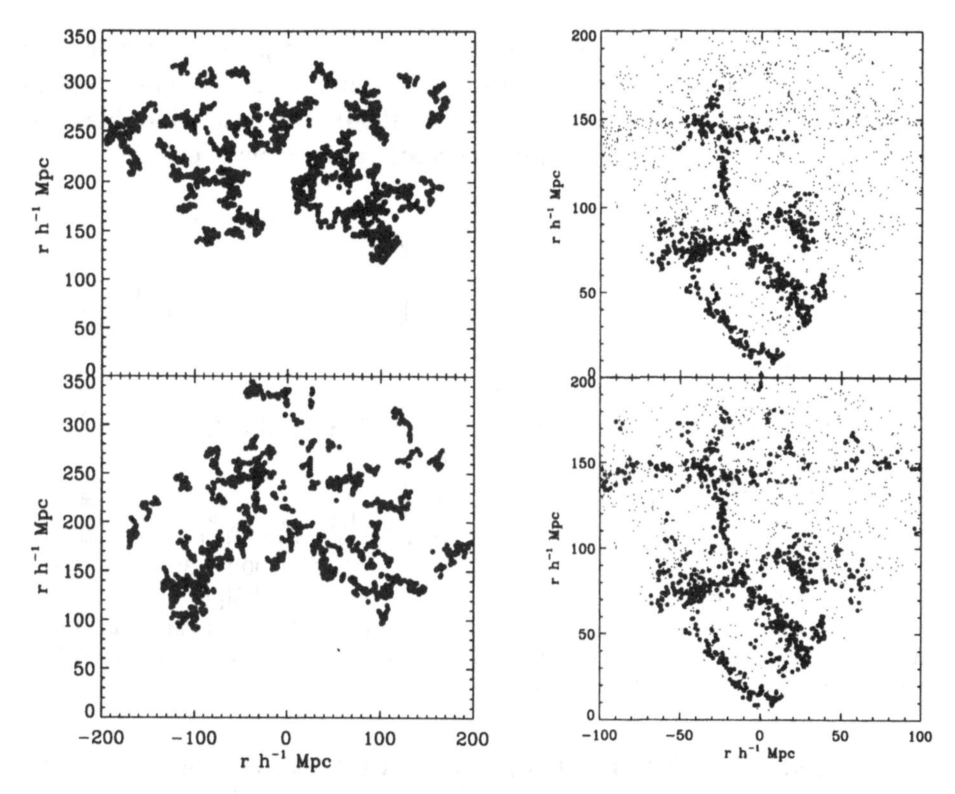

Figure 1. Left panel: Spatial distribution of the richer clusters in the northern (top panel) and southern (bottom panel) slices of LCRS. Right panel: Spatial distribution of the richer clusters in the DURS for two threshold richness of clusters.

that allows to connect various measure of initial amplitude of perturbations. However, at $z \ll 1$ the power spectrum is far from the CDM-like one and this expression cannot connect the really measured τ_0 and σ_8. σ_8 can be expressed through the two point autocorrelation function, $\xi(r)$, (Pebbles 1993).

3. Walls in observed galaxy distribution

At small redshifts rich wall-like superclusters of galaxies similar to the Great Wall (Ramella, Geller & Huchra 1991) are the most conspicuous and distinct structure elements. The representative samples of such elements can be extracted for the LCRS and DURS using the cluster (friend-of-friend) analysis. The examples of such walls in the LCRS and DURS are presented in Fig. 1. In all cases ~40 – 50% of galaxies occupy ~ 5 – 7 per cent of the volumes at overdensities of about a factor of 7 – 10 above the mean. The galaxy distributions in one slice of the LCRS along three radial and three transverse lines and in the Canada - France Redshift Survey (CFRS,

Crampton et al. 1995) are plotted in Fig. 2. Higher peaks correspond to the selected clusters. Similar 1D galaxy distribution was observed in deep pencil beam surveys (e.g., Broadhurst et al. 1990). The noticeable thickness of the peaks is caused by a random orientation of beams and walls.

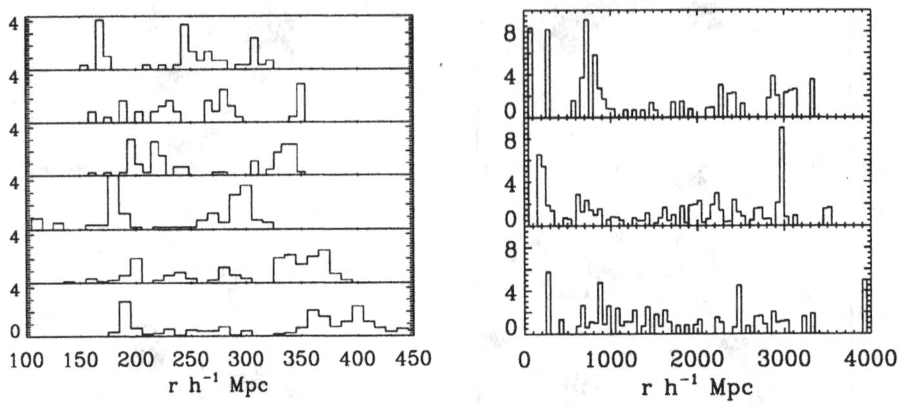

Figure 2. Left panel: The galaxy number density in the LCRS along three radial lines (3 top panels) and three transverse lines (3 bottom panels). Right panel: The galaxy number density in three fields of the CFRS.

The analysis of both the LCRS and mock catalogues (Cole et al. 1998) shown that there are essential variations of the density of galaxies and wall properties averaged over a volume $\sim 10^6 h^{-3}\mathrm{Mpc}^3$. The statistical meaning of these variations is in question but they demonstrate that the homogeneity of the universe can be significantly distorted even at so large scales. The analysis of richer and more representative catalogues such as 2dF, and SLOAN surveys, will allow to establish the actual characteristics of galaxy distribution in the universe.

4. Basic characteristics of walls

To obtain more reliable statistical characteristics of observed walls four samples of wall-like structure elements were selected in each slice of the LCRS. Selection of walls was performed with the observed sample and the influence of only usual selection effect was taken into account.

The basic averaged parameters of walls such as the wall thickness, $\langle h_w \rangle$, the wall separation, $\langle D_{sep} \rangle$, the surface density of walls, m_w, dimensionless surface density of walls, $q_w = m_w/l_0/\langle n_{gal} \rangle$, and the velocity dispersion of galaxies within walls, $\langle w_w \rangle$, were found. Some of them are summarized in Table 1. The averaging was performed separately for the three northern and three southern slices of the LCRS. Results averaged over all six slices and obtained for the DURS are also listed in Table 1.

TABLE 1. Mean characteristics of walls

	LCRS − N	*LCRS − S*	LCRS	DURS
$\langle q_w/\Gamma\rangle$	3.7 ± 0.9	$5.3 \pm 1.$	$4.5 \pm 1.$	5.3 ± 1.2
$\langle \tau_m/\sqrt{\Gamma}\rangle$	0.8 ± 0.1	0.9 ± 0.1	0.8 ± 0.1	0.9 ± 0.1
$\langle h_w\rangle$	8.8 ± 1.0	8.7 ± 0.8	8.7 ± 0.9	11.1 ± 0.6
$\langle w_w\rangle$	250 ± 27	250 ± 22	250 ± 25	320 ± 20
$\langle D_{sep}\rangle$	61 ± 6	66 ± 10	64 ± 8	46 ± 6

$\langle h_w\rangle$, $\langle D_{sep}\rangle$ & $\langle w_w\rangle$ are measured in h^{-1}Mpc and km/s, respectively.

These wall parameters can be compared with both properties of simulated walls, and with predictions of Zel'dovich theory. Such comparison confirms that the observed walls are partly virialized (quasi)stationary objects, and their life time is probably defined by the small scale clustering of matter compressed within walls. This comparison allows to estimate the fundamental cosmological parameters such as the factor Γ, the amplitude of perturbations, and the possible large scale bias between the spatial distributions of dark matter and galaxies.

The fundamental characteristic of walls is their surface density, m_w, defined as an amount of galaxies at unity of wall surface, for example, at $1h^{-2}\mathrm{Mpc}^2$. The approximate expression for the expected probability distribution function (PDF) of m_w has obtained in Demiański & Doroshkevich (1999). For Gaussian initial perturbations with the CDM-like power spectrum it can be written as follows:

$$N_m = \frac{1}{\sqrt{2\pi}\tau}\frac{1}{\sqrt{q_w}}\exp\left(-\frac{q_w}{8\tau^2}\right)\mathrm{erf}\left(\sqrt{\frac{q_w}{8\tau^2}}\right), \tag{4.1}$$

$$q_w = \frac{m_w}{l_0\langle n_{gal}\rangle}, \quad \langle q_w\rangle = 8(0.5 + 1/\pi)\tau^2 \approx 6.55\tau^2,$$

where $\langle n_{gal}\rangle$ is the mean density of galaxies in a sample, l_0 was defined by (2.1) and τ characterizes the amplitude of perturbations. The expression (4.1) connects τ with the mean surface density of walls and allows us to estimate $\tau = \tau_m$ for a measured $\langle q_w\rangle$.

The matter compressed within walls is found to be partly virialized and the PDF of *reduced* velocity dispersion of galaxies

$$\omega = \frac{w_w}{\langle w_w\rangle}\sqrt{\frac{m_w}{\langle m_w\rangle}}, \tag{4.2}$$

is close to Gaussian

$$N_\omega \approx \frac{1}{\sqrt{2\pi}\sigma_\omega}\exp\left(-\frac{(\omega - \langle\omega\rangle)^2}{2\sigma_\omega^2}\right), \ \langle\omega\rangle \approx 1, \ \sigma_\omega \approx 0.2. \tag{4.3}$$

The distribution of measured surface density of walls, $N_m(q_w/\langle q_w\rangle)$, and reduced velocity dispersion within walls, $N_\omega(\omega)$, are plotted in Fig. 3 for the LCRS together with the expected fits (4.1) and (4.3). This is an evidence in favor of Gaussianity of initial perturbations, because the PDF (4.1) was derived for such distribution of initial perturbations. The distribution of the wall separation also plotted in Fig. 3 for the LCRS is well fitted by the exponential law that is consistent with a 1D Poissonian-like wall distribution along a random straight line.

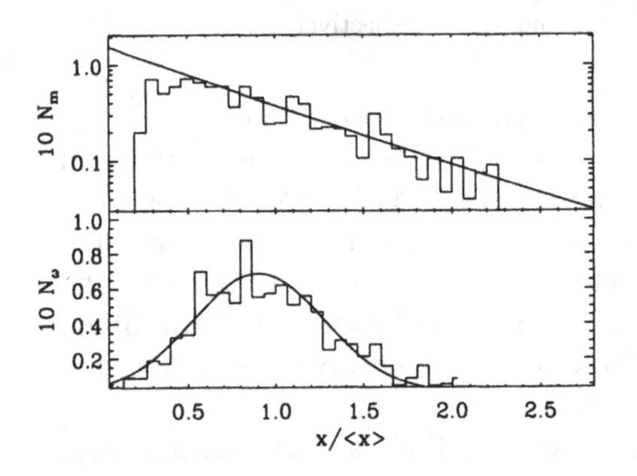

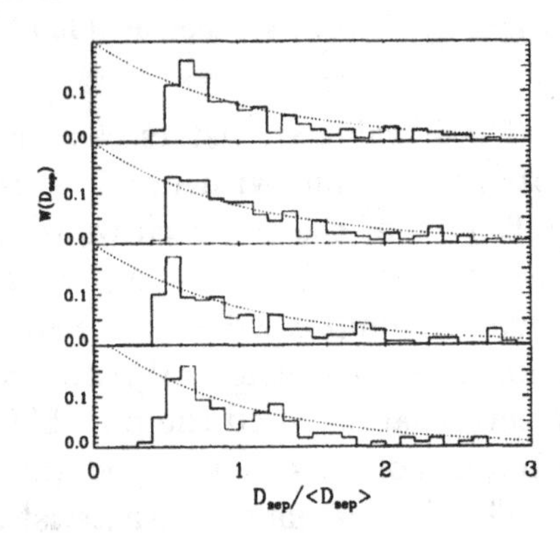

Figure 3. Left panel: PDFs $N_m(q_w/\langle q_w\rangle)$, and $N_\omega(\omega/\langle\omega\rangle)$ averaged over a sample of walls in the LCRS. Fits (4.1) and (4.3) are shown by solid lines. Right panel: The PDFs for the wall separation in the LCRS for several samples of walls. Exponential fits are plotted by dotted lines.

Results obtained for the DURS are not so reliable due to smaller number of galaxies in the catalogue. In the range of errors they agree with parameters found for the LCRS. Wall parameters listed in Table 1 are approximately consistent with those found in Doroshkevich et al. (1996, 1999 b, 1999 c) with other methods for the LCRS and DURS and with parameters compiled by Oort (1983) for the nearest superclusters of galaxies.

5. The amplitude of perturbations and the large scale bias

The measured mean surface density of walls allows us to estimate the amplitude of perturbations, τ_m, as follows:

$$\tau_m = \sqrt{\frac{\langle q_w\rangle}{6.55}} \approx (0.36 \pm 0.04)\sqrt{\frac{\Gamma}{0.2}} \tag{5.1}$$

The independent estimates of the amplitude of perturbations can be obtained from the autocorrelation function using relation (2.3). In the redshift space the two points autocorrelation functions found for the LCRS in Tucker et al. (1997), and for the DURS in Ratcliffe et al. (1998) are well consistent with that obtained for the Stromolo – APM redshift survey (Loveday et al. 1995). In all the cases we have

$$r_\xi \approx 6.3h^{-1}\text{Mpc}, \quad \gamma \approx 1.5, \quad r_0 \approx 40h^{-1}\text{Mpc}, \quad \tau_\xi \approx 0.2\left(\frac{\Gamma}{0.2}\right), \qquad (5.2)$$

with a formal precision ~10% that is well consistent with estimates (2.2).

Difference between τ_m and τ_ξ implies that the *expected* wall richness is smaller then actually measured one by a factor

$$\tau_\xi^2/\tau_m^2 \approx 0.3(\Gamma/0.2).$$

This difference can be caused by possible large scale bias between the spatial distributions of DM and luminous matter. Comparison with results obtained for mock catalogues (Cole et al. 1998) shows that it can be influenced by the insufficient representativity of observed sample of walls. More reliable estimates can be obtained with larger samples, such as the forthcoming 2dF and SLOAN surveys.

ACKNOWLEDGEMENTS

This paper was supported by Denmark's Grundforskningsfond through its support for an establishment of Theoretical Astrophysics Center.

References

1. Babul A., Starkman G.D., (1992), A quantitative measure of structure in the three-dimensional galaxy distribution - Sheets and filaments, *ApJ*, **401**, 28 – 33
2. Bardeen J.M., Bond J.R., Kaiser N., Szalay A., (1986), The statistics of peaks of Gaussian random fields, *ApJ.*, **304**, 15 – 37
3. Barrow J.D., Bhavsar S.P., Sonoda D., (1985), Minimal spanning trees, filaments and galaxy clustering, *MNRAS*, **216**, 17 – 35
4. Broadhurst T.J., Ellis R.S., Koo D.C., Szalay A.S., (1990), Large-scale distribution of galaxies at the Galactic poles, *Nature*, **343**, 726–730.
5. Bunn E.F., and White M., (1997), The 4 Year COBE Normalization and Large-Scale Structure *ApJ.*, **480**, 6 – 12.
6. Cole S., Hatton, S., Weinberg D.H. & Frenk C.S., (1998), Mock 2dF and SDSS galaxy redshift surveys, *MNRAS*, **300**, 945 – 956
7. Coles P., (1992), in Feigelson E.D., Babu G.J., eds. Statistical Challenges in Modern Astronomy, Springer, New-York, p.57–69
8. Coles P., Davies A.G., & Pearson R.C., (1996), Quantifying the topology of large-scale structure, *MNRAS*, **281**, 1375-1382
9. Crampton D., Le Fevre O., Lilly S.J., & Hammer F., (1995), The Canada-France Redshift Survey. V. Global Properties of the Sample *ApJ.*, **455**, 96 – 108.
10. de Lapparent V., Geller M.J., Huchra J.P., (1988), The mean density and two-point correlation function for the CfA redshift survey slices, *ApJ*, **332**, 44 – 51
11. Dekel A. & Silk J., (1986), Dwarf galaxies, cold dark matter and biased galaxy formation, *ApJ.*, **303**, 39 – 46.
12. Dekel A. & Rees M.,J., (1987), Physical mechanisms for biased galaxy formation, *Nature*, **326**, 455 – 463.

13. Demiański M. & Doroshkevich A., (1999), Statistical characteristics of formation and evolution of structure in the Universe, *MNRAS*, **306**, 779 - 791
14. Demiański M. & Doroshkevich A., Müller V., & Turchaninov V.I., (1999), Statistical characteristics of simulated large scale matter distribution, *MNRAS*, submitted
15. Doroshkevich, A.G., Tucker, D.L., Oemler, A.A., Kirshner, R.P., Lin, H, Shectman, S.A., Landy, S.D., Fong, R., (1996), Large- and superlarge-scale structure in the LCRS, *MNRAS*, **283**, 1281 - 1299
16. Doroshkevich A.G., Müller V., Retzlaff J., & Turchaninov V.I., (1999a), Superlarge-scale structure in N-body simulations, *MNRAS*, **306**, 575 - 588.
17. Doroshkevich A.G., Tucker D.L., Fong R., Turchaninov V., Lin H, (1999b), Large Scale Galaxy Distribution in the LCRS, *MNRAS*, submitted.
18. Doroshkevich A., Fong R., McCracken G., Ratcliffe A., Shanks T., & Turchaninov V.I., (1999c), Superlarge-Scale Structure in the DURS, *MNRAS*, submitted.
19. Efstathiou G., (1996), in Schaeffer R., Silk J., Spiro M., & Zinn-Justin J., eds., Les Houches, Session LX, Cosmology and Large Scale Structure, Elsvier Science B.V
20. Gott J.R., Melott A.L., Dickinson M., (1986), The sponge-like topology of large-scale structure in the universe, *ApJ.*, **306**, 341 - 353
21. Kirshner R.P., Oemler A.J., Schechter P.L., Shectman S.A., (1983), A deep survey of galaxies, *Astron.J.*, **88**, 1285–1297
22. Loveday J., Maddox S.J., Efstathiou G., Peterson B.A., (1995), The Stromolo-APM redshift survey, *MNRAS*, **442**, 457-468
23. Mecke K.R., Buchert T., Wagner H., (1994), Robust morphological measures for large-scale structure in the Universe, *A&A*, **288**, 697 - 707
24. Oort J.H., (1983), Superclusters, *ARA&A*, **21**, 373 - 401
25. Pearson R.C., Coles P., (1995), Quantifying the geometry of large-scale structure, *MNRAS*, **272**, 231 - 244.
26. Pebbles, P.J.E., (1993), Principles of Physical Cosmology, Princeton University Press.
27. Ramella M., Geller M.J., Huchra J.P., 1992, The distribution of galaxies within the Great Wall, *ApJ*, **384**, 396 - 403
28. Ratcliffe A., Shanks T., Parker Q.A. & Fong R., (1998), Large-scale structure via the two-point correlation function, *MNRAS*, **296**, 173 - 190
29. Sahni V., Sathyaprakash B.S., Shandarin S., (1997), Probing Large-Scale Structure Using Percolation and Genus Curves, *ApJ.*, **476**, L1 - L5.
30. Schaeffer R., (1996), in Schaeffer R., Silk J., Spiro M., & Zinn-Justin J., eds., Les Houches, Session LX, Cosmology and Large Scale Structure, Elsvier Science B.V.
31. Schmalzing J., Gottlöber S., Klypin A., Kravtsov A., (1999), Quantifying the evolution of higher-order clustering, *MNRAS*, in press, astro-ph/9906475.
32. Shandarin S., Yess C., (1998), Detection of Network Structure in the Las Campanas Redshift Survey, *ApJ.*, **505**, 12 - 17.
33. Shectman S.A., Landy S.D., Oemler A., Tucker D.L., Lin H., Kirshner R.P., Schechter P.L., (1996), The Las Campanas Redshift Survey, *ApJ*, **470**, 172 - 191
34. Tucker D.L., Oemler A.,Jr., Kirshner R.P., Lin H., Shectman S.A., Landy S.D., Schechter P.L., Müller V., Gottlöber S., Einasto J., (1997), The LCRS galaxy-galaxy correlation function, *MNRAS*, **285**, L5-L9
35. van de Weygaert R., (1991), Ph.D. Thesis, University of Leiden
36. Vishniac E.T., (1986), in Kolb E.W., Turner M.S., Lindley D., Olive K., Seckel D., eds., Inner Space/ Outer Space. University Chicago Press, Chicago, p. 190 - 199
37. Zel'dovich Ya.B., Einasto J., Shandarin S.F., (1982), Giant voids in the universe, *Nature*, **300**, 407 - 411

GRAVITATIONAL LENSING FOR COSMOLOGY

FRANCIS BERNARDEAU
Service de Physique Théorique, C.E. de Saclay
F-91191 Gif-sur-Yvette Cedex, France

1. Introduction

The understanding of the formation of the large-scale structures of the Universe is a fundamental problem of cosmology because of its close links to the physics of the early Universe. The investigations that started in the mid-eighties that aimed at mapping the large-scale mass distribution in the Universe have been done mainly with galaxy surveys (either angular or 3D redshift surveys). Important results have been obtained with those surveys on the structuration of the Universe up to 100 h^{-1}Mpc scale. Further extensions of these results are expected from the next generation of surveys that are should be completed soon (2dF and SDSS). However, as our ambitions in falsifying our theories of structure formation are progressing, we are more and more hampered by the so called "bias problem": galaxies may not be faithful tracers of the underlying matter field.

One way to circumvent this problem[1] to map the large-scale structures with the use gravitational lens effects. The observed quantities are indeed directly proportional to the total mass fluctuations no matter what the light distribution is.

The aim of these lectures is to discuss the physics of gravitational lensing with a focus on the application of gravitational lensing to cosmology. More general considerations on lensing can be found in,

- "Gravitation" by Misner, Thorne and Wheeler (1973) for General Relativity and in particular for the presentation of the geometric optics.
- "Large-scale Structures of the Universe" by Peebles (1993) for the description of the large-scale structures of the Universe.
- "Gravitational lenses" by Schneider, Ehlers & Falco (1992) for a general (but rather mathematical) exhaustive presentation of the lens

[1] not to mention the CMB anisotropies, the physics of which is well understood

R.G. Crittenden and N.G. Turok (eds.), Structure Formation in the Universe, 351–370.

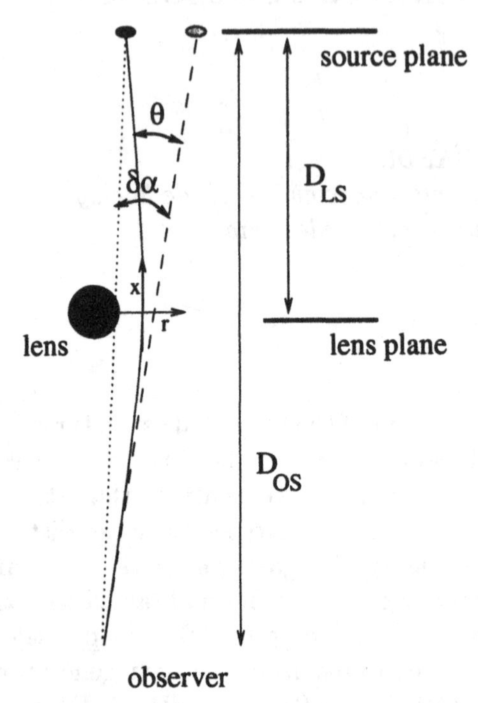

Figure 1. The geometrical relationship between the deflection angle θ and the displacement angle $\delta\alpha$.

physics.

Recent reviews and lecture notes can also be fruitfully read,

- Mellier (astro-ph/9812172) to appear in ARAA.
- Mellier (astro-ph/9901116) and Bernardeau (astro-ph/9901117) to appear in the proceedings of the Cargèse summer school "Theoretical and Observational Cosmology", August 1998, ed. M. Lachièze-Rey

The content of these notes is the following. In the first section I describe of the basic mechanisms of gravitational lenses, techniques and approximations that are usually employed. The last two sections are devoted to cosmological applications, from galaxy cluster to the mapping of the large-scale structures of the Universe, and their interests.

2. Physical mechanisms

The physical mechanisms of gravitational lenses are well known since the foundation of General Relativity. Any mass concentration is going to deflect photons that are passing by with a fraction angle per unit length, $\delta\theta/\delta s$, given by

$$\frac{\delta\vec{\theta}}{\delta s} = -2\,\vec{\nabla}_x \frac{\phi}{c^2} \tag{1}$$

where the spatial derivative is taken in a plane that is orthogonal to the photon trajectory and ϕ is the Newtonian potential[2].

2.1. BORN APPROXIMATION AND THIN LENS APPROXIMATION

In practice, the total deflection angle is at most about an arcmin. This is the case for the most massive galaxy clusters. It implies that in the subsequent calculations it is possible to ignore the bending of the trajectories and calculate the lens effects as if the trajectories were straight lines. This is the Born approximation.

Eventually, one can do another approximation by noting that in general the deflection takes place along a very small fraction of the trajectory between the sources and the observer. One can then assume that the lens effect is instantaneous and is produced through the crossing of a plane, the lens plane. This is the thin lens approximation.

2.2. THE INDUCED DISPLACEMENT

The direct consequence of this bending is a displacement of the apparent position of the background objects. This apparent displacement depends on the distance of the source plane, D_{OS}, and on the distance between the lens plane and the source plane D_{LS}. More precisely we have (see Fig. 1),

$$\vec{\beta} = \vec{\alpha} - \frac{2}{c^2}\frac{D_{LS}}{D_{OS}\,D_{OL}}\,\vec{\nabla}_\alpha\left(\int \mathrm{d}s\;\phi(s,\alpha)\right) \tag{2}$$

where $\vec{\alpha}$ is the angular position (it is a thus 2D vector) in the image plane, $\vec{\beta}$ is the position in the source plane. The gradient is taken here with respect to the angular position (this is why a D_{OL} factor appears). The total deflection is obtained by an integration along the line of sight, assuming the lens is thin.

2.3. THE AMPLIFICATION MATRIX

Actually the image of a background object is not only displaced. It is also deformed. This effect is due to the variations of the displacement field with respect to the apparent position. These variations induce a change of both the size and shape of the background objects. To quantify this effect one

[2]We will see in section 2.4 what is its meaning in a cosmological context.

can compute the amplification matrix $\mathcal{A}$ which describes the linear change between the source plane and the image plane,

$$\mathcal{A} = \left(\frac{\partial \alpha_i}{\partial \beta_j}\right). \tag{3}$$

Its inverse, $\mathcal{A}^{-1}$, is actually directly calculable in terms of the gravitational potential. It is given by the derivatives of the displacement with respect to the apparent position,

$$\mathcal{A}^{-1} \equiv \frac{\partial \beta_i}{\partial \alpha_j} = \delta_{ij} - 2\frac{D_{LS}}{D_{OS}\,D_{OL}}\phi_{,ij}. \tag{4}$$

2.4. GRAVITATIONAL LENSES IN COSMOLOGY

The extension of the lens equations to a cosmological context raises some technical difficulties because the background in which the objects are embedded is not flat. The aim of the following paragraphs is to raise possible ambiguities. Justifications of the results can be found in the textbooks quoted in the beginning.

2.4.1. *The angular distances*

The first remark is that the distances to be used are the angular distances. If one considers an object of proper size l at redshift z seen under an angle α, then by definition,

$$\alpha = \frac{l}{\mathcal{D}_0(z)}, \tag{5}$$

where $\mathcal{D}_0$ is the angular distance. This is the true distance at which this object would be in an Euclidean metric. At low redshift the curvature of the Universe has no effects and we always have,

$$\mathcal{D}_0(z) \approx \frac{c}{H_0} z. \tag{6}$$

In general however the relation between D_0 and z depends on the cosmological parameters. It shows a significant dependence, in particular with λ_0, at redshift beyond unity.

2.4.2. *Geometric optics in a weakly inhomogeneous Universe*

The second ambiguities is the source term for the deflection angle. The gravitational potential to be used is actually the cosmological gravitational potential ϕ defined with the Poisson equation,

$$\Delta\phi = 4\pi\, G\,\overline{\rho}\,a^2\,\delta_{\mathrm{mass}}. \tag{7}$$

where the Laplacian is taken with respect of the comoving angular distances. Sachs (1961) in particular gave a sound derivation of the equations of geometric optics, that justifies this point.

3. Galaxy clusters as gravitational lenses

The study of galaxy clusters has become a very active field since the discovery of the first gravitational arc by Soucail et al. (1988) in Abell cluster A370. Galaxy clusters give the most dramatic example of gravitational lens effects in a cosmological context.

3.1. THE ISOTHERMAL PROFILE

For an isothermal profile we assume that the local density $\rho(r)$ behaves like,

$$\rho(r) = \rho_0 \left(\frac{r}{r_0}\right)^{-2}. \tag{8}$$

With such a density profile the total mass is not finite. So this is not a realistic description but it is a good starting point for the central part of clusters. It is actually more convenient to parameterize the depth of a potential well with the velocity dispersion it induces. The velocity dispersion is due to the random velocities that particles acquire when they reach a sort of thermal equilibrium. Such a dispersion is in principle measurable with the observed galaxy velocities along the line of sight. The velocity dispersion is related to the mass $M(<r)$ of the potential well that is included within a radius r,

$$\sigma^2(r) \sim \frac{G\,M(<r)}{r}. \tag{9}$$

In case of a isothermal profile, the velocity dispersion is *independent* of the radius and we have

$$\sigma^2 = 2\pi\, G\, \rho_0\, r_0^2. \tag{10}$$

The integrated potential along the line of sight is given by,

$$\varphi(r) = 2\pi\, \sigma^2\, r. \tag{11}$$

As a consequence the amplitude of the displacement is independent of the distance to the cluster center and

$$\vec{\beta} = \vec{\alpha} - \frac{4\pi}{c^2}\frac{D_{LS}}{D_{OS}}\sigma^2\frac{\vec{\alpha}}{\alpha}. \tag{12}$$

For $\vec{\beta} = 0$ there is a degenerate solution for,

$$\alpha = R_E = \frac{4\pi}{c^2}\frac{D_{LS}}{D_{OS}}\sigma^2, \tag{13}$$

corresponding to the position of the so-called Einstein ring. The value of R_E depends both on the velocity dispersion and on the angular distances. The number of images depends in this case on the value of the impact parameter. If it is too large (i.e. larger than R_E) then each background object has only one image. For a galaxy cluster of a typical velocity dispersion of 500 km/s, and for a source plane situated at twice the distance of the lens, the size of the Einstein ring is about 0.5 arcmin. It is interesting to note that the size of the Einstein ring is directly proportional to the square of velocity dispersion (in units of c^2) and to the ratio D_{LS}/D_{OS}.

In this case, the amplification matrix reads,

$$\mathcal{A}^{-1} = \begin{pmatrix} 1 & 0 \\ 0 & 1 - \frac{1}{x} \end{pmatrix}, \tag{14}$$

where we have,

$$x = \frac{r}{R_E}. \tag{15}$$

As a result the amplification is given by,

$$\mu = \frac{x}{1 - x}. \tag{16}$$

Once again the amplification becomes infinite when $x \to 1$ that is, close to the critical line.

3.2. THE CRITICAL LINES FOR A SPHERICALLY SYMMETRIC MASS DISTRIBUTION

In this paragraph I am interested in a more general situation where the only assumption is that we have a spherically symmetric profile. The displacement is then given by the radial derivative of the projected potential,

$$\vec{\beta} = \vec{\alpha} - \frac{\mathrm{d}\varphi}{\mathrm{d}r}\frac{\vec{\alpha}}{\alpha}. \tag{17}$$

It is interesting to visualize this relation with a graphic representation. This is proposed in Fig. 2. The number and position of the images of a given background object are given by the number of intersection points between the curve and a straight line of slope unity. This is a direct consequence of the relation,

$$b - a = -\frac{\mathrm{d}\varphi(a)}{\mathrm{d}a} \tag{18}$$

when the potential is computed along a given axis that crosses the cluster through the center and a and b are the abscissa on this axis of one given object in respectively the source and the image plane.

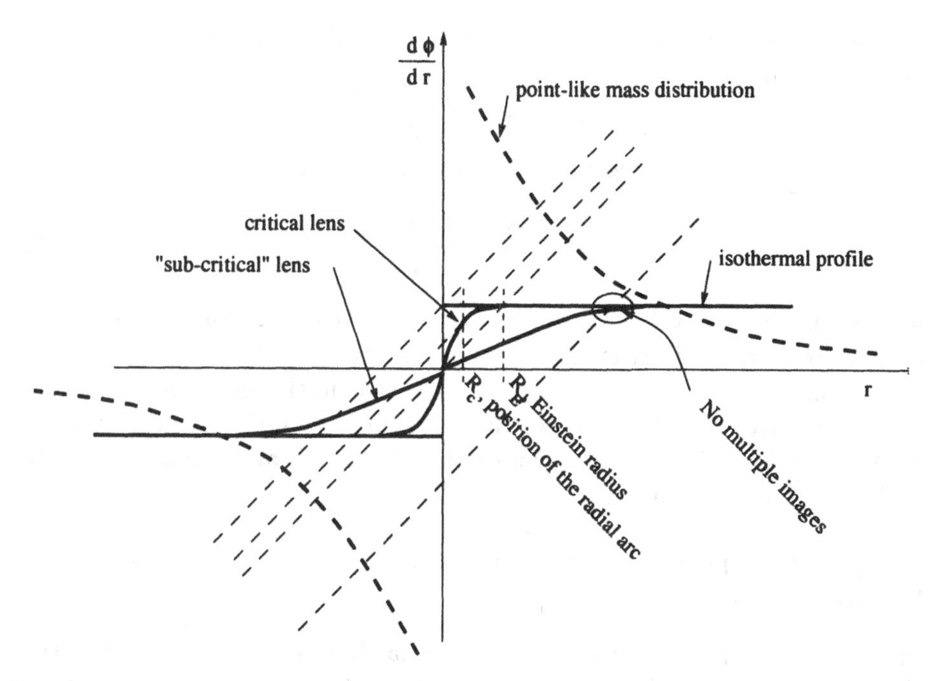

Figure 2. Graphical determination of the position and number of images from the shape of the potential.

The interesting quantity is also the amplification matrix that indicates the position of the critical lines. In general this matrix reads,

$$\mathcal{A}^{-1} = \begin{pmatrix} 1 - \frac{\partial^2 \varphi}{\partial r^2} & 0 \\ 0 & 1 - \frac{1}{r}\frac{\partial \varphi}{\partial r} \end{pmatrix}, \tag{19}$$

when it is written in the basis $(\vec{e}_r, \vec{e}_\theta)$. Then the amplification is infinite in two cases, when

$$\frac{\partial^2 \varphi}{\partial r^2} = 1 \ \text{ or } \ \frac{1}{r}\frac{\partial \varphi}{\partial r} = 1. \tag{20}$$

The second eigenvalue corresponds to the same case as for a singular isothermal profile. At this particular position the source forms an Einstein ring. The first eigenvalue, however, is associated with an eigenvector that is along the first direction, that is along the radial direction. It means that the "arc" which is thus formed is radial. It graphically corresponds to the case of two merging roots. It is therefore directly associated with the behavior of the potential near the origin.

3.3. THE ISOTHERMAL PROFILE WITH A CORE RADIUS

Let us consider a simple case where the projected potential is made regular near the origin,

$$\varphi(r) = \varphi_0 \sqrt{1 + (r/r_c)^2}. \tag{21}$$

The constant φ_0 is related to the velocity dispersion with

$$\varphi_0 = \frac{4\pi\sigma^2}{c^2} \frac{D_{LS}\, D_{OL}}{D_{OS}\, r_c} \tag{22}$$

where σ is here the velocity dispersion at a radius much larger that r_c (the velocity dispersion decreases to zero at the origin in this model). This is a more realistic case. It is interesting to note that in this case the potential is not necessarily critical (there may be no region of multiple images, see Fig. 2). When it is critical the discovery of a radial arc is an extremely precious indication for the value of the core radius.

3.4. CRITICAL LINES AND CAUSTIC LINES IN REALISTIC MASS DISTRIBUTIONS

In realistic reconstructions of lens potential however, it is very rare that the lens is circular. Most of the time the mass distribution of the lens is much more complicated. It induces complex features and series of multiple images.

The simplest assumption beyond the spherically symmetric models is to introduce an ellipticity ϵ in the mass distribution (Kassiola & Kovner 1993),

$$\varphi = \varphi_0 \sqrt{1 + r_{\rm em}^2/r_c^2} \;\text{ with }\; r_{\rm em}^2 = \frac{x^2}{(1-\epsilon)^2} + \frac{y^2}{(1+\epsilon)^2}. \tag{23}$$

To understand the physics it induces one should introduce the caustics and critical lines. The *critical lines* are the location on the image plane of the points of infinite magnification. The *caustic lines* are the location of these points on the source plane. These points are determined by the lines on which $\det(\mathcal{A}^{-1}) = 0$. It means that arcs are along the critical lines and that they are produced by background galaxies that happen to be located on the caustics. On Fig. 3 one can see the shapes of the critical lines and caustics for different depth of the potential (23) (or equivalently for different positions of the source plane).

Eventually the reconstruction of galaxy cluster mass maps requires the use of more complicated models and it can be necessary to perform non-parametric mass reconstructions. Recent results have been obtained by AbdelSalam et al. (1998) for few clusters.

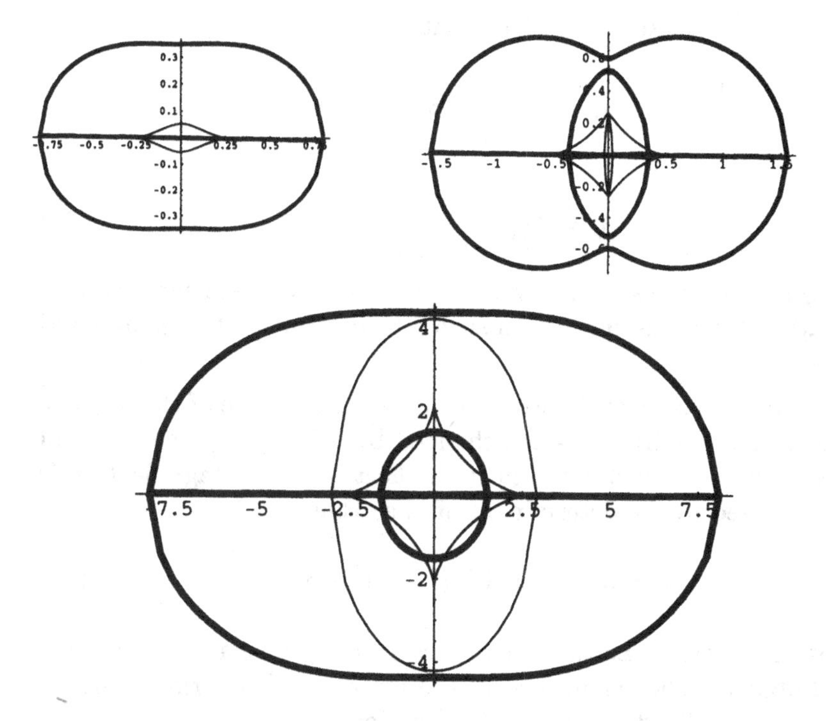

Figure 3. Shape of the caustic lines (thin lines) and critical lines (thick lines) for an elliptic potential and for different values of the central potential.

4. The weak lensing regime

In this section the possibility of using the lens effects to probe the large-scale structures of the Universe is considered. The difficulty is here that the distortion induced by the lenses can be very small. The projected potential should then be reconstructed with a statistical analysis on the deformations measured on a large amount of background objects.

4.1. THE MATHEMATICAL DESCRIPTION OF THE WEAK LENSING REGIME

In the weak lensing regime the deformation of the background objects can be described by the amplification matrix, the components of its inverse are in general written,

$$\mathcal{A}^{-1} = \begin{pmatrix} 1-\kappa-\gamma_1 & -\gamma_2 \\ -\gamma_2 & 1-\kappa+\gamma_1 \end{pmatrix}, \tag{24}$$

taking advantage of the fact that it is a symmetric matrix. The components of this matrix are expressed in terms of the convergence, κ, (a scalar field) and the shear, γ (a pseudo vector field) with

$$\kappa = \frac{1}{2}\nabla^2\psi; \quad \gamma_1 = \frac{1}{2}(\psi_{,11} - \psi_{,22}) \ ; \quad \gamma_2 = \psi_{,12}, \tag{25}$$

with

$$\psi = 2\frac{D_{LS}}{D_{OS}\,D_{OL}}\phi. \tag{26}$$

The convergence describes the linear change of size and the shear describes the deformation. The consequences of such a transform can be decomposed in two aspects:

- The magnification effect. Lenses induce a change of size of the objects. As the surface brightness is not changed by this effect, the change of surface induces a direct magnification effect, μ. This magnification is directly related to the determinant of $\mathcal{A}$ so that,

$$\mu = \det(\mathcal{A}) = 1/\left[(1-\kappa)^2 - \gamma^2\right]. \tag{27}$$

- The distortion effect. Lenses also induce a change of shape of the background objects. The eigenvalues of the matrix $\mathcal{A}^{-1}$ determine the direction and amplitude of such a deformation.

4.2. THE GALAXY SHAPE MATRICES TO MEASURE THE DISTORTION FIELD

The distortion effects change the shape of the background objects. The objects appear elongated along the eigenvalues of the amplification matrix. When background objects are only moderately extended (this excludes the case of arcs), their shapes can be described by the matrix,

$$\mathcal{S} \equiv \int \mathrm{d}^2\theta\,\theta_i\,\theta_j\,\mathcal{I}(\theta). \tag{28}$$

It is easy to relate the shape matrix in the source plane to the one in the image plane. This is obtained by a simple change of variable that uses the fact that the surface brightness of the objects is not changed. It implies

$$\mathcal{S}^S = \mathcal{A}^{-1}\cdot\mathcal{S}^I\cdot\mathcal{A}^{-1}. \tag{29}$$

By averaging over the shape matrices in the source plane, assuming the intrinsic shape fluctuations are not correlated, one can eventually get[3] the

[3]See the Mellier references for a more detailed presentation of the data analysis techniques.

value of $\mathcal{A}^{-1}/\sqrt{\det \mathcal{A}^{-1}}$. The combination we have access to is totally independent of the amplification factor. As a consequence, the quantity which is measurable is the reduced shear field,

$$\mathbf{g} = \vec{\gamma}/(1-\kappa). \tag{30}$$

This quantity identifies with γ only in the limit of very weak lensing (i.e. when $\kappa \ll 1$).

4.3. THE CONSTRUCTION OF THE PROJECTED MASS DENSITY

The elaboration of methods for reconstructing mass maps from distortion fields is not a trivial issue. In a pioneering paper, Kaiser & Squires (1993) showed that this is indeed possible, at least in the weak lensing regime. It is indeed not too difficult to show that (in the single lens approximation),

$$\nabla\kappa = -\begin{pmatrix} \partial_1 & \partial_2 \\ -\partial_2 & \partial_1 \end{pmatrix} \cdot \begin{pmatrix} \gamma_1 \\ \gamma_2 \end{pmatrix} \tag{31}$$

when $\kappa \ll 1$ and $\gamma_i \ll 1$. By simple Fourier transforms it is then possible to recover κ from a distortion map. An extension of this relation has been given by Kaiser (1995) beyond the linear approximation that relates the reduced shear to κ.

The first reconstruction of a mass map of a galaxy cluster has been done on MS1224 by Fahlman et al. (1994). Many other reconstructions have now been done or are under preparation (see Mellier 1998).

5. The weak lensing as a probe of the Large-Scale Structures

5.1. THE LARGE-SCALE STRUCTURES

The idea of probing the large-scale structures with gravitational effects is very attractive. The gravitational survey offers indeed a unique way of probing the mass concentrations in the Universe since, contrary to galaxy survey, it can provide us with mass maps of the Universe that are free of any bias. Its interpretations in terms of cosmological parameters would then be straightforward and independent on hypothesis on galaxy or cluster formation theories.

The physical mechanisms are the same in the context of large-scale structures and the source for the gravitational effects is the gravitational potential ϕ given by Eq. (7). It is important to remember that the source term of this equation is $\overline{\rho}(t)\,\delta_{\text{mass}}(t,\mathbf{x})$. The density contrast $\delta_{\text{mass}}(\mathbf{x})$ is usually written in terms of its Fourier transforms,

$$\delta_{\text{mass}}(\mathbf{x}) = \int \frac{\mathrm{d}^3\mathbf{k}}{(2\pi)^{3/2}}\,\delta_{\text{mass}}(\mathbf{k})\,D_+(t)\,\exp(\mathrm{i}\mathbf{k}.\mathbf{x}). \tag{32}$$

The density field is then entirely defined by the statistical properties[4] of the random variables $\delta_{\rm mass}(\mathbf{k})$. At large enough scale the field is (almost) Gaussian (at least for Gaussian initial conditions which is the case in inflationary scenarios). The amplitude of the fluctuations grows with time in linear theory in a known way $D_+(t)$. This function is simply proportional to the expansion factor for an Einstein-de Sitter Universe.

The variables are then entirely determined by the power spectrum $P(k)$,

$$\langle \delta_{\rm mass}(\mathbf{k})\delta_{\rm mass}(\mathbf{k}')\rangle = \delta_{\rm Dirac}(\mathbf{k}+\mathbf{k}')\,P(k). \tag{33}$$

The cosmological model is therefore completely determined by the power spectrum, Ω and λ as long as the the dark matter distribution is concerned.

5.2. THE RELATION BETWEEN THE LOCAL CONVERGENCE AND THE LOCAL DENSITY CONTRAST

The relation between the convergence and the local density contrasts in the local universe can be derived easily from Eq. (7),

$$\kappa(\gamma) = \frac{3}{2}\Omega_0 \int {\rm d}z_s\, n(z_s) \int {\rm d}\chi\, \frac{\mathcal{D}(\chi_s,\chi)\,\mathcal{D}(\chi)}{\mathcal{D}(\chi_s)}\,\delta_{\rm mass}(\chi,\gamma)\,(1+z). \tag{34}$$

In this relation the redshift distribution of the sources in normalized so that,

$$\int {\rm d}z_s\, n(z_s) = 1. \tag{35}$$

All the distances are expressed in units of c/H_0. The relation (34) is then totally dimensionless.

5.3. THE EFFICIENCY FUNCTION

It is convenient to define the efficiency function, $w(z)$, with

$$w(z) = \frac{3}{2}\Omega_0 \int {\rm d}z_s\, n(z_s)\, \frac{\mathcal{D}(\chi_s,\chi)\,\mathcal{D}(\chi)}{\mathcal{D}(\chi_s)}\,(1+z) \tag{36}$$

On Fig. 5 one can see the shape of the efficiency function for different hypothesis for the source distribution. Obviously the further the sources are the more numerous the lenses that can be detected are, and the larger the effect is.

[4]For a detailed introduction to large-scale structure formation theory and phenomenology see lecture notes of Bertschinger, 1996

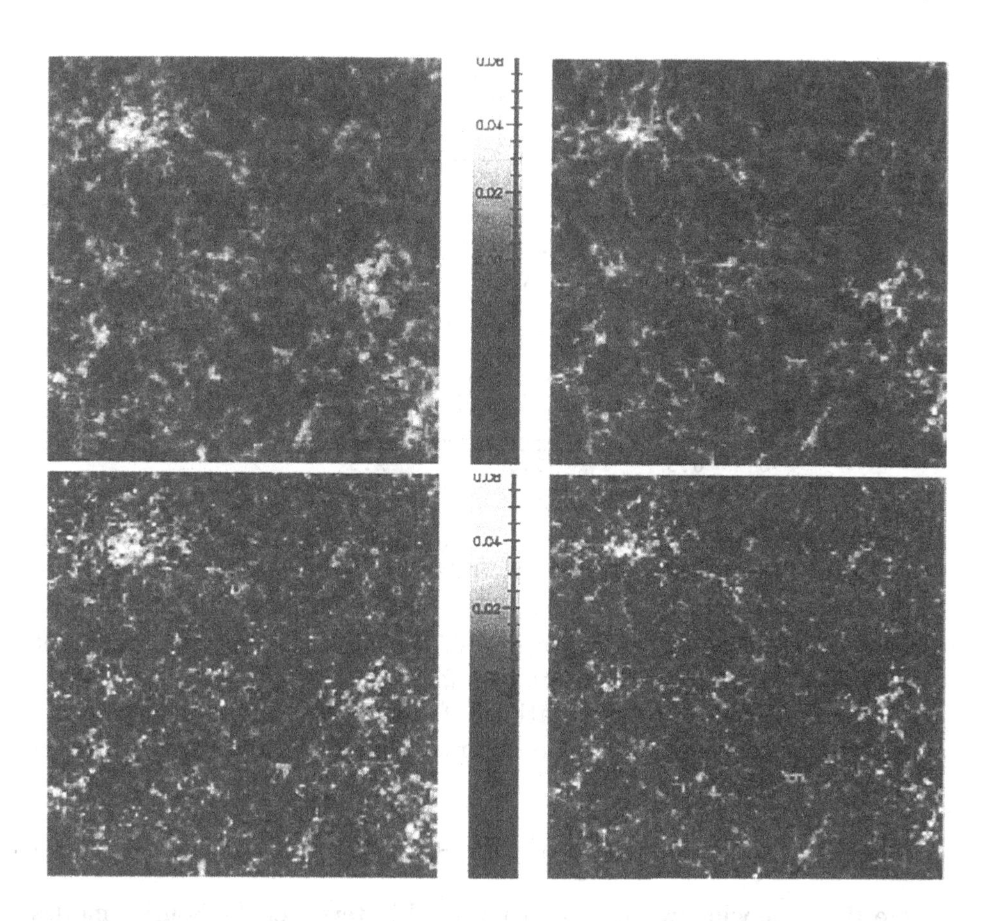

Figure 4. Example of reconstructions of projected mass maps. The top panels show the initial noise-free κ map for either $\Omega = 1$ (left panel) or $\Omega = 0.3$ (right panel) with the same underlying linear random field and the same rms distortion. The bottom panels show the reconstructed κ maps with noise included in the shear maps. The maps cover a total area of 25 degrees^2. Each pixel has an angular size of 2.5 arcmin^2 and averages the shear signal expected from deep CCD exposures (about 30 galaxy/arcmin^2). The sources are assumed to be all at redshift unity and to have a realistic intrinsic ellipticity distribution. Such a survey is easily accessible to MEGACAM at CFHT. The precision with which the images can be reconstructed and the striking differences between the two cosmological models demonstrate the great interest such a survey would have.

5.4. THE AMPLITUDE OF THE CONVERGENCE FLUCTUATIONS

From this equation it is obvious that the amplitude of κ is directly proportional of the density fluctuation amplitude and that the two point correlation function of the κ field is related to the shape of the density power spectrum. We usually parameterize its amplitude with σ_8 which is the r.m.s. of the density contrast in a sphere of radius $8\,h^{-1}$Mpc. The relation (34) also shows that κ depends on the cosmological parameters. There is a sig-

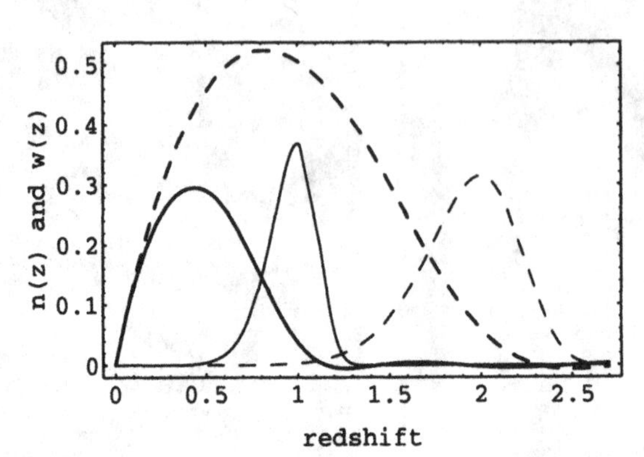

Figure 5. shape of the efficiency function, $w(z)$ (thick lines), for two different hypothesis on the shape of the redshift distribution of the sources (thin lines).

nificant dependence in the expression of the distances but the dominant contribution comes from the overall Ω_0 factor.

The amplitude of the fluctuations of κ depends on the angular scale at which the convergence map is filtered. We can introduce the filtered convergence κ_θ, with

$$\kappa_\theta(\gamma) = \int \mathrm{d}^2\gamma' \, \kappa(\gamma + \gamma') \, W_\theta(\gamma'), \tag{37}$$

where W is the window function. Expressed in terms of the Fourier modes it reads,

$$\begin{aligned} \kappa_\theta(\gamma) = & \int \mathrm{d}\chi \, w(z) \int \frac{\mathrm{d}^2\mathbf{k}_\perp}{2\pi} \frac{\mathrm{d}k_r}{(2\pi)^{1/2}} \, \delta(\mathbf{k}) \, D_+(z) \, \times \\ & \exp\left[\mathrm{i}\mathbf{k}_\mathrm{r}\chi(\mathrm{z}) + \mathrm{i}\mathbf{k}_\perp.\gamma\mathcal{D}(\mathrm{z}0)\right] \, W\left[k_\perp\theta\mathcal{D}(z)\right], \end{aligned} \tag{38}$$

where the wave vector $\mathbf{k}$ has been decomposed in two parts k_r and $\mathbf{k}_\perp$ that are respectively along the line of sight and perpendicular. The computation of the r.m.s. of κ_θ is analytic in the small angle approximation only. It eventually reads,

$$\langle\kappa_\theta^2\rangle = \int \mathrm{d}\chi \, w^2(\chi) \int \frac{\mathrm{d}^2\mathbf{k}}{2\pi} \, P(k) \, W^2(k_\perp\theta\mathcal{D}). \tag{39}$$

For realistic models of the power spectrum (e.g. Baugh and Gaztañaga, 1996), the numerical result is (Bernardeau et al. 1997),

$$\langle\kappa_\theta^2\rangle^{1/2} \approx 0.01 \; \sigma_8 \; \Omega_0^{0.8} \; z_s^{0.75} \left(\frac{\theta}{1\,\mathrm{deg}}\right)^{-(n+2)/2}. \tag{40}$$

One can note the strength of the dependence on the redshift of the sources. This was first stressed by Villumsen (1996) who pointed out that the Ω_0 dependence is roughly given by the Ω value at the redshift of the sources. These results are slightly affected by the introduction of the non-linear effects in the shape of the power spectrum (Miralda-Escudé 1991, Jain & Seljak 1997).

5.5. THE EXPECTED SIGNAL TO NOISE RATIO

Are the effects from large-scale structures measurable? It depends on the number density of background objects for which the shape matrices are measurable. In current deep galaxy survey the typical mean number density of objects is about 50 arcmin^{-2}. The precision of the measured distortion at the degree scale is then about,

$$\Delta_{\text{noise}}\kappa = \frac{0.3}{\sqrt{50\ 60^2}} \approx 10^{-3}, \tag{41}$$

for an intrinsic ellipticity of sources of about 0.3. This number is to be compared with the expected amplitude of the signal coming from the large-scale structures, about 1% according to Eq. (40) (see also earlier computations by Blandford et al. 1991, Miralda-Escudé 1991, Kaiser 1992). This makes such detection a priori possible with a signal to noise ratio around 10 (provided the instrumental noise can be controlled down to such a low level!).

5.6. SEPARATE MEASUREMENTS OF Ω_0 AND σ_8

In Eq. (40) one can see that the amplitude of the fluctuations depend both on σ_8 and on Ω_0. A question that then arises is whether it is possible to separate the amplitude of the power spectrum from the cosmological parameters. A simple examination of the equation (34) shows that it should be the case, because, for a given value of σ_κ, the density field is more strongly evolved into the non-linear regime when Ω_0 is low.

The consequences of this are twofold. The nonlinearities change the angular scale at which the non-linear dynamics starts to amplify the growth of structures. This effect was more particularly investigated by Jain & Seljak (1997) who showed that the emergence of the nonlinear regime is apparent in the shape of the angular two-point function. This effect is however quite subtle since it might reveal difficult to separate from peculiarities in the shape of the initial power spectrum.

The other aspect is that nonlinear effects induce non-Gaussian features due to mode couplings. These effects have been studied extensively in Perturbation Theory. Technically one can write the local density contrast as an expansion with respect to the initial density fluctuations,

$$\delta_{\text{mass}}(\mathbf{x}) = \delta^{(1)}_{\text{mass}}(\mathbf{x}) + \delta^{(2)}_{\text{mass}}(\mathbf{x}) + \dots \tag{42}$$

where $\delta^{(1)}_{\text{mass}}(\mathbf{x})$ is proportional to the initial density field (this is the term we have considered so far), $\delta^{(2)}_{\text{mass}}(\mathbf{x})$ is quadratic, etc. Second order perturbation theory provides us with the expression of $\delta^{(2)}_{\text{mass}}(\mathbf{x})$ (there are many references for the perturbation theory calculations, Peebles 1980, Fry 1984, Goroff et al. 1986, Bouchet et al. 1993 for the Ω dependence of this result),

$$\delta^{(2)}_{\text{mass}}(t,\mathbf{x}) = \int \frac{\mathrm{d}^3\mathbf{k}_1}{(2\pi)^{3/2}} \frac{\mathrm{d}^3\mathbf{k}_2}{(2\pi)^{3/2}} D_+^2(t)\, \delta_{\text{lin.}}(\mathbf{k}_1)\delta_{\text{lin.}}(\mathbf{k}_2) \exp[\mathrm{i}(\mathbf{k}_1+\mathbf{k}_2)\cdot\mathbf{x}] \times \left[\frac{5}{7} + \frac{\mathbf{k}_1\cdot\mathbf{k}_2}{k_1^2} + \frac{2}{7}\frac{(\mathbf{k}_1\cdot\mathbf{k}_2)^2}{k_1^2 k_2^2}\right], \tag{43}$$

where $\delta_{\text{lin.}}(\mathbf{k})$ are the Fourier components of the *linear* density field. It behaves essentially as the square of the linear term, with a non-trivial geometric function that contains the non-local effects of gravity.

Equivalently it is possible to expand the local convergence in terms of the initial density field,

$$\kappa(\gamma) = \kappa^{(1)}(\gamma) + \kappa^{(2)}(\gamma) + \dots \tag{44}$$

The apparition of a non-zero $\kappa^{(2)}$ induces non-Gaussian effects that can be revealed for instance by the computation of the skewness, third moment, of κ_θ (Bernardeau et al. 1997),

$$\langle\kappa_\theta^3\rangle = \langle \left(\kappa_\theta^{(1)}\right)^3 \rangle + 3 \langle \left(\kappa_\theta^{(1)}\right)^2 \kappa_\theta^{(2)} \rangle + \dots \tag{45}$$

The actual dominant term of this expansion is $3 \langle \left(\kappa^{(1)}\right)^2 \kappa^{(2)} \rangle$ since the first term vanishes for Gaussian initial conditions. For the computation of such term one should plug in Eq. (34) the expression of $\delta^{(2)}_{\text{mass}}$ in Eq. (43) and do the computations in the small angle approximation (and using specific properties of the angular top-hat window function, Bernardeau 1995).

Eventually perturbation theory gives the following result for a realistic power spectrum (Bernardeau et al. 1997),

$$s_3(\theta) \equiv \frac{\langle\kappa_\theta^3\rangle}{\langle\kappa_\theta^2\rangle^2} = 40\ \Omega_0^{-0.8}\ z_s^{-1.35}. \tag{46}$$

The origin of this skewness is relatively simple to understand: as the density field enters the non-linear regime the large mass concentrations tend

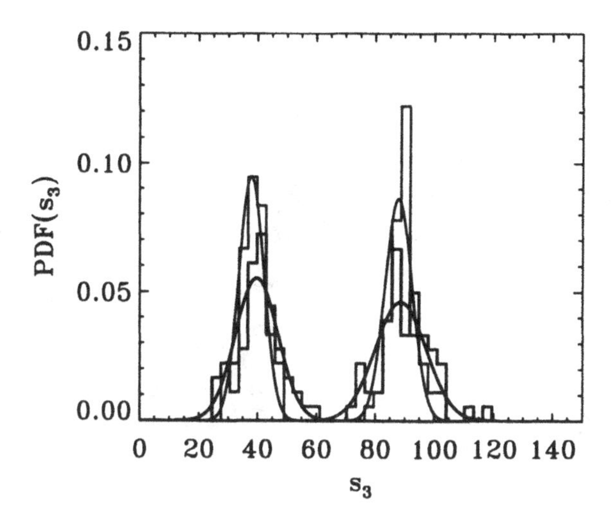

Figure 6. Histograms of the values of s_3, top-hat filter, for $\Omega = 1$ and $\Omega = 0.3$ for a 5×5 degree survey (thick lines) and a 10×10 degree survey (thin lines).

to acquire a large density contrast in a small volume. This induces rare occurrences of large negative convergences. The under-dense regions tend on the other hand to occupy a large fraction of the volume, but can induce only moderate positive convergences. This mechanism is clearly visible on the maps of figure (4). When the mean source redshift grows the skewness diminishes since the addition of independent layers of large-scale structures tend to dilute the non-Gaussianity.

What the Eq. (46) demonstrates is that distortion maps can be used to determine the cosmic density parameter, Ω_0, provided the redshift distribution of the sources is well known. The hierarchy exhibited in this relation is also a direct consequence of the hypothesis of Gaussian initial conditions. Such a hierarchy has been observed for instance in galaxy catalogues (see Bouchet et al. 1993 for results in the IRAS galaxy survey). It can be very effective in excluding models with non-Gaussian initial conditions (see the attempt of Gaztañaga & Mähönen 1996). To be more precise the actual histograms of the measured skewness in numerical simulations are presented in Fig. 6. They clearly demonstrate that the two cosmologies are easily separated. One can see that the scatter in s_3 is roughly the same in the two cases and that the difference in the relative precision is due to the differences in the expectation values.

The validity of Eq. (46) has been confirmed numerically by Gaztañaga & Bernardeau (1998), who showed it is valid for scales above a few tens of arcmins. A non-zero skewness has also been observed in numerical experi-

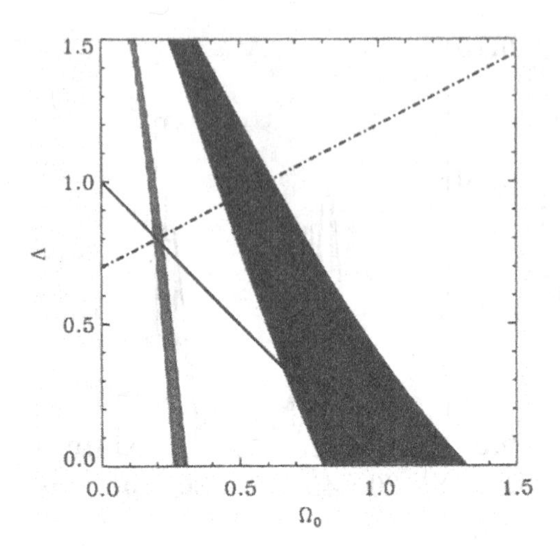

Figure 7. **Constraints that can be brought by a 10x10 square degree weak lensing survey in an $\Omega_0-\Lambda$ plane. The grey bands are the location of the 1 and 2-σ locations (respectively darker and lighter bands) allowed by a measured skewness that would be obtained with either $\Omega_0 = 0.3$ (left bands) or $\Omega_0 = 1$ (right bands). The solid straight lines corresponds to a zero curvature universe, and the dot-dashed lines to a fixed acceleration parameter, q_0.**

ments (van Waerbeke et al. 1999, Jain et al. 1999, White & Hu 1999). Large angular convergence maps can therefore provide new means for constraining fundamental cosmological parameters. Numerical results show that in maps of 10×10 square degrees it is reasonable to expect a precision[5] of a few percent on the normalization and about 2% to 5% on the cosmological density parameter depending on the underlying cosmological scenario (see Fig. 7).

5.7. PROSPECTS

From an observational point of view, the investigation of the large-scale structures of the Universe with gravitational lenses is in a very preliminary stage. After an early claim by Villumsen (1995), a direct evidence of the detection of a distortion signal of gravitational origin has been reported recently by Schneider et al. (1997).

There are at present many studies, either theoretical or numerical, that aim to examine all possible systematic errors (Bonnet & Mellier 1995, Kaiser et al. 1995), to optimize the data analysis concepts (such as the pixel

[5]Only the so called cosmic variance has been taken into account. Observational systematics, in particular due to possible anisotropic point spread functions, are neglected.

autocorrelation function by van Waerbeke et al. 1997) and the scientific interpretations of the resulting mass maps (Bernardeau 1998, Bernardeau et al. 1997, Seljak 1998, van Waerbeke et al. 1999). A few observational surveys are now emerging,

- the ESO key program jointly done by MPA and IAP;
- the DESCART project, part of the scientific program of the wide field CCD camera to be installed at the CFHT.

Acknowledgements

The author thanks Y. Mellier for innumerable discussions on the lens physics and IAP for its hospitality.

References

1. AbdelSalam, H.M., Saha, P. & Williams, L.L.R. 1998, *Mon. Not. R. astr. Soc.* **294**, 734
2. Baugh, C.M. & Gaztañaga, E. 1996, *Mon. Not. R. astr. Soc.* **280**, L37
3. Bernardeau, F. 1998, *Astr. & Astrophys.* **338**, 375
4. Bernardeau, F. 1995, *Astr. & Astrophys.* **301**, 309
5. Bernardeau, F., van Waerbeke, L. & Mellier, Y. 1997, *Astr. & Astrophys.* **324**, 15
6. Bertschinger, E. 1996, in "Cosmology and Large Scale Structure", Les Houches Session LX, August 1993, NATO series, eds. R. Schaeffer, J. Silk, M. Spiro, J. Zinn-Justin, Elsevier Science Press
7. Blandford, R. D., Saust, A. B., Brainerd, T. G., Villumsen, J. V. 1991, *Mon. Not. R. astr. Soc.* **251**, 600
8. Bonnet, H. & Mellier, Y. 1995 *Astr. & Astrophys.* **303**, 331
9. Bouchet, F., Juszkiewicz, R., Colombi, S. & Pellat, R., 1992, *Astrophys. J.* **394**, L5
10. Bouchet, F., Strauss, M.A., Davis, M., Fisher, K.B., Yahil, A. & Huchra, J.P. 1993, *Astrophys. J.* **417**, 36
11. Fry, J., 1984, *Astrophys. J.* **279**, 499
12. Fahlman, G., Kaiser, N., Squires, G., Woods, D. 1994, *Astrophys. J.* **437**, 56
13. Gaztañaga, E. & Bernardeau, F. 1998, *Astr. & Astrophys.* **331**, 829
14. Gaztañaga, E. & Mähönen, 1996, *Astrophys. J.* **462**, L1
15. Goroff, M.H., Grinstein, B., Rey, S.-J. & Wise, M.B. 1986, *Astrophys. J.* **311**, 6
16. Jain, B., Seljak, U. 1997, *Astrophys. J.* **484**, 560
17. Jain, B., Seljak, U., White, S., astro-ph/9901191
18. Kaiser, N. 1992 *Astrophys. J.* **388**, L72
19. Kaiser, N. 1995 *Astrophys. J.* **439**, 1
20. Kaiser, N. & Squires, G. 1993 *Astrophys. J.* **404**, 441
21. Kaiser, N., Squires, G. & Broadhurst, T. 1995 *Astrophys. J.* **449**, 460
22. Kassiola, A. & Kovner, I. 1993, *Astrophys. J.* ,, 417,450
23. Mellier, Y. 1998, astro-ph/9812172 to appear in ARAA
24. Miralda-Escudé, J. 1991 *Astrophys. J.* **380**, 1
25. Misner, C.W. Thorne, K. & Wheeler, J.A. 1973, Gravitation, San Francisco, Freeman.
26. Peebles, P.J.E. 1980; The Large–Scale Structure of the Universe; Princeton University Press, Princeton, N.J., USA;
27. Sachs, R. K. 1961, *Proc. Roc. Soc. London* A**264**, 309
28. Schneider, P., Ehlers, J., Falco, E. E. 1992, *Gravitational Lenses*, Springer.

29. Schneider, P., Van Waerbeke, L., Mellier, Y., Jain, B., Seitz, S., Fort, B. 1997, astro-ph/9705122.
30. Seljak, U. 1998, *Astrophys. J.* **506**, 64
31. Soucail, G., Mellier, Y., Fort, B., Mathez, G. & Cailloux, M. 1988 *Astr. & Astrophys.* **191**, L19
32. Villumsen, J. V., astro-ph/9507007
33. Villumsen, J. V. 1996, *Mon. Not. R. astr. Soc.* **281**, 369
34. van Waerbeke, L., Mellier, Schneider, P., Fort, B. & Mathez, G. 1997 *Astr. & Astrophys.* **317**, 303
35. van Waerbeke, L., Bernardeau, F. & Mellier, Y. 1999, *Astr. & Astrophys.* **342**, 15
36. White, M., Hu, W., astro-ph/9909165

THE SLOAN DIGITAL SKY SURVEY

A.S. SZALAY

Department of Physics and Astronomy
The Johns Hopkins University, Baltimore

for the SDSS Collaboration

1. Introduction

Astronomy is about to undergo a major paradigm shift, with data sets becoming larger, and more homogeneous, for the first time designed in the top-down fashion. In a few years it may be much easier to "dial-up" a part of the sky, when we need a rapid observation than wait for several months to access a (sometimes quite small) telescope. With several projects in multiple wavelengths under way, like the SDSS, 2MASS, GSC-2, POSS2, ROSAT, FIRST and DENIS projects, each surveying a large fraction of the sky, the concept of having a "Digital Sky", with multiple, TB size databases inter-operating in a seamless fashion is no longer an outlandish idea. More and more catalogs will be added and linked to the existing ones, query engines will become more sophisticated, and astronomers will have to be just as familiar with mining data as with observing on telescopes.

The Sloan Digital Sky Survey, hereafter the SDSS, is a project to digitally map about 1/2 of the Northern sky in five filter bands from UV to the near IR, and is expected to detect over 200 million objects in this area. Simultaneously, redshifts will be measured for the brightest 1 million galaxies. The SDSS will revolutionize the field of astronomy, increasing the amount of information available to researchers by several orders of magnitude. The resultant archive that will be used for scientific research will be large (exceeding several Terabytes) and complex: textual information, derived parameters, multi-band images, and spectra. The catalog will allow astronomers to study the evolution of the universe in greater detail and is intended to serve as the standard reference for the next several decades.

R.G. Crittenden and N.G. Turok (eds.), Structure Formation in the Universe, 371–382.

Figure 1. The view of the SDSS 2.5m telescope from the East.

2. The Sloan Digital Sky Survey

The Sloan Digital Sky Survey (SDSS) is a collaboration between the University of Chicago, Princeton University, the Johns Hopkins University, the University of Washington, Fermi National Accelerator Laboratory, the Japanese Promotion Group, the United States Naval Observatory, and the Institute for Advanced Study, Princeton, with additional funding provided by the Alfred P. Sloan Foundation and the National Science Foundation. In order to perform the observations, a dedicated 2.5 meter Ritchey-Chretien telescope was constructed at Apache Point, New Mexico, USA. This telescope is designed to have a large, flat focal plane which provides a 3 field of view. This design results from an attempt to balance the areal coverage of the instrument against the detector's pixel resolution.

The survey has two main components: a photometric survey, and a spectroscopic survey. The photometric survey is produced by drift scan imaging of 10,000 square degrees centered on the North Galactic Cap using five broad-band filters that range from the ultra-violet to the infra-red. The photometric imaging will use a CCD array that consists of 30 2K x 2K imaging CCDs, 22 2K x 400 astrometric CCDs, and 2 2K x 400 Focus CCDs. The data rate from this camera will exceed 8 Megabytes per second, and the total amount of raw data will exceed 40TB.

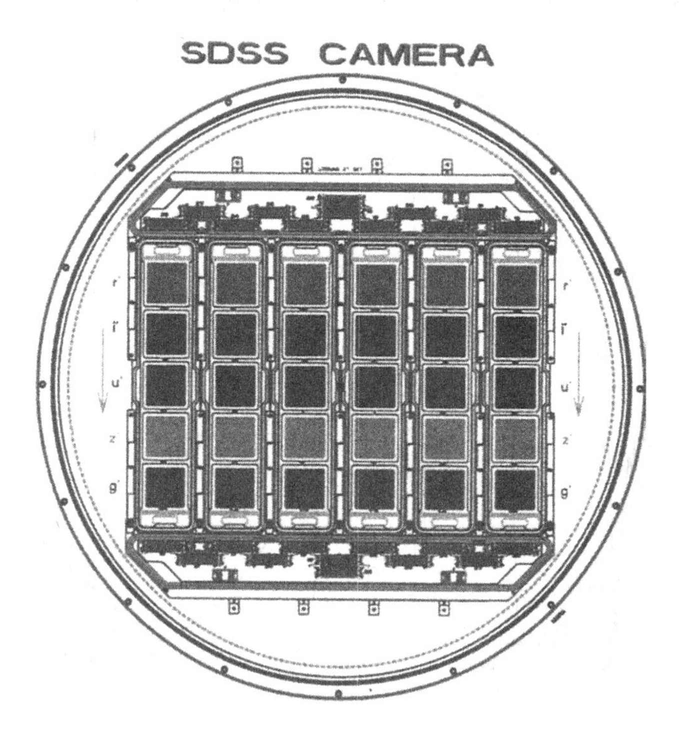

Figure 2. Front view of the SDSS camera assembly. This diagram shows the camera as it would be seen with the front cover and shutters removed, showing the 30 photometric and 24 astrometric/focus CCDs and their associated dewars and kinematic supports.

A considerable care has gone into selecting the filter system. The SDSS has designed a filter system similar to ugriz which should transform to and from it with little difficulty, in which the g , r and i filters are as wide as practicable consistent with keeping the overlap small. If we wish the u band to be a "good" one which is almost entirely contained between the Balmer jump and the atmospheric cutoff, it cannot be significantly wider than Thuan-Gunn u . We call the new system u' g' r' i' z' . The response curves and sensitivity data with the coatings and CCDs we will use are shown in Figure 3.

The spectroscopic survey will target over a million objects chosen from the photometric survey in an attempt to produce a statistically uniform sample. This survey will utilize two multi-fiber medium resolution spectrographs, with a total of 640 optical fibers, of 3 seconds of arc in diameter, that provide spectral coverage from 3900 - 9200 Å. The telescope will gather

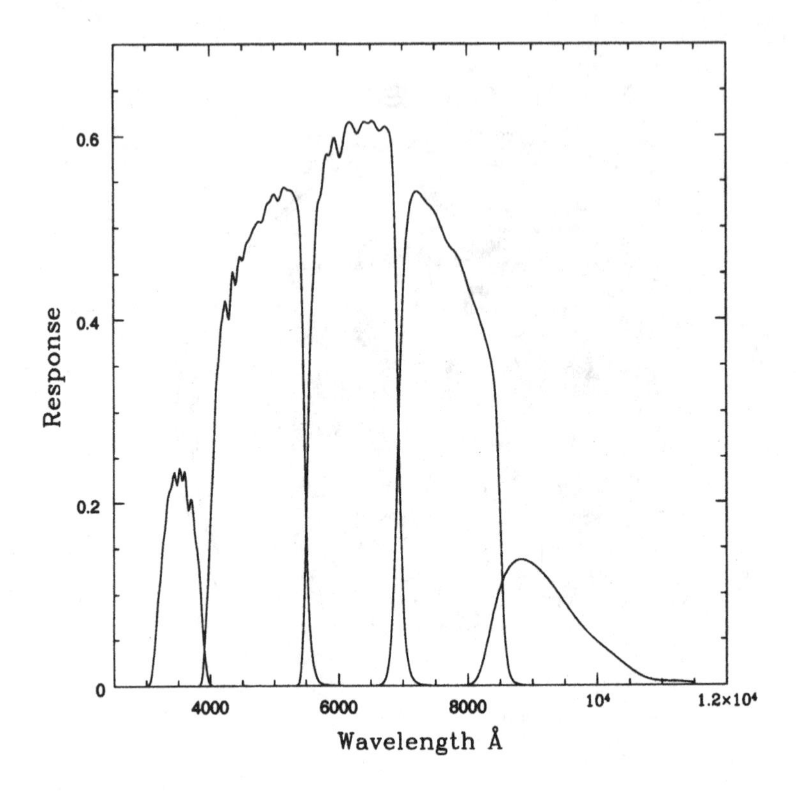

Figure 3. The SDSS system response curves. The responses are shown without atmospheric extinction. The curves represent expected total quantum efficiencies of the camera plus telescope on the sky.

about 5000 galaxy spectra in one night. The total number of spectra known to astronomers today is about 60,000 - only 12 days of SDSS data! Whenever the Northern Galactic cap is not accessible from the telescope site, a complementary survey will repeatedly image several areas in the Southern Galactic cap to study fainter objects and identify any variable sources.

2.1. THE DATA PRODUCTS

The SDSS will create four main data sets: a photometric catalog, a spectroscopic catalog, images, and spectra. The photometric catalog is expected to contain one hundred million galaxies, one hundred million stars, and one million quasars, with magnitudes, profiles, and observational information recorded in the archive. The anticipated size of this product is about 250GB. Each detected object will also have an associated image cutout ("atlas image") for each of the five filters, adding up to about 700GB. The spectroscopic catalog will contain identified emission and absorption lines,

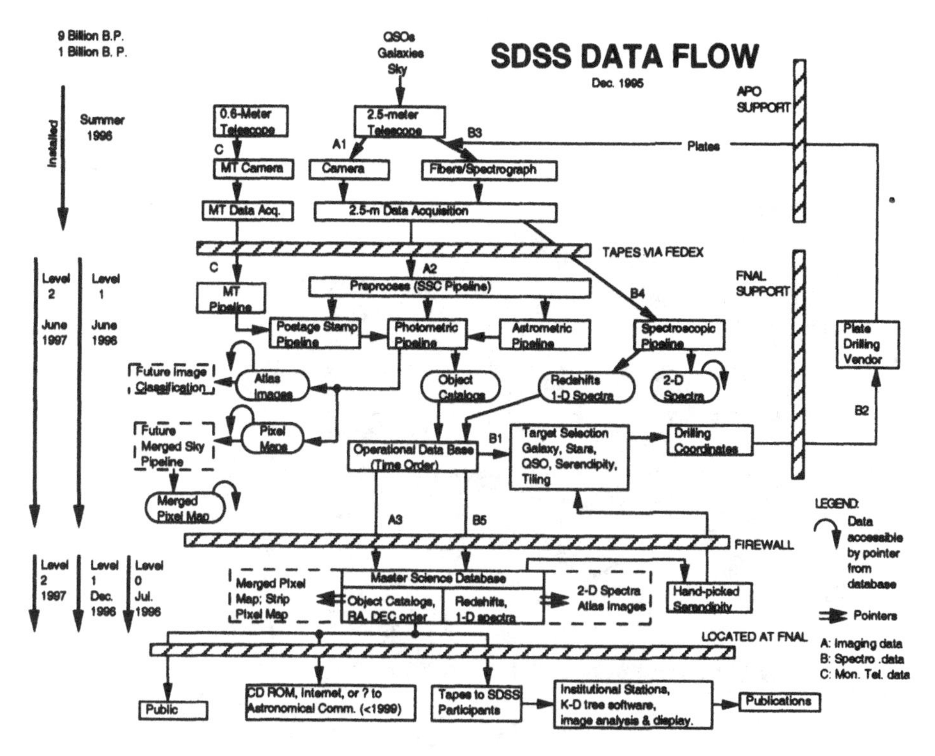

Figure 4. A diagram of the data reduction software for the SDSS project. Due to the large amount of data, a fully automated reduction system is necessary. The software is organized into pipelines which communicate via the Operational Database (ODB).

and one dimensional spectra for one million galaxies, one hundred thousand stars, one hundred thousand quasars, and about ten thousand clusters, totalling about 50GB. In addition, derived custom catalogs may be included, such as a photometric cluster catalog, or QSO absorption line catalog. Thus the amount of tracked information in these products is about 1TB.

The collaboration will release the data to the public after a period of thorough verification. The actual distribution method is still under discussion. This public archive is expected to remain the standard reference catalog for the next several decades, presenting additional design and legacy problems. Furthermore, the design of the SDSS science archive must allow for the archive to grow beyond the actual completion of the survey. As the reference astronomical data set, each subsequent astronomical survey will want to cross-identify its objects with the SDSS catalog, requiring that the archive, or at least a part of it be dynamic.

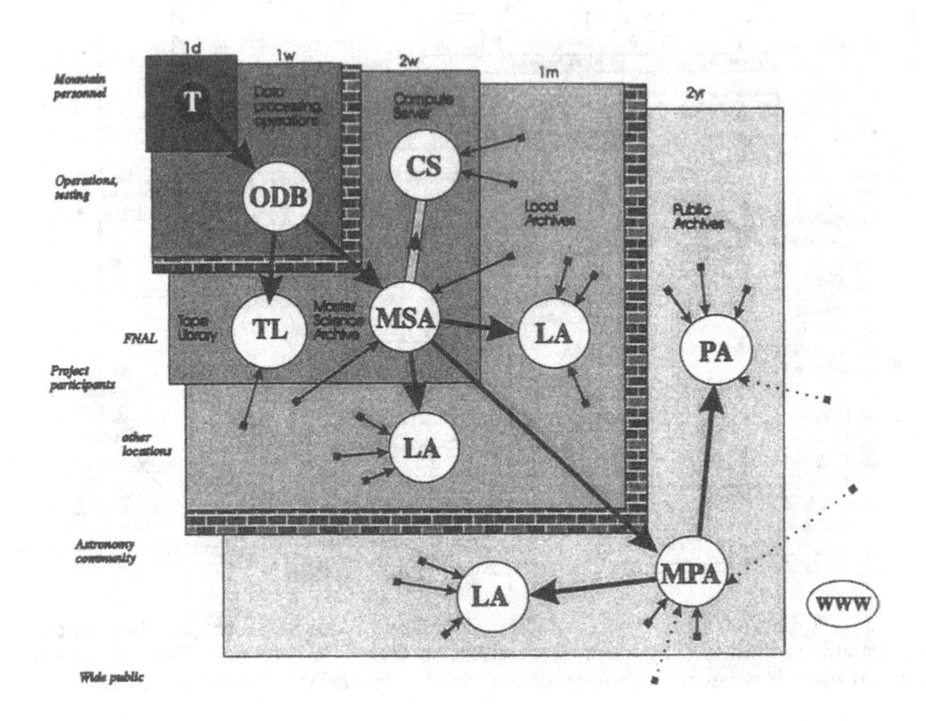

Figure 5. A conceptual data-flow diagram of the SDSS data. The data is taken at the telescope (T), and is shipped on tapes to FNAL, where it is processed within one week, and ingested into the operational archive (ODB), protected by a firewall, accessible only the personnel working on the data processing. Within two weeks, data will be transferred into the Master Science Archive (MSA). From there data will be replicated to local archives (LA) within another two weeks. The data gets into the public domain (MPA, LA) after two years of science verification, and recalibration, if necessary. These servers will provide data for the astronomy community. We will also provide a WWW based access for the wide public, to be defined in the near future.

3. The SDSS Archives

The survey archive is split into two orthogonal functionalities and the corresponding distinct components: an *operational archive*, where the raw data is reduced and mission critical information is stored; and the *science archive*, where calibrated data is available to the collaboration for analysis and is optimized for such queries. In the operational archive. data is reduced, but uncalibrated, since the calibration data is not necessarily taken at the same time as the observations. Calibrations will be provided on the fly, via method functions, and several versions will be accessible. The Science

Archive will contain only calibrated data, reorganized for efficient science use, using as much data clustering as possible. If a major revision of the calibrations is necessary, the Science Archive and its replications will have to be regenerated from the ODB. At any time the Master Science Archive will be the standard to be verified against.

High level requirements for the archive included (a) easy transfer of the binary database image from one architecture to another, (b) easy multi-platform availability and interoperability, (c) easy maintainance for future operating systems and platforms. In order to satisfy these, the Science Archive employs a three-tiered architecture: the user interface, the query support component, and the data warehouse. This distributed approach provides maximum flexibility, while maintaining portability, by isolating hardware specific features. The data warehouse, where most of the low-level I/O access happens is based upon an OODBMS (Objectivity/DB), where the porting issues depend mainly on the database vendor. Objectivity's database image is binary compatible, can be copied between different platforms, since the architecture is encoded on every page in the archive.

3.1. ACCESSING TERABYTES OF DATA

Today's approaches to accessing astronomical data do not scale into the Terabyte regime — brute force does not work! Assume a hypothetical 500 GB data set. The most popular data access technique today is the World Wide Web. Most universities can receive data at the bandwidth of about 15 kbytes/sec. The transfer time for this data set would be 1 year! If the data is residing locally within the building (access via Ethernet at 1 Mbytes/sec), the transfer time drops to 1 week. If the astronomer is logged on to the machine which contains the data, all of it on hard disk, then with SCSI bandwidth it still takes 1 day to scan through the data. Even faster hardware cannot support hundreds of "brute force" queries per day. With Terabyte catalogs, even small custom datasets are in the 10 GB range, thus a high level data management is needed.

4. Survey Goals

Both the photometric and the redshift survey will cover a larger fraction of the sky, a slightly elliptical region in the North, centered close to the NGP, and three strips at -10, 0 and +15 declinations. The length of these stripes is determined by extinction.

The design of the SDSS redshift survey is crucial to its effectiveness as a tool for studying large scale structure. The goal of this survey is not to measure a single statistical property of the distribution of galaxies, such as the power spectrum of density fluctuations, or the maximum size of non-

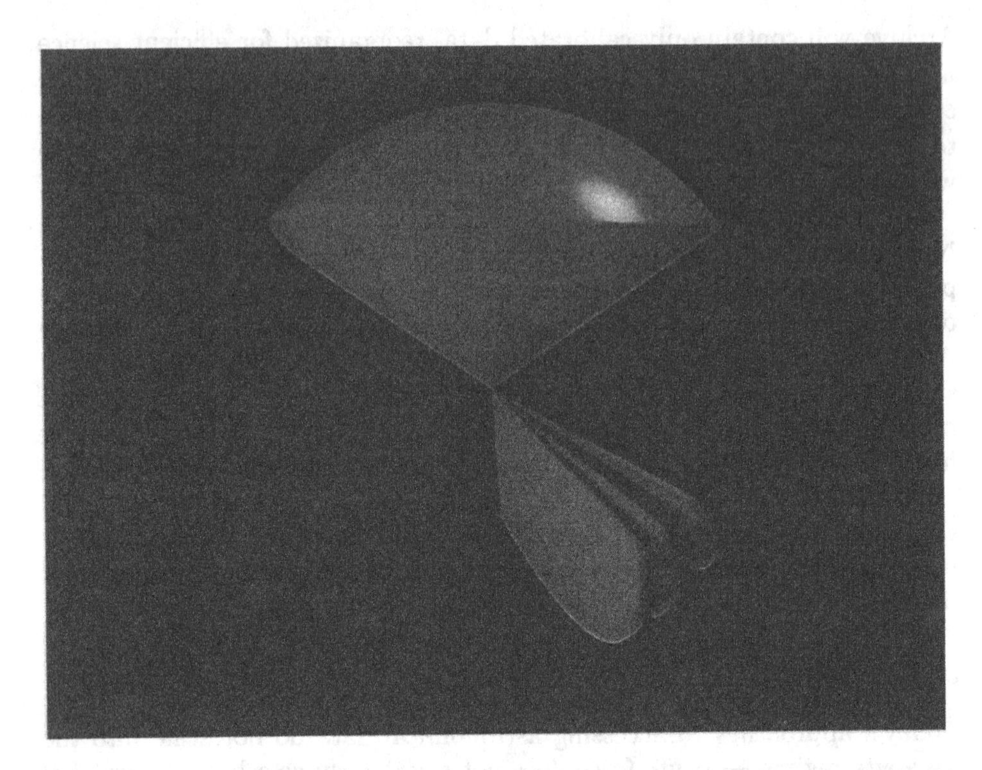

Figure 6. A ray-traced visualization of the volume covered by the SDSS survey: the upper cone is the Northern survey, while the three narrow fans below are the Southern stripes.

linear structures. Instead, we desire an accurate description of the galaxy distribution over a wide range of physical scales, in order to address the questions listed above. To meet this goal, we plan to do a wide-angle, deep redshift survey which fully samples the galaxy distribution.

Figure 7. displays a view of a simulated SDSS spectroscopic sample, drawn from a large N-body simulation of a low-density, cold dark matter Universe. For these simulations, the spectroscopic sample consisted of all galaxies with apparent magnitude $r' < 17.55$ and apparent half-light diameter greater than 2 arcseconds, selecting just under one million galaxies over the survey area. Since the time that this simulation was done, the details of the galaxy selection criteria have changed; see below, but the qualitative nature of this figure is not sensitively dependent on these details.

This figure shows the distribution of galaxies in redshift space in a 6 deg by 130 deg slice along the survey equator, which contains about 6% of the spectroscopic sample. The median depth of the sample is $300h^{-1}$ Mpc , and there is useful information on the density field to $600h^{-1}$ Mpc, with a few galaxies as far away as $1000h^{-1}$ Mpc.

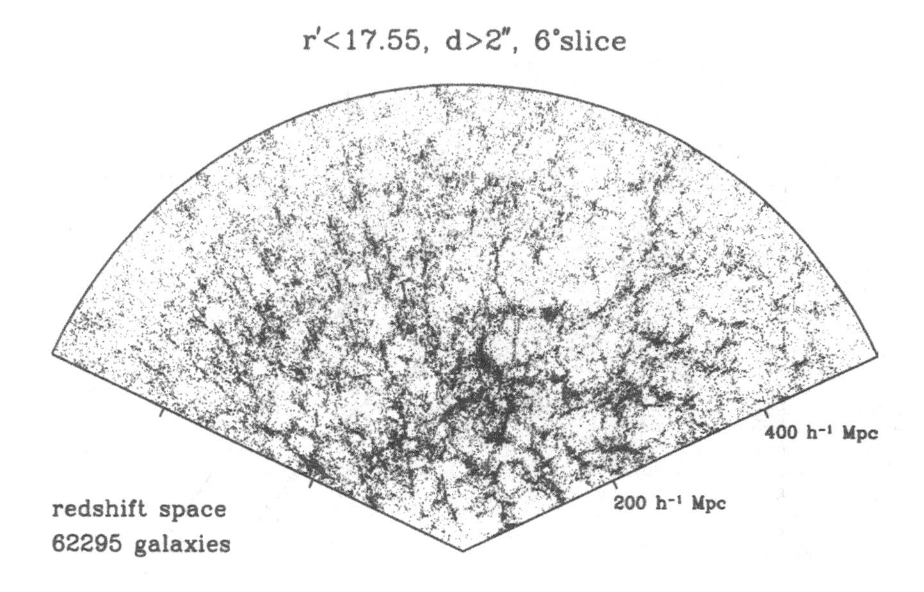

Figure 7. The redshift-space distribution of galaxies. The objects lie in a slice 6deg thick along the survey equator. Galaxies are plotted at the distance indicated by their redshift, hence the appearance of clusters as "fingers-of-God." This slice contains roughly 6% of the galaxy redshift sample. (D. Weinberg)

5. Large Scale Structure and Survey Strategy

A few examples from the recent history of redshift surveys illustrate how the detection of structure depends on both the survey geometry and the sampling rate. Observing in narrow pencil beams in the direction of the constellation Bootes, Kirshner et al. (1981) detected a $60h^{-1}$ Mpc diameter void in the galaxy distribution. The geometry and sampling of the Kirshner et al. survey was well suited to detection of a single void, but it could not answer the question of how common such structures are. De Lapparent, Geller and Huchra (1986, hereafter CfA2) measured redshifts of a magnitude-limited sample of galaxies in a narrow strip on the sky, and found that (1) large voids fill most of the volume of space and (2) these voids are surrounded by thin, dense, coherent structures. Demonstration of the first result was possible because of the geometry and depth of the survey, demonstration of the second because of the dense sampling. There were hints of this structure in the shallower survey of the same region by Huchra et al. (1983), but the deeper CfA2 sample was required to reveal the structure clearly. Sparse samples of the galaxy distribution, e.g. the QDOT (Saunders et al. 1991) survey of one in six IRAS galaxies, have pro-

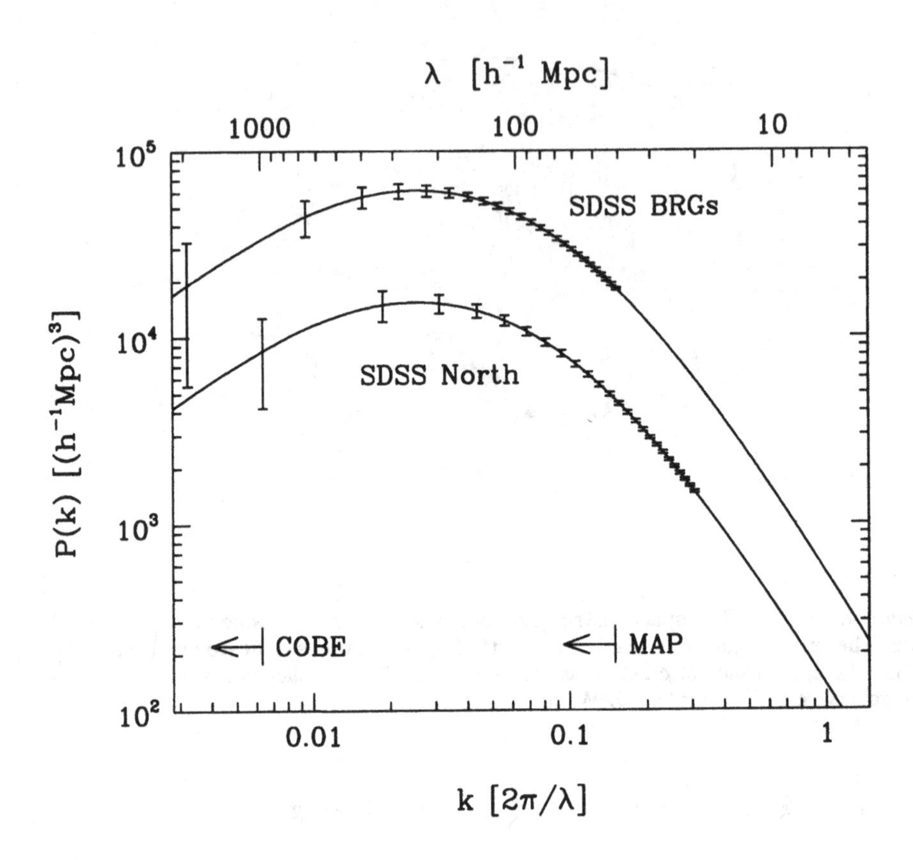

Figure 8. Model power spectra. The power spectra for the SDSS Northern galaxy survey and for the luminous red galaxies (BRGs) are shown. The latter are assumed to be biased by a factor of two with respect to the galaxies. The individual points and their error bars are statistically independent. On small scales, the errors are smaller than the smallest bars. Also shown are the smallest comoving wavelength scales accessible to COBE and to the upcoming Microwave Anisotropy Probe (MAP).

vided useful statistical measures of low-order clustering on large scales, but sparse surveys have less power to detect coherent overdense and underdense regions. Perhaps because of this limitation, members of the QDOT team have elected to follow up their one-in-six survey by obtaining a complete (one-in-one) survey of IRAS galaxies to the same limiting flux, now nearing completion.

In order to encompass a fair sample of the Universe, a redshift survey must sample a very large volume. We do not know a priori how large such a volume must be, but existing observations provide the minimum constraints. Our picture of the geometry and size of existing structures became clearer as the CfA2 survey was extended over a larger solid angle (e.g. Vogeley et al. 1994). The largest of the structures found is a dense region measuring $150h^{-1}$ Mpc by $50h^{-1}$ Mpc – the "Great Wall" (Geller and Huchra 1989). Giovanelli et al. (1986) find a similar coherent structure in a complete redshift survey of the Perseus-Pisces region, and da Costa et al. (1988) find another such "wall" in the Southern hemisphere (cf. Santiago et al. 1996). However, none of these surveys covers a large enough volume to investigate the frequency of these structures. In quantitative terms, the observed power spectrum of galaxy density fluctuations continues to rise on scales up to $> 100h^{-1}$ Mpc (Vogeley et al. 1992; Loveday et al. 1992; Fisher et al. 1993; Feldman et al. 1994; Park et al. 1994; Baugh and Efstathiou 1993; Peacock and Dodds 1994; da Costa et al. 1994; Landy et al. 1996). The observed power spectrum may be influenced by these few, largest, nearby structures. Thus a fair sample clearly must include many volumes with scale $100h^{-1}$ Mpc .

The sampling rate and geometry of a survey can strongly influence the largest structures seen (Szalay et al. 1991). While different survey geometries are appropriate for elucidating different features of large scale structure, a deep survey with a large opening angle and full sampling of the galaxy population is the only way to get a complete picture of the galaxy distribution. Coverage of the whole sky is extremely helpful if one wants to compare velocity and density fields in the nearby Universe, and redshift surveys of infrared-selected galaxies are invaluable for this purpose.

6. Summary

We are well along in the design and construction an extremely ambitious project, aiming to provide a useful tool for almost all astronomers in the world. The SDSS project has currently engineering quality imaging data for a few hundred square degrees and test spectra of over 24,000 objects. Some of the early discoveries cover high redshift quasars (Fan et al 2000), including one at $z = 5.8$, and the discovery of methane dwarfs (Fan et al 2000). These serve as a demonstration of the power of deep multicolor imaging.

This endeavour would not have been possible ten years ago, and even now we are pushing the limits of technology. We hope that our efforts will be successful, and the result will substantially change the way scientists do astronomy today. Having 200 million objects at our fingertips will undoubtedly lead to new major discoveries, and the spectroscopic followup of even a fraction of our fainter objects occupy astronomers for decades. The day when we have a "Digital Sky" at our desktop may be nearer than most astronomers think. Given the enormous public interest in astronomy, we hope that the resulting archive will also provide a challenge and inspiration to thousands of interested high-school students, and a lot of fun for the web-surfing public.

7. Acknowledgments

This material is based upon the Grey Book, the SDSS proposal to the National Science Foundation. The author would like to acknowledge helpful discussions with the SDSS participants, and emphasize the heroic efforts of the collaboration on making this data set the highest quality possible.

References

1. Baugh, C. M., and Efstathiou, G. 1993, MNRAS 265, 145.
2. da Costa, L. N., Pellegrini, P., Sargent, W., Tonry, J., Davis, M., Meiksin, A., Latham D., Menzies, J., and Coulson, I. 1988, ApJ 327, 544.
3. da Costa, L. N., Vogeley, M. S., Geller, M. J., Huchra, J. P. and Park, C. 1994, ApJL 437, 1.
4. de Lapparent, V., Geller, M.J., and Huchra, J.P. 1986, ApJL 302, 1.
5. Feldman, H., Kaiser, N., and Peacock, J. 1994, ApJ 426, 23.
6. Fisher, K.B., Davis, M., Strauss, M.A., Yahil, A., and Huchra, J.P. 1993, ApJ 402, 42.
7. Fisher, K. B., Huchra, J. P., Davis, M., Strauss, M. A., Yahil, A. and Schlegel, D. 1995, ApJSuppl, 100, 69.
8. Geller, M.J., and Huchra, J.P. 1989, Science 246, 897.
9. Giovanelli, R., Haynes, M.P., and Chincarini, G. 1986, ApJ 300, 77.
10. Huchra, J., Davis, M., Latham, D., and Tonry, J. 1983, ApJSuppl 52, 89.
11. Kirshner, R.P., Oemler, A., Schechter, P.L., and Shectman, S.A. 1981, ApJL 248, 57.
12. Landy, D.S., Shectman, S.A., Lin, H., Kirshner, R.P., Oemler, A.A., and Tucker, D. 1996, ApJL 456, 1.
13. Loveday, J., Efstathiou, G., Peterson, B., A., and Maddox, S. J. 1992, ApJ 400, 43.
14. Park, C., Vogeley, M.S., Geller, M.J., and Huchra, J.P. 1994, ApJ 431, 569.
15. Peacock, J. A., and Dodds, S. J. 1994, MNRAS 267, 1020.
16. Santiago, B. X., Strauss, M. A., Lahav, O., Davis, M., Dressler, A., and Huchra, J. P. 1996, ApJ 461, 38.
17. Saunders, W., Frenk, C. S., Rowan-Robinson, M., Efstathiou, G., Lawrence, A., Kaiser, N., Ellis, R. S., Crawford, J., Xia, X.-Y., and Parry, I. 1991, Nature 349, 32.
18. Shectman, S.A., Landy, S.D., Oemler, A., Tucker, D.L., Lin, H., Kirshner, R.P., and Schechter, P.L. 1996, ApJ 470, 172.
19. Strauss, M.A. and Willick, J.A. 1995, Physics Reports, 261, 271.
20. Szalay, A.S., Broadhurst, T.J., Ellman, N., Koo, D.C., and Ellis, R. 1991, Proc. Natl. Acad. Sci., 90, 4858.
21. Vogeley, M.S., Park, C., Geller, M.J., Huchra, J.P., 1992, ApJL 391, 5.
22. Vogeley, M.S., Park, C., Geller, M.J., Huchra, J.P., and Gott, J.R. 1994, ApJ 420, 525.
23. Fan,X. et al 2000a, AJ, 119, 1.
24. Fan,X. et al 2000b, AJ, 119, 928.